Major Impacts and Plate Tectonics

'Snowball', prior to explosion, formed an inverted hemisphere of 500 tons of TNT. When exploded, the TNT gave rise to a central uplift crater of the form to be seen on the moon. Such features can be attributed to the initial configuration of the TNT blocks, which approximate to the impact effects of a major comet.

The 'Prairie Flat' bomb was constructed to form a sphere resting on the test site and supported in the lower half by blocks of polystyrene. On explosion, this sphere of TNT gave rise to a crater comprising a series of concentric ridges, which can be compared with the impact features seen in the N African Sahara, which resulted from an 'air-burst' (see Chapter 5).

Major Impacts and Plate Tectonics

A model for the Phanerozoic evolution of
the Earth's lithosphere

Neville J. Price
B.Sc., Ph.D., D.Sc., Sen.F.G.S.

London and New York

First published 2001
by Routledge
11 New Fetter Lane, London EC4P 4EE

Simultaneously published in the USA and Canada
by Routledge
29 West 35th Street, New York, NY 10001

Routledge is an imprint of the Taylor & Francis Group

© 2001 Neville J. Price

Typeset in Goudy by
HWA Text and Data Management, Tunbridge Wells
Printed and bound in Great Britain by
St Edmundsbury Press, Bury St Edmunds, Suffolk

British Library Cataloguing in Publication Data
A catalogue record for this book is available from the British Library

Library of Congress Cataloging in Publication Data
Price, Neville J.
    Major impacts and plate tectonics : a model for the Phanerozoic evolution of the Earth's
    lithosphere / Neville Price.
        p. cm.
    Includes bibliographical references and index.
    1. Plate tectonics. 2. Catastrophes (Geology) 3. Impact. I. Title.
    QC511.4 .P74 2000
    551.1'36–dc21   00-037717

ISBN 0-415-23899-4

Dedicated to
Dr G.H.S. Jones
[1924–97]

of whom was said in a radio broadcast
by
Eugene Shoemaker

"While we, down in the States, were making
holes in the ground – the Jones boy, up in
Canada was making Moon craters on Earth".

This remark not only indicates Gareth Jones' standing,
but also reveals the late Eugene Shoemaker, arguably the
most famous of workers in this field in the last 50 years,
as a man of humour and generosity of spirit.

Contents

Preface

I have had the good fortune of carrying out post-graduate research for over 50 years. Consequently I have been influenced by many people. Many of these I grew to know personally and many more remained known only through their publications. Some influenced my thinking when I was still at school. Even when still a sixth-former the writings of Wegener and DuToit convinced me that the continents must have drifted apart. The evidence supporting their thesis was abundant, but for lack of a feasible mechanism, their arguments were all too readily dismissed.

Another early influence was the book by Arthur Holmes, entitled *Principles of Physical Geology*, the earliest edition of which I bought in Singapore, where, with thousands of other sailors of the Pacific Fleet of the Royal Navy, I waited impatiently to be sent home. The very much larger second edition, published in 1965, covered a wide range of topics and remains, in my view, the best book on Physical Geology ever written.

As an undergraduate from 1947–50, I found the subjects taught in Geology to be compartmentalised, heavily biased to taxonomy and not terribly exciting. My University College, at Aberystwyth, was situated on the shore of Cardigan Bay with its many magnificent cliff sections, with fold forms in thick sequences of interbedded grits and shales quite different from the fold geometries hidden in 'problem maps'. How such sequences of grits and shales developed was a mystery until about 1950 when the initially dubious concept of 'turbidity currents' was introduced.

Fortunately, half-way through my undergraduate studies Gilbert Wilson [from Imperial College], who was at that time the leading structural geologist in Britain, visited our Geology Department and presented lectures. In his lectures and in trips along the cliffs in the vicinity of Aberystwyth he explained how many of the structures developed. His influence upon me was so profound that my subsequent PhD study was based on mapping and interpreting the structures in a 40-mile-long section of the Aberystwyth Grits as revealed in the cliffs and foreshore of central Cardigan Bay. I was fortunate for, during the middle year of my postgraduate studies I transferred to Imperial College where, to quote his words, 'I sat on his feet'. [He was a gentleman of the old school and I benefited greatly from his guidance, not only in that one year, but also much later, when I joined the staff at Imperial College.]

I followed my PhD studies with almost ten years at the Mining Research Establishment in Isleworth, Middlesex. where I was mainly concerned in studying aspects of the elastic and strength characteristics of sedimentary rocks. I found that I had joined a small but prestigious band of researchers which included David Griggs, John Handin, Hugh Heard and Bill Brace, in the USA and J.C. Jaeger in Australia. In addition, through the work by M.K. Hubbert on

'Scaling' and also the application of stress functions to faulting problems, I found myself on a steep learning curve.

One of my colleagues, Gareth Jones [a former Captain in REME, where, among other things he learned how to handle explosives] and I became friends. However, within a year he emigrated to Canada and soon became a Scientific Civil Servant at Suffield Research Establishment, where, because of his war-time experience with explosives, he rapidly became a senior researcher involved with the effects of explosions in the range of 5–500 tons of TNT. We met from time to time and discussed the results of the research at Suffield, so that I gradually began to see its significance as regards natural major impacts. I encouraged him to write an 'overview' which he completed a few months before he died. Much of chapter 5 is based on this unique overview.

I joined the staff of the Geological Department at Imperial College in 1964 and encountered several people who influenced me, but of these, only John Norman was fired with enthusiasm for impacts. Like Gareth Jones, his war-time experiences pre-disposed him to the subject. He volunteered for the army on the outbreak of war. He fought with the Monmouthshire Regiment from the Normandy Beaches to Berlin during which trip he was 'mentioned in dispatches' [indeed, of the members of his officers mess, he was the only one to survive that trip] and in so doing created and observed a lot of craters. After the war he trained at Imperial College, became a 'photo-geologist' with Bird and company and later in Hunting's Aero-Survey. He eventually joined the staff at Imperial College in 1962. His ability to see faint traces of fault patterns on aerial photographs was amazing. Several students from Saudi Arabia joined him to learn the techniques of aerial [and by then] satellite imagery. Initially, even the satellite images of that desert area were lacking in contrast. On several occasions he showed me the original image, with his interpretation of what was there. I was politely sceptical – so he then showed me the *enhanced* version of the satellite image which completely vindicated his original interpretation. Thereafter, we had many happy times interpreting the patterns he established. Indeed, as a result of one of our collaborations, together with Eric Peters we were awarded the Consolidated Goldfields Medal of the Institution of Mining and Metallurgy. However, his masterpiece, much edited by pragmatic colleagues, is shown in Figure 5.27.

In 1984 I moved to the Department of Geological Sciences in University College London and a new group of colleagues. I have had long productive discussions with Claudio Vita-Finzi, which I hope will continue, for he has moved into the impact field. A lecture by Judith Milledge on the surface features of diamonds which can be attributed to blast effects resulting from major impacts was particularly entertaining and enlightening. The support rendered by Ron Dudman and his staff was and continues to be greatly appreciated. In addition, I benefited from words of wisdom from Alan Smith relating to the Atlas 3.3 programme. William Napier generously allowed me to cite his work before it went to press. I was greatly impressed by the analysis of Verne Oberbeck et al, regarding the impact flux. However, from my research I reached the conclusion that during the Upper Phanerozoic the flux was at least twice that established by these authors. I wrote to Verne and received a very civil response to the effect that their analysis was correct. In the event, honours were even, for as the reader will see, we were both right!

As regards the writing of this book I must express my gratitude to several people, namely Jorge Skarmeta, August Gudmundsson and my former Head of Department, Mike Audley-Charles, for reading and re-reading early versions of my manuscript. They were critical, in the best meaning of that word, supplied references and were long-suffering in the way that good friends are.

Some of the diagrams and photographs relating to experimental and natural impact features were over 40 years old, my own sketches were often rudimentary, so that much work was required to bring them up to the standards required by publishers. This was a task that was willingly, cheerfully and enthusiastically embraced by Janet Baker who, near the end of this phase was helped by Toby Stiles, Simon Tapper and Dominic Fortes, to whom I express my gratitude; and here I must also thank Kate Bruin for her constant support.

Closer to home, I received tremendous help from my immediate family. My wife Joan has long been responsible for smoothing my style and rectifying my cavalier treatment of the humble comma. However, I am recidivist, so I am wholly responsible for any recurring errors.

I depended heavily on the knowledge of my son David [and current Head of Department] regarding the literature relating to the deeper levels in the Mantle and Core [as discussed in Chapter 1]. In addition, he read, re-read and critically assessed my arguments and presentation in the devastating manner that can only be approached by a beloved son.

Finally, I feel I cannot finish without expressing my indebtedness to Ian Dalziel, without whose comments it is possible this book may never have been written.

Acknowledgements

The Geological Society of London has permitted me to quote widely from their excellent collection of papers in 'Magnetism and the causes of continental break-up' (eds. Storey, B., Alabaster, T. and Pankhurst, R.), *Geol. Soc. of London*, Special Publication 68.

The AGSO *Journal of Australian Geology and Geophysics* has also permitted me to draw upon the various excellent papers by R.S. Dietz, E.M. Shoemaker and C.S. Shoemaker, R.A.F. Grieve and M. Pilkington, and other leaders in their various fields. The general editor was A.Y. Glickson who, in addition, presented interesting papers. Unfortunately several of the 'elder statesmen' at this conference failed to last out the century.

Many of the diagrams in the text relate to arguments presented by various workers. Wherever possible, I have obtained permission from the author(s). However, many of the cited authors have died, sometimes decades ago. Hugh Heard, a close friend, was one such who fell by the wayside.

The Geological Survey of Canada have mapped wonderful structures in the northern regions of that country. Many of the structures are best seen in mosaics compiled from earlier photographs. The Geological Survey of Canada generously permit such photographs to be published, with acknowledgement.

An unusual situation exists regarding a coloured photograph of a 50 km diameter impact structure in the Sahara. This photograph is shown on page 25 of *An Encyclopedia of Geology* (1977), edited by A. Hallam. Without doubt, the photograph of this wonderful structure, which must be attributed to NASA, was taken sometime prior to 1977.

Chapter 1

Earth and the solar system

1.1 Introduction

Throughout the history of the study of the Earth, geologists have attempted to understand the factors and forces which shape the surface of our globe. In this book, we shall attempt to summarise the current theories of how the gross surface layer of the Earth evolves tectonically. In this introductory chapter, however, we shall outline the historical evolution of the theory of plate tectonics and then review the three-dimensional nature of the Earth in order to understand the factors which drive the evolution of the plates. Finally, we shall compare the tectonics of the Earth with other terrestrial planets, highlighting the similarities and differences between the nature and evolution of the Earth and its neighbours. This will lead us to conclude that the Earth is likely to have experienced a greater number of impacts from meteorites and comets than is generally accepted by the geological community. In subsequent chapters, we shall investigate what effect these impacts may have had – and will continue to have – on the evolution of our planet.

1.2 Plate tectonics

Plate Tectonics is a unifying concept in the Earth Sciences. It was formulated in the 1960s and provides a partial, mechanistic basis for the previously established model of *Continental Drift*. Although he was not the first to suggest that continents move relative to one another, the concept was mainly developed by Wegener in the 1910s and 1920s. He suggested that the continents had moved relative to each other, based on a study of palaeoclimatology, continental geometry, palaeontological provinces and structural correlations. His ideas, however, were criticised (e.g. Jeffreys, 1942) and were not completely accepted by the geological community, because he was unable to present a convincing mechanism for continental movement. Nevertheless, as a student, the present writer was convinced by his arguments, and especially by those of DuToit (1937), that the continents had indeed parted and drifted one from another.

After the Second World War, new fields of geophysics opened up. Palaeomagnetic studies in particular contributed greatly in reinforcing the evidence for relative continental motion from apparent polar wandering curves (Figure 1.1). In parallel, oceanographers had established the existence of continuous ridges and spurs of submarine mountain chains that usually coincided with mid-oceanic areas which were also areas of concentrated seismic activity, while marine geologists were becoming aware that the ocean floors were youngest at the ridges and became gradually older as one traversed normal to the ridges. Thus, considerable evidence was accumulated to indicate that the continents had, indeed, drifted one from

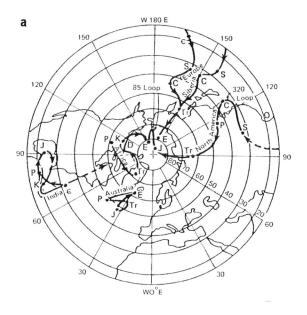

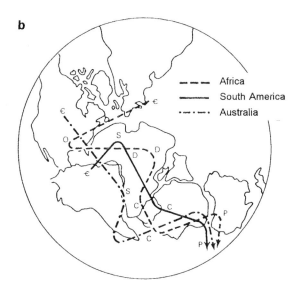

Figure 1.1 (a) Apparent polar wandering curves for N America, Europe (with offshoot for Siberia), Africa, Australia and India; compiled from many sources. E=Eocene, K=Cretaceous, J=Jurassic, Tr=Triassic, P=Permian, C=Carboniferous, S=Silurian and ε=Cambrian.
(b) Reconstructed polar wandering paths for Australia, S America and Africa. Continental fit around Antarctica (after Smith and Hallam, 1970). D=Devonian, O=Ordovician, other symbols as shown in (a).

another. Hess (1962) suggested that the ocean floor may be moving, possibly as the result of a form of convective motion. Raff and Mason (1961) and Menard (1969) showed that the ocean floor off the west coast of N America exhibited variations in strength of magnetic intensity, with a surprisingly regular degree of striping, the stripes being off-set by strike-slip faults or shear zones (Figure 1.2). Vine and Matthews (1963) argued that the striping on either side of a mid-ocean ridge was symmetrically arranged, which led them to put forward the idea of *ridge spreading*, which was in complete accord with the idea proposed by Hess (Figure 1.3).

One of the other important elements in bringing about the acceptance of plate tectonics was the seismic data that had been accumulated in the first 50–60 years of this century. Holmes (1965), for example, showed that the sites of the epicentres of earthquakes could occur anywhere in the world, but, when viewed in plan, were mainly concentrated in relatively narrow zones (Figure 1.4). Benioff (1954) (and Wadeti, who published only in Japanese) showed that, in the vicinity of deep oceanic trenches, where seismic activity was intense, earthquake data, when viewed in a vertical section, defined an inclined plane of activity

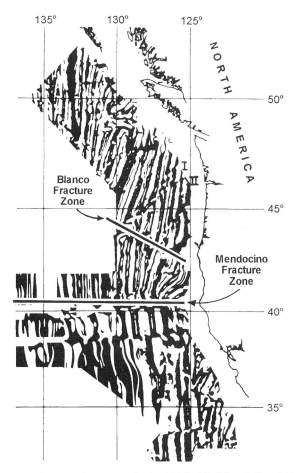

Figure 1.2a Linear magnetic anomaly patterns in the NE Pacific (after Raff and Mason, 1961).

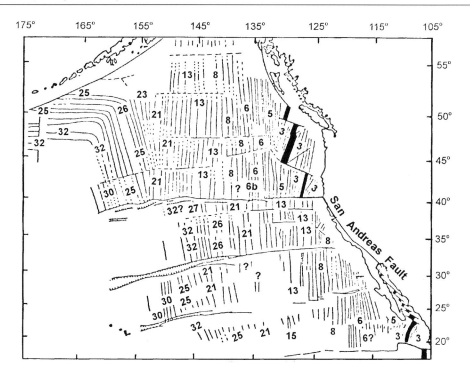

Figure 1.2b Magnetic anomaly map of the NE Pacific Basin (after Menard, 1969).

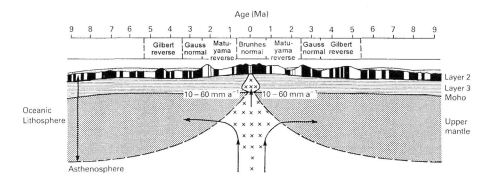

Figure 1.3 Sea floor spreading and the generation of magnetic lineation by the Vine–Matthews hypothesis.

which descended from the surface to a depth of 600–700 km (Figure 1.5). They inferred that the seismic data were the result of the oceanic lithosphere which was flexed, fractured and transported beneath the continental lithosphere.

These various conclusions were incorporated into a schematic three-dimensional diagram (Figure 1.6) by Isaaks *et al.* (1968), which combined the main features of sea-floor spreading and subduction. It illustrated how such spreading gave rise to the oceanic plate becoming older as it moved away from the ridge until, eventually, it was subducted to form a down-going slab. From this description, it can be inferred that plate tectonics is essentially a

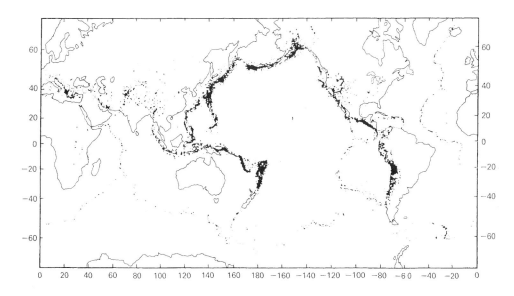

Figure 1.4 Distribution of world earthquakes 1961–69 (after National Earthquake Information Center Map NEIC-3005).

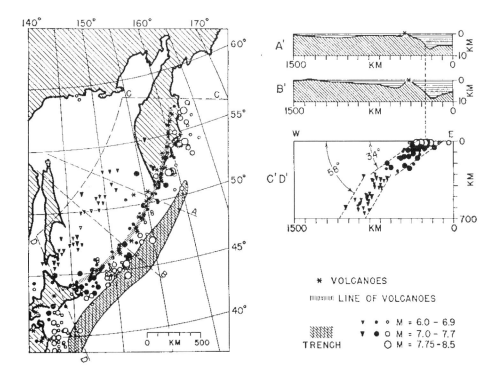

Figure 1.5 Map of the Kamchatka-Kurile volcanic arc with topographic profiles and a composite seismic profile (after Benioff, 1954).

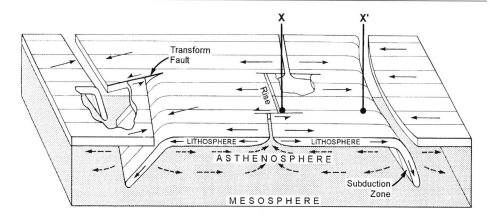

Figure 1.6 Schematic three-dimensional section diagram showing the major features of sea-floor spreading (after Isaaks *et al.*, 1968).

convective process, with hot, magmatic material being intruded at spreading centres, and old, cold slabs sinking back into the Earth's interior. The whole process is fundamentally driven by the cooling of the Earth, as it loses its primordial and radiogenic heat.

It was found that the vast majority of the known geologic and seismic data could be explained by defining 12 major plates (Figure 1.7) which interacted at their boundaries. There are three types of margin between plates: constructive margins (oceanic ridges), destructive margins (subduction zones or trenches) and conservative margins (e.g. transform faults). The motion of the plates can be described in terms of a pole of rotation via *Euler's 'fixed point' theorem*, which states that 'the most general displacement of a rigid body over the surface of a sphere can be regarded as a rotation about a suitable axis which passes through the centre of that sphere'. The motion of plates can be determined in a number of ways. These include palaeomagnetic reconstructions, as described by Smith *et al.* (1981) and available in software packages such as Atlas 3.3. Contemporary motions can also be determined by satellite laser ranging and very-long-baseline interferometry (VLBR) which uses the signals from quasars and terrestrial radio telescopes as receivers. Moreover, by assuming that certain volcanic features were associated with *stationary* hotspots (generally viewed as being generated by plumes upwelling from the mantle), the *absolute* plate motions were determined.

It can be seen from Figure 1.7 that plates range widely in size from the small Cocos and Nazca Plates to the huge Pacific Plate. These plates are viewed as being parts of the Earth's rigid lithosphere (discussed in more detail below) which moves over a more plastic (or viscous) deformable asthenosphere. The lithosphere contains the crust (both oceanic and continental) and part of the upper mantle. The areas of the twelve plates and the proportions of lithosphere containing continental crust are given, in alphabetical order, in Table 1.1.

The plate tectonic model is now generally accepted, and is routinely used to describe the evolution of oceanic regions. In addition, with this model, the general distribution of the continents over the past 500 Ma or so can now be inferred with some accuracy. Full and more comprehensive reviews of the foundations of the plate tectonic model can be found in many excellent texts by, for example, Kearey and Vine (1990) and Cox (1988). From these and other works, it is apparent that the plate tectonic model is good at explaining *what* happens to the Earth's lithosphere. However, it must be noted that, as Molnar and Stock

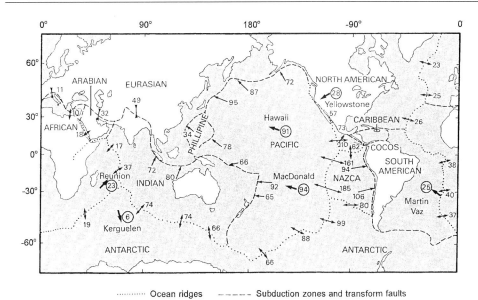

Figure 1.7 Plate movement vectors relative to fixed hotspot frame of reference (after Forsyth and Uyeda, 1978 and Morgan, 1972).

Table 1.1 Areas of the plates (in millions of km²).

Plate	Total area	Continental area
Africa	78.4	35.4
America N	58.8	35.0
America S	42.7	25.6
Antarctica	59.9	35.0
Arabia	4.9	4.2
Caribbean	3.5	1.4
Cocos	3.1	0.0
Eurasia	68.0	59.4
India	61.0	21.7
Nasca	16.4	0.0
Pacific	108.0	0.0
Philippines	5.7	0.0

(1987) says, 'the basic tenet of plate tectonics, rigid-body movements of large plates of lithosphere, fails apply to continental interiors, where buoyant continental crust can detach from the underlying mantle to form mountain ranges and broad zones of diffuse tectonic activity'. Furthermore, the theory behind plate tectonics is less well developed when it comes to explaining *how* or *why* plates move. These latter questions are fundamental to the paradigm and, of course, various attempts have been made to explain the manner in which various features of plate tectonics come into being. It is to these mechanistic problems that this book is addressed.

1.3 The structure and dynamics of the Earth

It is well established from seismology that the Earth can be divided into a crust (with depths of 10–70 km) that overlies the mantle (itself subdivided into an upper, transitional and lower part), which is separated at a depth of about 2880 km from the underlying core. Within the mantle, it appears that seismic velocity invariably increases with depth, except for a zone which may extend between about 75 and 200 km beneath the surface, known as the low velocity zone, where seismic velocities exhibit a local minimum. The zone is not ubiquitous, however, and is absent beneath old, cold continental cratonic regions. Throughout its depth, P and S waves can travel through the mantle (indicating that it is crystalline throughout), in contrast to the core, the outer part of which is liquid and unable to sustain shear waves.

The study of the composition and structure of the continental crust has occupied geologists for generations, and so will not be discussed further here. Cosmogenic, petrological and geophysical considerations enable the composition of the mantle to be estimated to be less silica-rich than the crust, with the upper mantle being peroditic (rich in olivine, pyroxene and garnet). Pressure-induced phase transformations in these minerals are responsible for the seismic discontinuities which mark the boundaries of the transition zone and the lower mantle. Experimental mineral physics leads us to believe that the lower mantle is predominantly composed of magnesium silicate perovskite ($MgSiO_3$), which may, in fact, depending on the exact compositional model of the planet one adopts, account for 40 per cent by volume of the entire Earth. The core is largely made of iron, alloyed with Ni and lighter elements such as O, S, Si or C. It is proposed that a crystalline inner core is slowly growing from the liquid outer core as the Earth cools, and indeed it is this cooling (coupled with the loss of radiogenic heat from the mantle) which drives the whole convective, plate tectonic evolution of the Earth's surface.

From the above, it is evident that the Earth can be divided into layers, which may be defined either seismically or chemically. In plate tectonics, however, we are greatly concerned with the rheological behaviour of our planet. That is to say, the model defines a rigid outer shell, or lithosphere, divided into a network of plates, that move over the underlying, plastic asthenosphere. We shall discuss these rheological units of the Earth in the following sections.

1.3.1 The lithosphere

The exact details of the nature of a tectonic plate are complex, and the meaning of the term lithosphere is not well defined. As we have already mentioned, the plate is formed at, and thickens as it cools during its movement away from, the spreading-ridge. From seismic studies of the oceanic plates, the boundary between the lithosphere and asthenosphere is often taken as the 4.3 km s^{-1} S-wave velocity contour. This boundary has often been considered to be the point at which partial melting of the mantle may start, and at which the low velocity zone (LVZ) begins. In this context therefore, the base of the lithosphere can be defined by an isotherm of about 1300°C. Recently, work by Kohlstedt *et al.* (1995), Hirth and Kohlstedt (1995) and others has shed doubt on the validity of viewing the LVZ as a zone of partial melting. Instead, they interpret the seismological (and rheological) behaviour of this zone to be owing to the presence of water dissolved into the mantle, forming olivines.

The thickness of the lithosphere in old, cold oceanic regions is about 100 km. The base of the lithosphere under continental areas is more variable and less distinct. Indeed, as already noted, the LVZ is not a globally ubiquitous phenomenon and is notably absent beneath Precambrian shield areas. As a result, defining the thickness of continental lithosphere is difficult. Estimates of the thickness of continental lithosphere have come, however, from the analysis of the elastic rebound from unloading, associated with the last deglaciation event (e.g. Peltier, 1984), which sometimes permit one to infer a thickness in excess of 200 km. Such deep roots to old continental areas can also be inferred from seismic studies (Archambeau *et al.*, 1969; and Hales, 1972) (see Figure 1.8). Tomographic analyses have also been conducted (Dziewonski and Woodhouse, 1987).

Anderson (1994), however, points out one basic confusion in the use of the term 'lithosphere'. The lithosphere is normally assumed to be the same as the strong, rigid outer shell of the Earth, but rock physics shows that silicate minerals have little strength at temperatures in excess of 700°C. He emphasises that only the relatively shallow, colder parts of the plate can be considered rigid. Thus, the lithosphere has a complex strength structure that is characterised by having a strong central layer, which is the part of the plate that can act as a stress guide. As we will see in later chapters, this complex, lithospheric structure is important in defining the response of the plates to changes in tectonic stress, and also is partly the reason why plate tectonics, in its simplest form, does not apply to continental regions. Further discussion of this problem may be found in Jackson (1998).

1.3.2 The asthenosphere

This is the non-rigid part of the Earth, which readily undergoes viscous flow, albeit at very slow strain-rates. As will be inferred from the above discussion, the detailed definition of

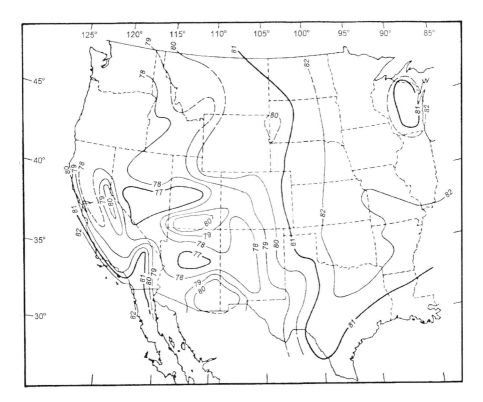

Figure 1.8a P_n velocity distribution in the western and central USA (after Archambeau *et al.*, 1969).

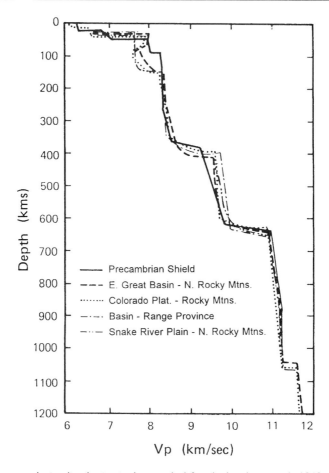

Figure 1.8b P-wave velocity distribution in the mantle (after Archambeau *et al.*, 1969 and Hales, 1972).

the extent of the asthenosphere is very poor. Some associate it with the low velocity zone, others with the upper mantle not within the plate, and some with the whole mantle not within the plate. Seismic tomographic studies permit one to infer that the entire mantle beneath the lithosphere is dynamic, and so perhaps this latter definition is to be preferred.

The problems with the definition of the terms lithosphere and asthenosphere reflect the problems in defining boundaries between parts of a rheological spectrum. Whether a material behaves as a rigid or plastic body depends on its Maxwell relaxation time, which in turn depends on its viscosity (η) and its elastic modulus (M). Viscosity is a measure of how easily flow occurs when a material is subjected to shear stress, and is defined by:

$$\eta = t/(d\varepsilon/dt)$$

where t is the applied shear stress and $d\varepsilon/dt$ is the strain rate, while the Maxwell relaxation time (T_r), is given by:

$$T_r = \eta/M$$

The relaxation time is a convenient approximation to the timescale of deformation, where the transition from purely viscous behaviour to purely elastic behaviour occurs.

Typical viscosities for liquids such as water, or high-temperature basaltic lava at the surface are:

$$H_2O = 10^{-3} \text{ Pa s}$$
$$\text{Lava} = 10^2 \text{ Pa s},$$

but the viscosity of the mantle is in the range $10^{21} - 10^{22}$ Pa s, as estimated from glacial rebound studies. More detailed results indicate that the LVZ has a viscosity of about 4×10^{19} Pa s, while the rest of the mantle has a viscosity that increases slightly with depth from 10^{21} to about 10^{22} Pa s. Given that the shear moduli at the base of the lithosphere are about 60 GPa, its Maxwell relaxation time is about 10 to 100 years.

The viscosity depends on the mechanism of flow. Water is a liquid with no long-range atomic order, but the mantle is made of crystalline material. At high temperatures, once a crystal is stressed past the elastic limit, crystal creep occurs, either by dislocation glide and/or climb or by diffusional flow involving movement of vacancies. As both processes need atomic motion, they are thermally activated, and so the viscosity of rock or crystals is highly temperature dependent. The hotter, lower lithosphere and underlying mantle has a large but finite viscosity, and so can be plastic at geological strain rates, but the upper, cold, oceanic crust covered lithosphere must have a much greater viscosity, since the basic assumption of plate tectonics requires it to behave as a rigid elastic material with a relaxation time in excess of 10^8 years. It should be noted, however, that it is obvious that the continental crust undergoes deformation on a much shorter time scale than this, which has led some workers (for example, see England, 1996) to treat continental crustal material also as a fluid.

1.3.3 Convective instability and mantle dynamics

As we have emphasised, plate tectonics is a convective process, thermally driven by the negative bouyancy of old, cold, lithospheric slabs which sink into the interior of the hot mantle at subduction zones. To accommodate this sinking, there must be motion below the surface too. This dynamical behaviour of the mantle has been discussed at great length in several major texts (e.g. Jackson, 1998), and the effect of mantle dynamics on plate motion has recently been reviewed by Lithgow-Bertelloni and Richards (1998). But in order to discuss how these mantle processes affect, or are affected by, plate process, we must briefly consider the fluid dynamical behaviour of the Earth's mantle.

Many descriptions of plate tectonics assume that the mantle convects actively ('like a pan of soup') and plates are driven by this motion. Most of the results outlined below have been derived from the analysis of simple systems, and although we expect that the general principles will always hold true, we must recall that the Earth is not 'like a pan of soup'. Important factors which cannot be ignored and which are still the focus of fluid dynamical research include (a) the spherical geometry of the Earth, (b) the role of internal radioactive heating, (c) the effects of pressure and temperature on density, and (d) the non-Newtonian nature of crystalline creep.

The behaviour of a homogeneous fluid heated uniformly from below was studied experimentally by Benard, in the 19th century, who found three stages in the evolution of heat flow in the system. With a small thermal gradient, there is no convection, and heat passes through the system by conduction. As the thermal gradient increases, stable convection is initiated, with well-defined convection cells, with an aspect ratio typically of 1 : 1. Finally, with very large thermal gradients across the system, chaotic, turbulent, time-dependent convection develops, and the regular pattern of convection cells is broken up.

The theory to describe this system was developed by Rayleigh, also in the 19th century, and the behaviour can be described in terms of the dimensionless Rayleigh number, Ra. This number reflects the balance between buoyancy forces which promote convection, and conduction effects which inhibit it. Ra can be defined as

$$Ra = a.dT.d.g.z^3 / \eta.k$$

where a is the coefficient of volumetric expansion, dT is the temperature gradient across the system, d is the density, g is the acceleration due to gravity, z is the depth of the convecting fluid, η is the viscosity and k is the thermal diffusivity. Theory and experiment show that systems will convect only if Ra is greater than about 10^3, which is known as the critical Rayleigh number, Ra_{Crit}. If one evaluates Ra for the mantle, one finds that it has a value in the range 10^5 to 10^7, which permits one to infer that mantle convection is turbulent.

In a convective system, the average temperature in the central part of the cell is relatively constant (or at least adiabatic), but there are upper and lower thermal boundary layers, where fluid flow is essentially horizontal and in which heat flow is locally dominated by conduction. The thickness, t, of the thermal boundary layer (in a system of constant viscosity) is given by

$$t \approx 1/2.z(Ra_{crit}/Ra)^{1/3}$$

For z of about 2800 km, Ra in the range 10^6–10^7, and Ra_{Crit} about 10^3, then t is approximately 100 km. Thus, from such a simple model of mantle convection, one could semi-quantitatively predict the thickness of the lithosphere, if this is seen as the upper thermal boundary layer in the convective system. Futhermore, with reasonable estimates of about 3000 K, this model would predict lithospheric temperature gradients of about 15 K km^{-1}, which again is in good accord with average global values. In this view of mantle convection, the D' zone could be identified with the lower thermal boundary layer. Stability analysis of such a lower thermal boundary layer suggests that it would be unstable, which has led to the suggestion that the D' zone is the source of the plumes inferred from the study of hotspots. The treatment of the orgin of plumes is discussed in greater detail in Jackson (1998).

One issue which is currently the subject of considerable debate is whether the mantle undergoes layered convection, in which the boundary between the transition zone of the mantle and the lower mantle acts as a barrier to convection. The need to have a layered mantle comes from the geochemical analysis of mantle-derived magmas, which leads one to infer that several different and chemically isolated mantle sources are needed to explain the Sr, Pb, Nd, etc. isotope data found. The argument is that if the mantle convected as a whole, as opposed to in layers, chemical mixing would occur and a homogeneous mantle with no isotopically distinct sources would be created.

Fluid dynamical modelling and mineral physics have combined to provide possible reasons why layered convection may occur. First, it should be recalled that our knowledge of mineral physics is still imprecise, and from seismic data alone, we cannot exclude the possibility that the lower mantle could be chemically distinct from the upper mantle and transition zone, and in particular it could be more Fe rich. Such a concentration of higher atomic weight material in the lower mantle could prevent effective mantle mixing. Furthermore, the phase change marking the 670 km boundary is endothermic which, as thermodynamic and fluid dynamical analysis shows, will inhibit convection across this interface. A full analysis of these factors of structural and chemical density effects is not easily made, and full computational models are needed. An early study is described by Christensen (1989), who found that the mantle lay very close to the transition between layered and whole-body convection. A more recent study by Tackley *et al.* (1993) was carried out using a spherical shell model. They found that a three-dimensional flow pattern was produced containing cylindrical plumes and linear decending sheets. The dynamics were dominated by the accumulation of downwelling, cold material above the 670 km boundary, which periodically avalanched into the lower mantle. Similar results have been presented by Solheim and Peltier (1994), who found that their simulation was *Ra* number dependent, and that the periodicity of the avalanches was controlled by instabilities which developed in the internal thermal boundary layer that develops when the convection is layered.

The question which now must be addressed is whether these models are supported by direct observation of the mantle. In recent years, seismic tomography (see, for example, Woodhouse and Dziewonski, 1989; Romanowicz, 1991) has given an increasingly resolved view of the internal thermal structure of the mantle, and it is now possible to correlate the observations with fluid dynamical models. Seismic tomography gives a three-dimensional image of seismic velocity of the Earth's interior, which can reasonably be interpreted in terms of the thermal structure of the Earth. At shallow depths, the mantle beneath ridges is hot and under continental shield areas it is cold, but these anomalies do not necessarily persist below about 300 km. It appears that the distribution of hotspots correlates strongly with anomalously hot regions at the core–mantle boundary (CMB), supporting the suggestion that at least a significant number of hotspots are the result of plume initiation in the unstable D' zone. There is a ring of high velocities extending through the lower mantle around the rim of the Pacific, apparently correlated with the circum-Pacific subduction zones. The geoid anomalies correlate with the general distribution of velocity highs and lows in the mantle. All of these support the view that tomography images mantle convection.

Woodward *et al.* (1994) have recently compared seismic tomographic and fluid dynamical flow models to investigate the role of the 670 km discontinuity. Their results are consistent with a picture of convective flow in the Earth's mantle, in which whole mantle style flow is dominant at present, but in which phase transition-induced, localised layering also exists. There are slow (i.e. hot) features that extend vertically through the 670 km boundary beneath, for example, the E Pacific, Indian Ocean and mid-Atlantic spreading centres, while local discontinuities exist where there are large downwellings, such as under the W Pacific, S America and Alaska. Thus, fluid dynamics and seismic tomography seem now to indicate that the upper and lower mantle do mix, but perhaps aspects of the circulation are layered. Recently, Allegre (1997) has suggested that the isotopic geochemical data and geophysical data might be reconciled, if there had been a transition from layered convection to the whole-mantle convection seen today, some time in the past 1.0 Ga. If the transition had been this recent, the geochemical reservoirs would still be relatively isolated. He points out that the geochemical signal reflects the whole history of mantle evolution, but seismic tomography is showing the convective regime as it is today.

1.4 Factors determining plate motion

There is no doubt that plates play a vital and integral part in mantle convection. What is less clear, however, is what actually controls the surface motion of the plates. Indeed, recent papers by Wen and Anderson (1997) and by Lithgow-Bertelloni and Richards (1998) both emphasise that the principal problem still facing geodynamics is how to explain global plate motions. The traditional view is that plates are pulled towards trenches by the sinking, cold lithosphere, and pushed away from ridges by the thickening oeanic lithosphere as it cools. Turcotte and Oxburgh (1967) showed that a simple model could be constructed which qualitatively described such a system, and subsequently, many workers established models involving forces such as slab-pull, ridge-push, etc., which were considered to be line forces which were acting along plate boundaries. However, more recent work by Wen and Anderson (1997) and by Lithgow-Bertelloni and Richards (1998) has attempted to integrate descriptions of mantle flow and predict and rationalise plate motion.

In the broadest sense, plates are part of the convective cycle. However, many researchers argue that their rigidity and the presence of a low-viscosity layer underlying most plates suggests that the motion of plates can, in many ways, be decoupled from some aspects of the convective style of the underlying mantle. Obviously the location and penetration of down-going slabs must couple with the flow directions of the mantle, but there is no strong evidence to support the suggestions that the distribution of ridges reflects the deep seated, first order upward flow of major mantle plumes. Thus, Davies (1988) observed 'Plate geometry and kinematics generally reflect the mechanical properties of the solid lithosphere rather than those of the fluid mantle underneath'. In which case, what are the forces which act on plates, and how do they affect tectonic processes? Traditionally these forces have been divided into *basal* and *edge* forces.

1.4.1 Basal force

The basal (F_B) force (either a driving or a retardational force depending on the differential motion) is the force acting on the base of the plate resulting from viscous coupling with the asthenosphere. The continental and oceanic lithosphere may behave differently, as the LVZ is less well defined below continents, and the continental lithosphere may also have a relatively deep, possibly undulating, 'keel' (Figure 1.9). If F_B is a retarding force, as one might expect if plate motion is independent of mantle flow directions, then it might be expected to resist plates containing a large proportion of continental material more than plates dominated by oceanic lithosphere. As a result, one might expect plates with a high percentage of continental crust to move slowly.

For current plate geometries this is generally true, for major plates see Table 1.2 from which it is reasonable to conclude that drag at the base of continental lithosphere is a major retarding force on plate motion. If F_B is a retarding force, what then are the driving forces?

1.4.2 Edge forces

There are a number of forces which act on the edge of plates including ridge-push, slab-pull, collisional force, suction forces, transform fault forces, etc. The roles of these forces have been analysed in a number of studies (e.g. Forsyth and Uyeda, 1978; Richardson, 1992; Bott, 1993). A number of early analyses favoured the role of slab pull in determining plate motion.

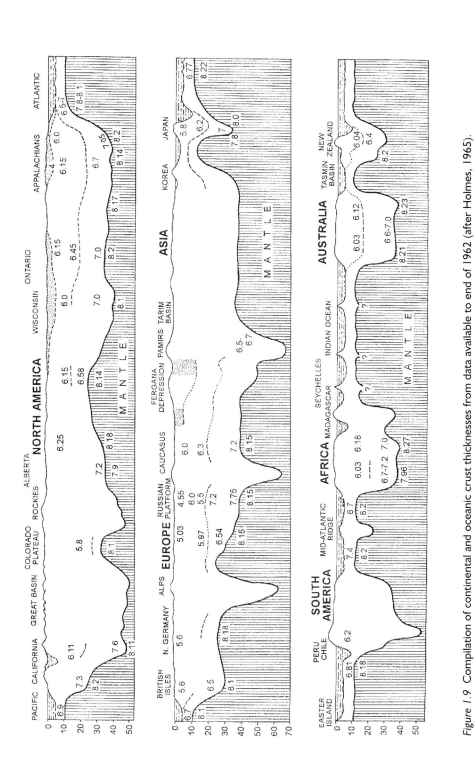

Figure I.9 Compilation of continental and oceanic crust thicknesses from data available to end of 1962 (after Holmes, 1965).

Table 1.2

	Absolute velocity (cm y^{-1})	Continent (per cent)
Eurasian	0.7	74
North American	1.1	60
South American	1.3	49
Antarctica	1.7	25
African	2.1	39
Arabian	4.2	90
Indian	6.1	25
Philippine	6.4	0
Nazca	7.6	0
Pacific	8.0	0

Table 1.3

	Absolute velocity (cmy^{-1})	Circumference (km)	Effective trench (km)	%
Eurasian	0.7	42100	0	0.0
North American	1.1	38800	1000	2.6
South American	1.3	30500	300	1.0
Antarctica	1.7	35600	0	0.0
African	2.1	41800	900	2.1
Arabian	4.2	9800	0	0.0
Indian	6.1	42000	8300	19.8
Philippine	6.4	10300	5200	27.8
Pacific	8.0	49900	11300	22.6

Thus, it was suggested that the stress guide of the lithosphere transmitted the force of the cold, sinking slab to produce the surface motion. Forsyth and Uyeda inverted observed surface velocity fields and concluded that slab-pull was an order of magnitude more important than any other force, and that oceanic plates attached to substantial amounts of down-going slab move with a near 'terminal velocity' at which the gravitational body force pulling the slab downwards is balanced by the resistive forces acting on the plate. Their inference is further supported by the simple observation of the high degree of correlation between absolute velocity and length of trench boundary of plate as a percentage of total circumference, see Table 1.3.

This would be all well and good, but more recent studies have suggested that in fact the story is much more complex that it appears. The conclusions of the World Stress Map Project as reported by Zoback (1992b) are that first order midplate stress fields are the result of compressional forces applied at plate boundaries, primarily ridge push and continental collisional forces, rather than the tensile force that would be expected from a predominantly slab-pull determined stress field. Furthermore, she reports no evidence of large, lateral stress gradients which would be expected across large plates, if simple resistive or driving basal drag (parallel or antiparallel to absolute motion) controlled the intraplate stress field.

Richardson (1992) analysed the torques acting on plates, and concluded that there was a strong correlation between ridge torque poles and the motion of the N American, S American, Pacific, Cocos and Eurasian plates (although Lithgow-Bertelloni and Richards, 1998 continue to contend that slab-pull effects are more dominant). Richardson (1992) also points out that the ridge torque directions agree well with the orientations of maximum horizontal stresses in large plates, and he concludes that slab forces must be largely balanced within the subducted slab itself, and that slab-pull forces contribute relatively little to the deformation or stress distribution in the surface plates. Thus it appears that ridge push is a very important mechanism in determining intraplate stress levels, and, if Richardson's conclusions are accepted, plate kinematics. We shall see how these stresses can be effectively developed in Chapter 3.

Work on the major problem of geodynamics – determining the causes of plate motion – is still continuing. The dichotomy between models that explain plate motion in terms of forces deriving from within the plates themselves, and those that attempt to describe plate kinematics in terms of mantle flow, still has to be resolved. The major problem probably lies in effectively describing the complex rheological changes that occur as the lithosphere first cools as it forms at a ridge, and then warms up after its subduction.

In addition to the problem of the factors which determine plate motion, there are several other regions where the theory behind the plate tectonic model is patchy or even non-existent. These include (i) the cause of the apparently sudden changes in plate motion which define so many geological stage boundaries (Lithgow-Bertelloni and Richards, 1998), as exemplified by the great bend in the Hawaiian-Emperor seamount chain, (ii) the reason and mechanism for continental rifting, which for example gave rise to the break up of Gondwana, and (iii) how subduction zones are formed. In this book we shall attempt to provide solutions to these problems.

If we are truly to understand how our planet evolves, it is essential for us to be able to explain and model these processes. Obviously, much information which will be vital to our understanding of such plate tectonic processes will come from closer observation of the Earth. However, some insight into how our own planet evolves may also come from the study of other bodies in our Solar System. Of particular interest are questions such as: is plate tectonics a ubiquitous mechanism for the evolution of terrestrial planets?, and: do other planets exhibit tectonic features not seen on Earth, and if so why? In the next section we shall review some of the observations which are key to answering these questions.

1.5 Tectonic evolution of terrestial planets

As we have seen, plate tectonics is now accepted as the major mechanism for the crustal evolution of the Earth, but the question still remains as to whether plate tectonic processes have played, or do still play, a role in the evolution of other terrestrial planets or satellites in the Solar System. Observation of the Earth's twin, Venus, reveals a very different planet, with a CO_2-dominated atmosphere generating surface temperatures of 477 °C. Notably, Venus has a considerably different global hypsometric distribution from that measured on Earth (Rosenblatt *et al.*, 1994). The Earth has an essentially bimodal hypsometric distribution (associated with oceanic and continental crusts) while Venus is essentially mono-modal, with its peak being coincident with the terrestrial oceanic peak. Unlike Earth, gravity anomalies on Venus correlate with high topography, and so the highlands of Venus can be best modelled, not necessarily in terms of compositionally compensated topography (as on Earth), but rather as dynamically supported topography (Bindschadler *et al.*, 1992). The equatorial highlands of Venus are thought to be associated with ascending mantle plumes, and on this basis Kiefer and Hager (1991) have successfully, simultaneously, modelled the topography and geoid associated with this feature. No extensive networks of ridge or trench systems have been identified, although locally such features can occasionally be seen.

As put forward by Arkani-Hamed (1993), it seems that the classical concept of plate tectonics – namely that rigid plates cover the surface of the planet and major motions only take place along their boundaries – does not seem to operate on Venus. The observation of intensive tectonism over the majority of Venus and the strong evidence for the coupling of Venusian mantle convection to the lithosphere permits one to suggest that the lithosphere of Venus undergoes significant deformation, and does not act as a stress guide as it does on Earth. A deformable mechanical upper boundary layer at the top of the mantle convective cell, with a crust being strongly deformed and thickened and the mantle part being dragged by mantle convection, seems the probable model for the tectonic regime on Venus (Lenardic et al., 1993). This difference in behaviour between Earth and Venus which could be explained by the high Venusian surface temperatures, would imply a much thinner and so weaker lithosphere on Venus relative to Earth, while the high temperature might also suggest a dehydrated planet giving rise to an anhydrous mantle and absence of a low viscosity layer under the Venusian lithosphere. (But see McKenzie et al., 1992 and Turcotte, 1993 for contrasting views.)

Another striking feature of the Venusian surface is the existence and preservation of a large number of impact craters. More than 900 craters with diameters ranging from 1.5 km to >300 km have been detected. They are apparently distributed evenly over the Venusian surface. It has been inferred that the average surface age of Venus is <500 Ma, and that there is no trace of crust any older than this (Strom et al., 1994). This has been explained both in terms of (a) 'catastrophic resurfacing hypotheses' involving planetary-wide, periodic lithospheric instability or time-variable mantle convection instabilities (Herrick and Parmentier, 1994), and (b) more tectonically evolutionary hypotheses, based, for instance, on gradual volcanic or tectonic resurfacing, which is, perhaps, understandable in the absence of significant 'continental material' and a weaker lithosphere (see Solomon, 1993, for a review).

Thus, although attempts have been made to relate certain features of the surface of Venus (McKenzie et al., 1992) and more recently of Mars (Sleep, 1994) to plate tectonic processes, it is evident that plate-like processes are either absent or much less important on other planets than they are on Earth. This may reflect the unique combination of size, temperature and, perhaps, especially the hydrated state of our planet. Observations of other planets, however, highlight one other important mechanism for tectonic evolution not commonly discussed in the context of terrestrial tectonics – namely impact related phenomena. Only recently has the role and significance of meteorite impacts on global evolution begun to be been taken seriously. Alvarez et al. (1980) postulated impacts at the K/T boundary to be responsible for the large number of extinctions observed at that time. Hildebrand et al. (1991) identified a 180 km diameter 65 Ma old impact crater at Chicxulub, which they suggested could have been produced by such an impact event. In fact, McLaren and Goodfellow (1990) estimated that the Earth will be hit by a meteoritic body of 5 km diameter more than once every 10 Ma. Such an impact would generate a significant crater (the size of which depends on the impact velocity, meteorite density, etc.) and greatly stress the environment. Over 150 impact craters have now been identified on Earth (the largest being about 800 km in diameter; Girdler et al., 1992). Indeed, some of the larger impacts have been invoked as being the cause of major tectonic events such as the break up of Gondwana (Oberbeck et al., 1993) and the trigger for the 65 Ma old continental flood basalt flows of the Deccan Traps (Negi et al., 1993). As we shall outline later in this book, there are definite mechanisms by which impacts on the Earth could be responsible for some of the outstanding, 'no-theory areas' in plate tectonics referred to above, namely the rapid changes in plate motion, the initiation of subduction and the rifting of continents. However, before we can assess the role that impacts may have played in the evolution of the Earth, we must be clear about the nature and origin of the possible impactors, which are thought likely to be either derived from the asteroid belt or comets from the Oort cloud.

1.6 Sources of impact bodies

In this section we shall follow a historical approach in that, until quite recently, it was assumed that the most important, indeed, almost only source of impacting bodies were asteroids that had originated in the belt of such bodies which are situated between Mars and Jupiter. Comets have been recorded for thousands of years and have been considered to be harbingers of famine and desease, but not usually of impacts. Their importance regarding the development of major impact structures has been a very recent innovation.

1.6.1 The asteroids

The asteroids, or minor planets, in the solar system occupy a zone, or belt, between Mars and Jupiter, which extends from 2.2 to 3.3 AU from the Sun (i.e. they are spread over a belt which differs in radius by about 1.5×10^8 km). They are thought to have developed as the result of collisions between originally larger bodies that circulated between Mars and Jupiter. Individual asteroids, in this belt, are non-luminous rocky bodies. Those with diameters of greater than 1.0 km probably total about 100,000. Only about 200 have a diameter exceeding 100 km. The largest is Ceres, which has a diameter of 1000 km. The majority of the planetoids in this belt appear to have a reasonable stable configuration relative to their neighbours. However, they may, from time to time, be forced out of this stable configuration by other asteroids, sometimes known as the Trojan Group, which have an elliptical orbit which extends from near the Sun to beyond the Asteroid Belt. The total number of asteroids in this group may be as many as 5000. A smaller number, of about 1200–1800, termed the Apollo Group, cross Earth's trajectory and are colloquially called 'Earth-crossers'. These, which we shall discuss later, are the asteroids which are most likely to cause impacts on Earth. The Trojan and Apollo asteroids are to some extent self-sustaining as regards numbers, in that they will tend continuously to disrupt the main asteroid belt.

From the relatively small meteorites believed to be derived from the asteroid belt that strike Earth at relatively low velocities, it can be inferred that there are three types of asteroids, namely (a) stony, (b) iron and (c) hybrids containing both iron and stony material. Stony meteorites, which comprise 93 per cent of those recorded on Earth, consist mainly of olivine and pyroxene. The third type of meteorite consists of a combination of iron plus the minerals found in the other two forms of meteorites, with an iron content which is usually less than 30 per cent.

The Apollo, or *Earth-crossing* asteroids, of which at least 110 are currently known (though the numbers are continually increasing) were named after the first asteroid with an Earth-crossing track, which was discovered in 1932. One of the closest observed approaches to the Earth occurred in 1937 when Hermes passed at a distance of 780,000 km. Because such Earth-crossers have only been discovered relatively recently, one may reasonably infer that there are a far greater number of undiscovered Earth-crossing asteroids. In 1997, Shoemaker estimated the number of such asteroids with a diameter of greater than 0.7 km to be 800 +/– 400. This latter figure has since been up-graded to over 2000, to accommodate newly discovered Earth-crossers.

Shoemaker (1997) has also estimated that the *average* impact velocities, from the first of these groups is 24.6 km s^{-1}, though, of course, the possible range of impact velocities is large.

Two large bodies are shown in Figure 1.10. These are asteroids Ida and Gaspra which were photographed from the spacecraft Galileo. The approximate dimensions of Ida are 56 km by 24 km by 21 km, (equivalent to a sphere 33.6 km in diameter). It will be noted that the cratering which, on Ida, is particularly well developed, is further testimony of the importance of impacts in the evolution of planets and satellites in the Solar System.

Scientists in the recently convened *Earthwatch* project have suggested that there is a second, hitherto undetected, asteroid belt (of which Earth is an element) concentrated with 'Earth-approaching' asteroids. Of the unknown total number, 15 are listed (Rabinowitz *et al.*, 1994). In general, it is likely that the majority of Earth-approaching asteroids are small and have a low differential velocity relative to Earth.

Chyba (1993) argues that recent observations show that Earth-approachers which have a diameter of less than 40 m are 10–100 times more likely than was originally inferred from earlier data. The explosions attendant upon these objects entering the atmosphere occur at 10–45 km above the Earth's surface. Only objects with an explosive energy in excess of 30 megatons are likely to break through to the Earth's surface. He demonstrates that 'spacewatch' objects with diameters of up to 50 m have kinetic energies of less than about 10 megatons high-explosive equivalent. Consequently, although collisions with the Earth's atmosphere are relatively frequent, they do not usually possess the potential to generate craters.

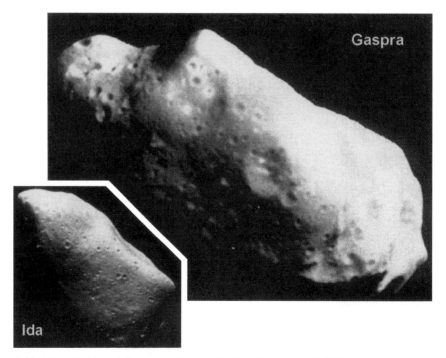

Figure 1.10 Asteroids Ida and Gaspra as photographed by the spacecraft Galileo.

In this context it is interesting to cite the evidence based on analyses of infra-red images, collected by United States spy satellites which were recently declassified by the Pentagon. Between 1975 and 1992, 136 *flashes* were detected in the upper atmosphere (Chyba, 1993). These flashes have been interpreted as being the result of small, potentially-impacting bodies burning up in the upper atmosphere. It has been estimated that the bodies had a diameter of about 60 m. Had one of these possessed a diameter of about 100 m or more, and had it been travelling at a relative velocity to the Earth of about 10 km s^{-1}, it would have survived passage through the atmosphere and hit the Earth's surface. Such an impact could have been sufficient to destroy a large city. Further, it can be concluded that the cited average rate of 8 per annum of these upper atmosphere events is conservative because the spy satellites picked up the events purely by chance. Of course, smaller bodies with a low relative velocity are successful in hitting the Earth's surface, but these are not 'destructive'.

A radar image of asteroid *4179 Toutatis*, was obtained by S. Ostro on 8 and 9 of December 1992, when the object(s) were at a distance of about 4,000,000 km (Figure 1.11). Like other near-Earth asteroids observed by Ostro and his collaborators, it can be seen in this figure that Toutatis is composed of two distinct objects in contact with one another. The two objects have a diameter of 2.5 and 1.5 km respectively, and both show impact craters upon their surfaces (Grieve, 1993). Were two such coupled bodies to make contact with the Earth's surface they could give rise to 'companion craters', examples of which are known to occur in the Canadian Shield.

1.6.2 Comets

A comet is thought to be comprised mainly of snow or ice with a relatively small nucleus consisting of a conglomeration of material that may range in size from dust to large blocks, the composition of which is comparable with that of the asteroids. Indeed, Clube and Napier (1982) suggest that as comets pass the Sun, significant melting or sublimation of the ice

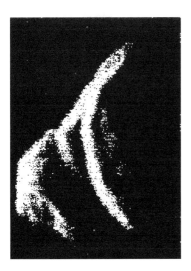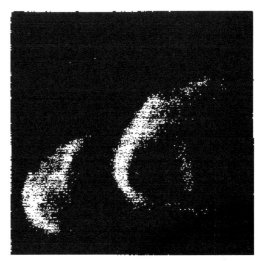

Figure 1.11 Radar images of 'Earth-approaching' duo-asteroid 4179 Toutatis, obtained by S. Ostro of the Goldstone Deep Space Communications Complex in California on 8/9 December, 1992. The objects, with diameters of 2.5 and 1.5 km respectively, were at a distance of about 4,000,000 km.

takes place and the volatile gas produced is blown away by the solar wind. After a number of cycles, the ice is eventually completely lost, so that the comet (if it is not destroyed by impact) evolves into a body which could closely resemble an asteroid.

Comets have been observed and recorded by man for well over a thousand years. Some are thought to be single-pass events, while others, such as Halley's Comet, return at well-regulated intervals. Delevan's Comet has the longest estimated return time of 24,000,000 years, with an aphelion of about 28×10^{12} km.

One of the sources for comets, particularly those which are non-returning or have a long return-time, is the *Oort cloud* which lies in a zone that starts beyond the planets, but which does not extend beyond 0.5 parsec (approximately 100,000 AU, where an AU is the radius of the Earth's orbit around the Sun). Originally, the existence of the Oort cloud was met with some scepticism, however, its existence is no longer in dispute.

The population of the Oort cloud was estimated from records of comets which were nearly, but not quite unbound (i.e. hyperbolic orbits) and is estimated at $2–4 \times 10^{12}$ proto comets. Hence, although as many as 200,000 such bodies may be 'shaken loose' from their orbits by the influence of extra-solar bodies or gas clouds, which may occur periodically, about every 30 Ma, the Oort cloud will continue to form a reservoir of potential Earth-crossing bodies to a time when the Earth can no longer sustain advanced forms of life.

It is possible that the comets which are dislodged from this cloud are generally single-pass events around the Sun, or returning comets of long periodicity. In addition, it is probable that the comets from the Oort cloud may also dislodge icy asteroids from a second source of comets; namely the Kuiper Belt which lies in the general orbit of Uranus and Neptune. Comets from this second source are likely to provide the majority of comets with relatively short-return times of the order of 100 years.

Only in recent years has it been realised how important comets are in generating impacts on Earth. For example, Shoemaker (1997), in his Fermor Lecture at a meeting of the Geological Society of London, indicated that 'the available evidence suggests that, late in geological time, comets have produced about half the impact craters on Earth >20 km diameter and nearly all >100 km diameter'. At the same Meeting, Bailey and Emel'yanenko presented results which support Shoemakers statement. They indicate that long-period comets are captured from the near-parabolic flux and also from orbits with the aphelia in the dense core of the Oort cloud. These authors predict that there are of the order of 4000 such comets. This, they point out, is at least 100 times more than the observed number of Halley-type comets. From this observation they infer that there should be many undiscovered inert Halley-type asteroids or compact streams of cometary debris still to be found in the inner solar system.

Clube and Napier (1982) suggested that Halley-type comets in many passages around the Sun were subject to devolatisation, so that the ice mantle was eventually stripped away, leaving what Bailey and Emel'yanenko termed Halley-type comets.

Fortunately for mankind, a really major impact on Earth appears not to have occurred during the period of recorded history. Subconsciously, or even consciously, therefore, such events are categorised as rare or remote occurrences and tend to be dismissed from our thoughts. As we shall see, impacts which produce a 1.0 km, or greater, diameter are only rare events when measured in terms of the recorded history of mankind. They are far from rare events when measured by more usual geological time-scales.

The two most important, practical reasons for the lack of acceptance of the importance of major impact structures were, firstly, the lack of a sufficiently large 'world view', so that extensive circular features were not readily recognised. (This difficulty has now, of course, disappeared with the arrival of satellite imagery.) Secondly, relatively small features, which have now been shown to be of impact origin, were, in the main, for many years attributed to volcanism, or *crypto-volcanism*, where *crypto* means secret, hidden or concealed, so that 'crypto-volcanism' really translates into 'unknown-mechanism'.

There is a well-documented history of events which demonstrates beyond reasonable doubt that Earth has experienced, and continues to experience, bombardment by meteoric and cometary material. Much of this material falls as *dust*. The arrival of relatively small iron meteorites at the Earth's surface during historic times is well documented (Clube and Napier, 1982; Lewis, 1996; Verschuur, 1966). Such events have, as yet, given rise to the recorded death of relatively few human beings. For example, on the 10 March 1987, a certain David Leisure was rendered unconscious by the air-blast of a passing meteorite before it broke; and a nearby horse had its skull crushed by a fragment of that same body after it shattered. Other, more recent events which have given rise to damage of houses and motor vehicles are also documented by Lewis.

A recent event in which an extra-terrestrial body penetrated deep into the Earth's atmosphere and exploded at about 7 km above the Earth's surface occurred in the tundra of Siberia on 30 June 1908 in the valley of the Tunguska River. The air-blast of this event flattened trees in a 'butterfly' pattern over an area of about 2000 km^2. Again, as far as is known, this event caused no deaths, though the tent belonging to two brothers, who were sufficiently near the event, was 'blown' away. We shall comment on this event in a later chapter.

In the *Introduction* to his book, Lewis presents a reconstruction of events which followed from a Tunguska-like event in the vicinity of Constantinople in AD 472. He concluded the reconstruction with the statement, 'Many thought it was the end of the world...'. Lewis also cites 94 different incidents between 1420 BC and 1992 and the property damage, injuries and deaths caused by the fall of meteorites. Early events are well spaced in time and become progressively more frequent as we approach the present day. One should not, of course infer from this that the rates of impact are increasing. Thousands of years ago, the Earth's population was much smaller than it is today and also, only events which occurred near well- populated sites were likely to be recorded, or entered into verbally repeated, local 'mythology'.

However, the most important event of this type, that surely must impress the Earth-scientists, occurred not on Earth but on Jupiter. We refer, of course, to the brilliant work that led to the discovery of the Shoemaker-Levy 9 meteoritic bodies and the predicted impacts that were eventually made with Jupiter and which were seen world-wide on television programmes (Figure 1.12), (Benka, 1994; Verschuur, 1996).

1.6.3 Impact flux

The amount of extraterrestrial material that falls to the Earth's surface has been assessed in a number of ways (Grady, 1997). These include the micrometeoric dust that collects in the ice-sheets of the polar regions and the small meteoritic pebbles that accumulate in some desert areas of the world, such as the Nullarbor plain of S Australia. Recently, assessments have been made on the micrometeoritic flux above the Earth's atmosphere by studying the small blemishes and pock-marks that develop on panels of communication satellites with panels which remain oriented to outer space. Iron meteorites occasionally fall to Earth, but these rarely exceed a few tens of centimetres in diameter. Lewis (1996) and Verschuur (1996) cite examples when such iron meteorites have caused social disturbances.

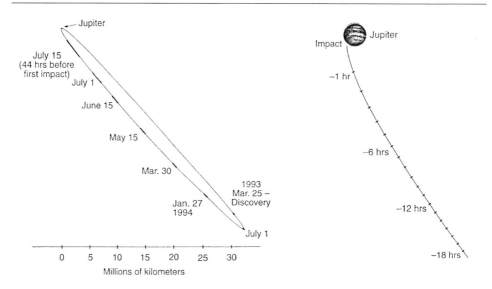

Figure 1.12 The 'once in a millennium' Shoemaker-Levy 9 impact event on Jupiter, as forecast by the named researchers and captured on television.

However, in this book we are concerned with major meteoritic impacts and the mildly sceptical reader will care little for the 'dust and small pebbles' that arrive from space, so we shall not deal further with such micro- or small-scale meteorites.

The obvious starting point in estimating the flux of moderate, to large-scale, meteorites and comets that make contact with the Earth is the cratering evidence which has come to light in the 20th century. At the time of writing, there are only 150 craters and impact features that are accredited with the rating *certain*. Of these, only three certain impact craters are in the range 100–200 km diameter. From the diameter of such craters, one can estimate the approximate energy of an impacting body generated. The energy (E) of a spherical meteorite is given by

$$E = m.V^2/2$$

where m is the mass of the body and V is its velocity. Obviously, this equation does not permit of a unique solution, for the velocity at impact may range from less than 10 km s^{-1} to more than 45 km s^{-1}.

The cratering database, although it is increasing annually, is far from adequate. Consequently, the search for and identification of major impact features should be one of our urgent tasks.

The other task relates to estimating the numbers of large meteoritic and comet bodies that exist in orbits within the Solar System which cross Earth's orbit, thereby constituting a serious hazard to Earth and its lifeforms.

Seyfert and Sirkin (1979) suggested that impact craters on Earth exhibited a succession of maxima which were at intervals of approximately 26 Ma. They also held that these 'impact epochs' could be correlated with a wide range of disturbances around the world, such as the product of large igneous provinces (LIPs), variations in sea-levels and orogenies, to which one could add widespread devastation or even extinction of some species, such as the dinosaurs.

Napier (1997) presented a time series analysis of impact craters, which indicates that such a periodicity does indeed seem to be present. It may be accounted for by the ticking of a well-known Galactic clock, whose mechanism has gradually become understood. This analysis supports a 'catastrophist' interpretation of Earth history.

1.6.4 The effects of atmosphere

When one looks at the Moon through even a low-powered telescope, one cannot but be impressed by the evidence of intense bombardment recorded on the surface of our satellite. The higher the degree of magnification used, the smaller are the diameters of craters that can be seen. Astronauts, from 'fly-by' and Moon landings, have recorded myriads of small craters with diameters of only a few metres. As the Moon is devoid of a significant atmosphere, erosion or covering of small craters will only result from subsequent, larger impacts, coverage by ejecta and the small degree of weathering that results from the monthly cycle of temperature changes. Many of the craters could have been in existence for 4000 Ma.

By comparison with the Moon, the cratering record of Earth is miniscule. This is in spite of the fact that the cross-sectional target area of Earth is 13 times greater than the Moon. Moreover, because of the much greater mass of Earth relative to that of its satellite, Earth should attract impacts by a further factor of about 20. Hence, Earth should have an impact record over 250 times greater than that of the Moon.

Unlike the Moon (see Figure 1.13), Earth has experienced extensive erosion and also widespread burial of older surfaces beneath extensive and sometimes thick sediments. In addition, if the impacts occurred in the ocean, obvious evidence of such impacts would not, in general, exist for more than about 150 Ma, for the oceanic lithosphere would be subducted, taking with it any evidence of impacts that may have existed. In addition, the Earth has a significant atmosphere (see Table 1.4). This atmosphere insures that smaller bodies will be heated to high temperatures and destroyed as it penetrates deeper and deeper into the atmosphere. Thus, the Earth has been saved from the massive bombardment of smaller missiles which continually take place on the Moon.

Table 1.4 Simple representation of the Earth's atmosphere.

Band	Height (km)	Atmosphere density (kg m³)	Temperature range (°K)
Exosphere	400–550	10^{-10}–10^{-12}	1500
Ionosphere	50–400	10^{-2}–0^{-10}	200–1500
Strato/Troposphere	0–50	1–10^{-2}	250–200

One may infer from this simple table that an incoming cometary body would encounter high temperatures in the Exosphere, but relatively low resistive pressure. In the Ionosphere the ambient temperature would fall as the body penetrated deeper into the Earth's atmosphere, but it would encounter a rapid increase in temperature engendered by passing through the progressively more dense Ionosphere. On entering the Stratosphere the air density increases dramatically, and frictional drag would engender high surface temperatures on the comet. If the comet entered at an angle of 30°, the path length before impact would be about 1100 km and if the average velocity of transit was 30 km/s⁻¹, the duration of heating prior to impact would be about 37 s.

Figure 1.13 Photograph of the 'full Moon' revealing the intense development of cratering on its Earth-facing surface.

1.6.5 Summary and conclusions

In the preceeding sections we have presented a brief outline of those known facts and reasonable inferences that relate to the mode of deformation of the main planets and satellites of the Solar System. Two main conclusions can be drawn.

(1) Although some of the inner planets may have exhibited in their early history a tendency towards a form of simple and elementary plate tectonics, the Earth is *unique* in the Solar System as regards the degree to which this form of deformation has developed. It is clear that this mode of behaviour has been made possible because Earth has retained water and has sustained active and plentiful life-forms.

(2) The Earth has experienced the same or (because of its size) an even greater degree of bombardment and deformation as is exhibited by the other Terrestrial Planets and our Moon. The only difference is that because of plate motions, most traces of the impacts that occurred in the oceans prior to 150 Ma have been swept down into the mantle by subduction, while those which have caused impact structures in continental areas are usually rapidly attacked by agents of weathering and denudation, so that they become progressively more difficult to recognise.

Chapter 2

Characteristics, stresses in, and strength of oceanic lithosphere

2.1 Introduction

In this and the following few chapters, we shall look at various elements of the plate tectonics paradigm. The overall concept, which is based on the relative motions of plates on a global scale over a period covering the Phanerozoic and even earlier times, cannot reasonably be challenged. Nevertheless, no-one is likely to claim that our understanding of the processes which give rise to plate movements is complete.

Relative plate motions over the recent past (i.e. about 0–50 Ma) have been established with a high degree of accuracy. The data used to establish such movements include the locations and azimuths of transform faults and the width and dates of magnetic anomalies of the ocean floor, while information regarding slip vectors and the relative motion of adjacent plates can be inferred from current earthquakes. Data from all plate pairs, when combined, result in a set of the best fitting angular velocity vectors which are globally self-consistent (Chase, 1978; Minster and Jordan, 1978; DeMets *et al.*, 1990).

The movement of the larger plates over the past 150 +/–50 Ma (which represents the *residence time* of some of the larger current oceanic plates) is made by compiling sea-floor spreading data which relate to plates separated by a spreading ridge. These data permit both the longitude and the latitude of the plates to be established.

As pointed out by Jurdy *et al.* (1995), further back in the Phanerozoic, evaluation of the movement mainly relies on the palaeomagnetic data obtained in continental regions. These data permit only the latitude to be known with reasonable precision, but longitudinal positions are not defined and must rely on geological constraints. Such geological constraints become progressively less precise until, in the Precambrian, the picture becomes progressively more hazy. Indeed, it has been argued (Davies, 1992) that the high heat flow and fast convection in the early Precambrian would tend to inhibit the development of a subducting oceanic plate. Consequently, he suggests, the plate tectonics mechanism has not always been in existence, but gradually evolved as the Earth cooled.

The accumulation of geological and geophysical data during the 20th century has been a remarkable and accelerating process. If this progress is maintained, the 21st century will see an even greater accumulation of pertinent data, which will permit plate movements to be more clearly defined than they are at present. However, one cannot reasonably expect that the early Precambrian will ever reveal a detailed picture of plate movements.

The *what and where* coupled with *when* have enabled the Earth scientist to build up the serial movement picture of plate tectonics as it is currently known. Much less effort has been expended on trying to answer the questions *why, or how?*

The mechanisms by which plate movements can be best understood are considered in the following chapter, with discussions in Chapters 3 and 4 of other problematic phenomena for which currently held concepts are somewhat less than completely satisfactory. In this chapter, we first outline, very briefly, the evolution of thought that gave rise to the plate tectonic paradigm. Then we indicate some of the more pertinent characteristics of oceanic and continental lithospheres. We note that, because oceanic plates are able to support oceanic islands for periods of many tens of millions of years, the oceanic lithosphere must, of necessity, contain a relatively thick, strong, elastic layer. We estimate the magnitude of the vertical stress that develops in the lithosphere as a result of gravitational loading, and the horizontal stresses that may occur in the elastic layer within the lithosphere. To ascertain the distribution of the maximum magnitude of stress that can develop within the lithosphere, we make brief reference to Rock Mechanics data.

In order to establish a restraint on the type of mechanisms which can be expected to influence plate motion, we consider a simple 'barge' model which may be *pulled,* or *pushed.* It is also reasonable to assume that plates may move as the result of both pull and push. It is argued that both mechanisms may operate on plates.

The orientation of the maximum horizontal stress has been determined, in many areas of the world, from direct measurements, or inferred from the orientation of various geological structures (Zoback, 1992a, 1992b, and others). From these data it can be inferred that, even if both *pull and push* mechanisms operate, the dominant mechanism is push. Finally, we discuss the strength and probable thickness of the *strong* layer in the oceanic lithosphere.

2.2 Characteristics of oceanic lithosphere

It is apposite here to outline some of the dominant and important characteristics of oceanic lithosphere and note their apparent simplicity and limited age-range. In Chapter 3, it will be argued that the generation of forces which cause plate motion is mainly related to the oceanic lithosphere, while the continental lithospheres are mainly passive, driven elements. Indeed, in its simplest form, plate tectonics fails to describe processes in the continental crust. As noted in Chapter 1, simple plate tectonics is based on Eulerian rotation on a sphere and assumes that plates act as stress guides, which are not readily deformable. This concept appears to work for oceanic crust, but obviously does not for continental crust.

Only when lithospheres with continental crusts collide, do they experience the most dramatic deformation, as, for example, when India was thrust against Asia. The continental crust, being too thick and having too low a density to be subducted, has caused tilting and thickening of the crust, thereby giving rise to the Himalayas. Alternatively, some mountain chains form without recourse to such continent/continent collision. For example, the northern and central portions of the Andes ride up over the subducting Pacific Oceanic lithosphere (Figure 2.1). Continental lithospheres can also experience internal, and usually less dramatic, deformation when they are subjected to the compression generated on opposite sides by opposing oceanic lithospheres. We shall discuss these latter aspects later in the book.

The tectonic consequences of such collisions of continental lithospheres, or of continental over-riding, subducting oceanic lithosphere are described in a host of texts and papers, so will not be considered here. Consequently, except for these collision phases, because continental components of plates are mainly but not, of course, completely passive, we shall concentrate on the characteristics of oceanic elements of lithospheric plates.

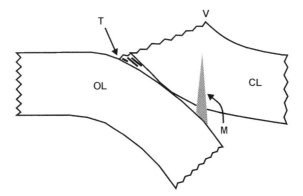

Figure 2.1 Section through continental lithosphere overriding oceanic lithosphere. OL = oceanic lithosphere; CL = continental lithosphere; T = trench; V = volcano; M = intrusive melt.

2.2.1 General morphology of oceanic crust and lithosphere

Oceanic crust is created at the spreading-ridges of the world, so that it is reasonable to start our description with them. These features are elongated regions of topographic highs that generally reach elevations of 3.0–4.0 km above the abyssal plains, though the occasional volcanic island on, or adjacent to, the spreading-ridge, may attain an elevation of more than twice this height. As can be inferred from Figure 2.2, these ridges interconnect and, in total, extend for more than 60,000 km. The importance of these ridges was initially recognised because of the relatively high level of seismic activity associated with them. However, interpretation of the seismic data reveals that the majority of the individual events are the result of normal faulting. The magnitude of these seismic events is commonly moderate (i.e. usually less than magnitude 5 on the Richter scale).

One may infer from bathometric maps or charts of the oceans, that the ridge makes well-marked features several hundreds of kilometres in width. However, these maps and charts tend to minimise the importance of the *ridge slopes*, which extend from the ridge, proper, to the oceanic abyssal plain. In the abyssal plains, the slopes may be as little as 1 : 1000 but, as the ridge is approached, the slopes will be closer to 1 : 50.

It has been established (Parsons and Sclater, 1977) that the depth (Z) to which a ridge subsides, relative to the depth of the ridge, is as shown in Figure 2.3 for the N Pacific and the N Atlantic. This relationship, holds generally, for all ridges. However, it is emphasised that the distance of any point, of a specific age *t*', from the ridge will depend upon the rate of spreading. For a fast spreading-ridge, the point associated with time *t*' will be a long way from the ridge, so that the slope from the ridge will be very gentle. When the rate of spreading is slow, the converse situation exists and the slope from the ridge is somewhat steeper.

Oceanic crust is formed in the vicinity of the spreading-ridges. It is comprised of three layers, the uppermost of which comprises extruded pillow lavas and the occasional intruded sill (Figure 2.4). Layer 2 is largely comprised of innumerable sheets of dykes. (Such dyke swarms are beautifully exposed in road-side exposures in the Troodos mountains, Cyprus, where oceanic crust has 'come ashore'.) The third and thickest layer consists mainly of gabbro. The combined group of the three layers usually have a total crustal thickness of about 7.0 km.

Below the crust, lies the mantle of the oceanic lithosphere, which is mainly composed of peridotite. Each and every oceanic lithospheric plate is underlain by a zone in the astheno-

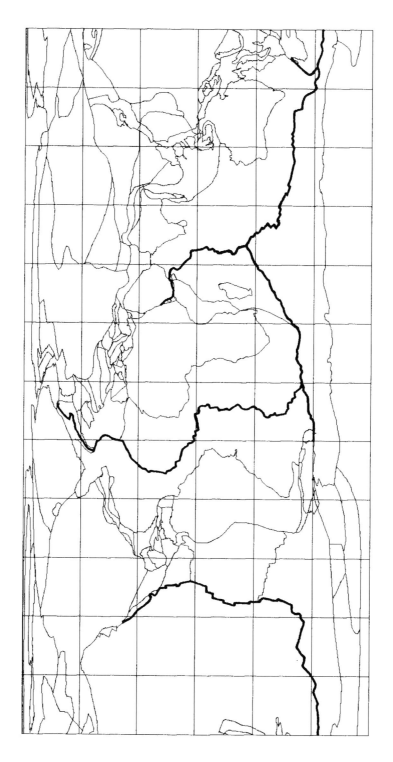

Figure 2.2 General outline of continents (including) shelves (thin lines) and the distribution of major oceanic ridges (thick lines).

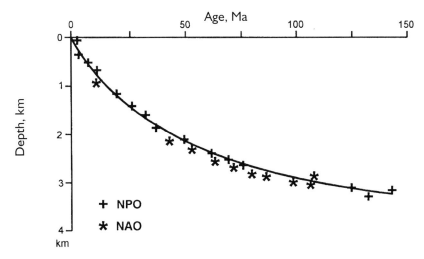

Figure 2.3 Depth/age relationship of oceanic floor relative to the spreading-ridge for the N Pacific and N Atlantic (after Parsons and Sclater, 1977).

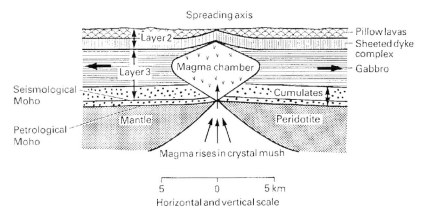

Figure 2.4 Section showing the main features of oceanic crust and upper mantle in diagrammatic cross-section. The somewhat schematic shape of the magmatic chamber, that supplies the eruptive rocks at the spreading-ridge, is based on a seismic reflection model of Detrick *et al.*, 1978.

sphere, which is perhaps 50 to 100 km thick (Turcotte and Schubert, 1982). This zone is situated at depths where the ambient conditions of temperature and pressure are close to the solidus of the mantle. This zone came to the notice of seismologists because of the attenuation of seismic signals, and the fact that the velocity of seismic waves in this layer is noticeably lower than in the mantle immediately above and below. Accordingly, this low-velocity zone was designated as the LVZ. The change in seismic velocity is attributed to the existence of small quantities of melt, water, or hydrous state within the zone, or to currently ill-defined, pre-melting phenomena. This pre-melting, or melt, phenomenon is also held to cause a reduction of the viscosity of the mantle material, relative to that above and below the zone, by about two orders of magnitude (Turcotte and Schubert, 1982). Consequently the abbreviation LVZ is used not only to allude to a low *velocity* zone, but also to a low *viscosity* zone.

At, or near the spreading-ridge, molten basic material is intruded or extruded at high initial temperatures (about 1300°C), but rapidly cools in, or adjacent to, oceanic water. This water, driven by thermal effects, circulates freely through the pillow lavas and also penetrates downward through the fractures which develop, as the result of cooling, in the sills and dykes. This convective circulation of water occurs throughout the uppermost layers of the ocean crust and according to Lister (1992) may even enter the uppermost part of the lower layer. However, hydration reactions occur between the circulating water and the rock. The hydrated minerals, thus formed, have a greater volume than the minerals they replace, so that eventually the rocks lower in the ocean crust, together with the mantle, are effectively sealed against further penetration of sea water. At only a few kilometres below the surface of the oceanic crust, the rock forming the ocean lithosphere probably contains little free water (Bergeron, 1997). (As we shall see, dry basic rock can be immensely strong.) Only in areas of subduction of oceanic lithosphere, where wet sediments are carried down into the astheno-sphere, below the lithosphere, can water migrate upwards and enter into the lower regions of the lithosphere. However, in these circumstances, it is usually the lower regions of continental lithosphere which are invaded by the upwelling water.

Because the zones, which define the oceanic crust, were initiated at the ridge and are continually transported from their site of origin, they extend, virtually unchanged in thickness, from the newest part of the ocean floor, currently being generated at the active ridges, to the oldest parts which are just about to be subducted.

The rate of sedimentation in the open ocean areas, far distant from continental shores, will tend to be reasonably constant. Hence, the thickness of the superficial sediments will be related to the age of the ocean floor upon which they rest. The ocean floor is being generated at the spreading-ridge, so, as one would expect, the thickness of the sediments at the crest of the ridge is virtually zero. Away from the spreading-ridge, the pillow lavas are the first to become covered with sediments. Further away from this ridge, any minor volcanic cones which may have been generated at, or near, the spreading-ridge are progressively covered by sediments until they too disappear beneath the sedimentary cover where, at the abyssal plains, the thickness of sediment builds up to about 300 m.

Beneath the crust, a more homogeneous and isotropic rock unit develops. This lower rock unit was originally part of the fluid asthenosphere and became 'rigid' on cooling, and thereby formed the main part of the lithosphere, as the appropriate temperature and pressure conditions for the transition were encountered.

Because the lithosphere becomes thicker away from the spreading-ridge, the geothermal gradient changes from several hundred degrees centigrade per kilometre of depth near the ridge to as little as $11°C$ km^{-1} at the most distant parts, in the oceanic lithosphere, away from the spreading-ridge. This sub-crustal rock unit becomes progressively thicker away from the ridge-spreading, until, in the larger plates, the total thickness of the lithosphere can approach a maximum of about 100 km. The temperature gradient depends not only upon the thickness of the lithosphere, but also on the gradient inherited from previous positions; hence the distribution of these gradients is related to the velocity of spreading of the oceanic plate. The distribution of temperatures in the N Pacific Plate, based upon a cooling model proposed by Parsons and Sclater (1977), was expressed graphically by Kirby (1980) and is shown in Figure 2.5.

As regards the surface morphology of the spreading-ridge, on their flanks, the slopes of the ridge are relatively small. Thus, taking the average relief of ridges relative to the 'level' abyssal ocean floor to about 4–5 km, and the abyssal areas to be 1000 km or more from the

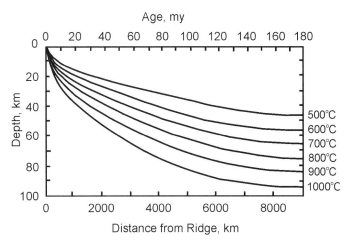

Figure 2.5 Temperature distribution and general shape of a section through the Pacific Plate (after Parsons and Sclater, 1977).

ridge, it follows that the average slope is less than 0.25° and is, therefore, barely perceptible. Only near the crest of the spreading-ridge is the average dip of the ocean floor likely to exceed a few degrees.

A generalised section across any spreading-ridge will show the simple form of Figure 2.6. However, in a limited and detailed profile, which extends for only 200–500 km across the ridge, the features are seen to be of two types. Either they form a reasonably well-defined peak, comparable with that shown in Figure 2.6a, or else define a central graben (Figure 2.6b and c). The profiles shown in these figures greatly exaggerate the relief of the features.

The East Pacific Rise is typical of the single-peak type of section, while the Mid-Atlantic Ridge is representative of the central, or axial, graben type. By inspection of a suitable map showing more detailed section of these ridges and their spreading rate, in various parts of the world, it can be inferred that those ridges with a central graben are associated with a relatively slow spreading rate (<5.0 cm a^{-1}), while the single-peak ridges occur where the spreading rate is relatively fast (i.e. >5.0 cm a^{-1}).

The morphology of the simple section, exhibited by the East Pacific Ridge, extends throughout the length of this spreading-ridge. The Atlantic-type ridge is much less uniform. Although grabens are a common feature, they are not of uniform size. Along the Mid-Atlantic Ridge, for example, the floor of the graben is as much as 1.5 km below its rim; it is often between 10–40 km in width and devoid of sediments. Such dimensions are comparable with some of the major terrestrial grabens.

The detail of the morphology of the submarine graben floor is similar to parts of Iceland, with minor fault scarps which run approximately parallel to the trend of the ridge, though in Iceland, because of the infilling by sediments, the depth of the grabens is small compared with those in the Atlantic Ridge proper. Some of the fractures are the loci of fissure eruptions, or may determine the location of a range of small volcanic cones.

Sections through the Atlantic and the Carlsbad Ridges reveal that, unlike the Rhine Graben, the floor of the graben is usually inclined rather than flat. However, the flat floor of the Rhine Graben can be attributed to erosional and depositional effects of the Rhine. Such an influence is, of course, absent in oceanic spreading-ridges.

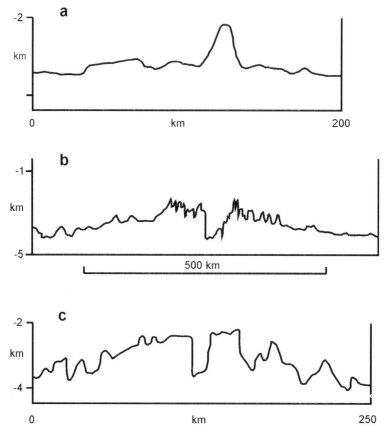

Figure 2.6 Sections through spreading-ridges showing (a) simple peak; (b) and (c) graben.

In addition, oceanic lithosphere is cut by a large number of vertical planes, or zones of shear displacement, which sometimes off-set spreading-ridges by tens or hundreds of kilometres and which usually trend at high angles to these ridges, and, indeed, tend to trend in the direction of plate motion. These features which are known as *transform* faults, are not readily explained in terms of mechanics usually applied to the interpretation of shear fractures.

One final and over-ridingly important characteristic is that oceanic lithosphere is more dense than the underlying asthenosphere. Consequently, oceanic lithosphere is always intrinsically unstable and, given the opportunity, will sink into the depths of the mantle.

2.3 Response of lithosphere to loading

It was estimated by Darwin (1882), who assumed the Earth's topography to be uncompensated, that the stress difference required to support the differences in topographic elevation could be as much as 600 bar. Barrell (1914) considered the departures from isostatic equilibrium to be of such magnitude that it required the support of a strong lithosphere. Later, Jeffreys (1942) included the effect of isostasy and showed that, when one considered such features as the Himalayas and deep ocean trenches, the inferred stress differences exceeded 1.6 kb. More recent evidence for the existence of considerable strength in continental lithosphere, relates to loads resulting from extensive cover of thick ice-sheets for periods in excess of a million years.

It has also been recognised for decades that oceanic lithosphere is strong (Barrel, 1914; Vening-Meinesz, 1941; Gunn, 1944). Such strength is required in order to satisfy the observation that oceanic plates are translated, in some instances by many thousands of kilometres, during which time they are able to support large surface loads such as volcanoes and seamounts. The depression of the sea-floor by such loads permits one to infer that a significant part of the lithosphere behaves elastically and that the underlying asthenosphere is weak.

By assuming reasonable values for the elastic constants of the rocks in the lithosphere, and the downward elastic displacement brought about by loading of the lithosphere by the mass of a volcano or seamount, the *approximate* thickness of the elastic layer can be inferred. Such elastic models have been used to explain a wide range of geological and geophysical observations. For example, Hanks (1971) used simple mathematical modelling to explain the large-magnitude, shallow-earthquake events associated with the Kuril Trench–Hokkaido Rise system: Watts and Cochran (1974) and Bodine *et al.* (1981) studied the gravity anomalies and flexure of the lithosphere along the Hawaiian-Emperor Chain, while Turcotte and Schubert (1982) give an overview of flexure and rheology of oceanic lithosphere.

The simple geometry of oceanic floors represented in Figure 2.3 is sometimes marred by the presence of *oceanic plateau basalt provinces,* which may range in size from masses with areas of many tens of thousands of square kilometres, and which stand as much as 4.0 km above the abyssal plain, to single volcanic islands, or chains of such islands and seamounts. When the province is extensive, it is mainly in isostatic equilibrium, though it may be surrounded by a depressed zone of the ocean-floor, which marks elastic behaviour of the lithosphere. Volcanic islands or chains exhibit comparable adjacent depressions. However, before we consider these latter structures, it is necessary to outline how the magnitude of lithospheric stresses can be assessed.

2.3.1 Lithospheric stress

One of the fundamental elements in understanding the manner in which the lithosphere may deform is the magnitude of the stresses within it, and the parameters which control the limits of deformation. The simplest component of stress to ascertain is the vertical stress (S_z), which is brought about by gravitational loading, so that at a depth from the surface (z)

$$S_z = \delta.g.z \tag{2.1}$$

where δ is the density of the material, in this instance, basic rock, where we take $\delta = 3.3$ g cm^{-3} (or 3.3 tonnes m^{-3}). Thus, in oceanic lithosphere, the vertical pressure increases linearly with depth by 330 bar km^{-1} (33.0 MPa km). This simple equation enables one to calculate the vertical stress for a rock mass, whether it behaves as a simple elastic or viscous body. Here, we make the simplifying assumption that oceanic lithosphere has a constant density. As regards continental lithosphere, the situation is rendered more complex, in that crustal rock densities range from less than 2.0 g cm^{-3} (2.0 tonnes m^{-3}) to more than 3.0 g cm^{-3} (3.0 tonnes m^{-3}).

This vertical stress, induced in the rock mass by gravitational loading, has a tendency to spread a hypothetical vertical column of rock in the horizontal direction. However, as may be inferred from Figure 2.7, any such hypothetical column is adjacent to other such columns, each trying to spread laterally. If the lateral strains are held at zero, the lateral stresses which develop depend upon the rheological response of the body. If we are dealing with hot rocks or minerals at considerable depth, they will behave as a fluid, so that the horizontal stress will equal the vertical stress (i.e. the stresses are 'hydrostatic').

In the upper 40–60 km of oceanic lithosphere, much of the rock mass away from the spreading-ridge will behave, to a very close approximation, as an elastic body, even when the stresses act for very long periods (i.e. 1.5×10^8 years). Consequently, it can be shown (Price, 1966) that when the horizontal strain (e_h) is zero, the horizontal stress (S_h) is given by:

$$S_h = S_z / (m - 1) \tag{2.2}$$

where m is Poisson's number, the reciprocal of Poisson's ratio.

The value of m depends upon rock type. Moreover, it changes, in real rock, with the magnitude of ambient stresses (Price, 1966). However, for convenience, we shall deal only with linear-elastic theory, in which m does not change with ambient stress. Also, we shall assume that $m = 4.0$ (a reasonable approximation for many basic rock types). Hence, when the conditions of zero horizontal strain are met, then, from Equation 2.2, it follows that the horizontal stress, at any specific depth, is about one third the vertical pressure.

2.3.2 Horizontal stress

It is usual to refer to the vertical direction as the z axis and the two orthogonal horizontal axes as the x and y directions, where, commonly, the x direction is attributed to the greatest horizontal stress, and the y direction indicates the direction of action of the least horizontal stress. These stresses may, or may not, be principal stresses.

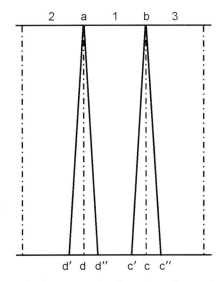

Figure 2.7 Hypothetical vertical column extending from the surface to depths in the lithosphere. If each column were free to expand laterally, d would move to d′ and d″ and c would move to c′ and c″. However, lateral stresses develop and increase with depth so that the tendency of every column is constrained by the tendency if adjacent columns are to expand laterally so that horizontal strain is preserved.

The stresses which result from a shortening of a plate parallel to the direction of movement are given by a combination of the horizontal stresses induced by gravity, plus the increment of compressive stress caused by the compressive strain, so that (Price, 1966) the maximum horizontal stress S_x is given by

$$S_x = \{S_z/(m-1)\} + S_c \tag{2.3a}$$

The compressive stress S_c causes strains to develop in the directions z and x. In the vertical direction (z), the surface can move upward as the result of this strain. However, as elastic strains are usually very small, the increase in vertical stress can conveniently be ignored.

The increase in stress in the y direction, assuming that strain $e_y = 0$, is equal to S_c/m, so that

$$S_y = \{S_z/(m-1)\} + S_c/m \tag{2.3b}$$

If the plate is being pulled, there will be extension in the direction of plate movement so that Eq. 2.3a and b, respectively, will become:

$$S_x = \{S_z/(m-1)\} - S_c \tag{2.4a}$$

and

$$S_y = \{S_z/(m-1)\} - S_c/m \tag{2.4b}$$

These equations[1] confirm our simple model, i.e. that if the plates are being *pushed*, then at or near the surface the axis of maximum horizontal stress should be aligned parallel to the direction of *absolute plate motion*. However, if the plates are being *pulled*, it is the axis of least horizontal stress that should be aligned parallel with the direction of absolute plate motion.

If a hypothetical plate had a rectangular shape and uniform boundary stresses, these equations would hold throughout much of the plate. However, since real plates are irregular in plan, boundary conditions vary from place to place. These equations indicate the type of stress conditions which are likely to exist in the central region of the plate which includes the axis of the resultant force.

Let us now ascertain to what extent evidence of stress orientations and their relative magnitudes in real plates may be inferred from studies using various methods.

2.4 Orientation of horizontal stresses

The orientation of principal stresses in the continental crust can be inferred by using a number of different techniques. The more accurate of these, which are best used at outcrops or in mines, and have only become available during the last few decades (Jaeger and Cook, 1969), are based on the *over-coring* method. The orientation and magnitude of the least horizontal stress can be determined by using the *hydraulic-fracturing technique*. The orientation of the axis of greatest horizontal stress can also be obtained by ascertaining the orientation of the elliptical cross-section of a deep, unlined, originally circular-section bore-hole. The elliptical form is caused by the process often referred to as wall *break-out*. Other methods used to infer stress orientation can sometimes be obtained from seismic data and also from the theoretical analysis of geological structures.

2.4.1 Over-coring

In the over-coring method of stress analysis, a small, planar surface (of the required orientation) is selected or prepared. Strain-gauges, usually set at 60° to each other are affixed to this surface by adhesive (Figure 2.8). When the adhesive has set, the network of gauges gives the apparent zero-strain condition in the rock. The rock surface, on which the gauges are placed, is then isolated from the surrounding rock by over-coring. The free surface around the gauges permits the rock stresses in the cylindrical stub, beneath the strain-gauged surface, to induce strains which are recorded by the strain-gauges. The elastic moduli (*Young's modulus E and Poisson's number m*) for the rock are subsequently ascertained in the laboratory from the detached stub, or a sample taken from an adjacent site. The strain-data obtained by over-coring can then be translated into magnitudes and orientation of *in situ*, principal two-dimensional stress. By carrying out this exercise with sets of gauges arrayed at right angles to each other, the three-dimensional stress array can be inferred.

An interesting feature of the data obtained from such measurements is that, even in non-tectonic areas, down to a depth of 300 m, the horizontal stresses are always larger than the vertical stress (Figure 2.9) and may be larger than the vertical stress at depths of several kilometres. It has been argued (Price and Cosgrove, 1990) that the increase in relative magnitude of the horizontal stresses as the surface is approached can be attributed to the fact that, with uplift and erosion, the vertical stress is reduced more rapidly than the horizontal stresses. Consequently, the stress data obtained at, or near, the surface does not permit one to infer that the maximum horizontal stress, at some depth in the plate, is also the maximum principal stress. However, one can infer that the maximum principal stress, as determined at or near the surface, may well approximate to the orientation of the maximum horizontal stress at depth. Even this, however, is open to question, for it has been established that deep stress orientations may change, quite abruptly, with the age of the rock unit (Cowgill, 1994). Such changes presumably reflect earlier, different movement directions of the plate.

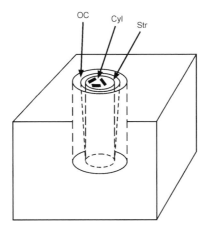

Figure 2.8 Strain gauges are cemented to a flat surface and wires from the gauges are taken up the barrel of a core-drill to circuits determining the expansion of the core. Over-coring to several inches (OC – dashed lines) enables the *in situ* stress to be released and gives rise to expansion of the 'core'. The dashed lines show, in exaggerated form, the strain release (Str).

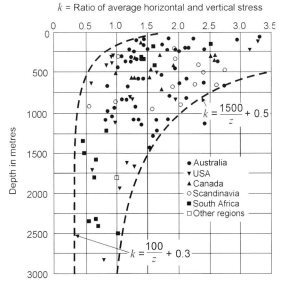

Figure 2.9 Ratio of horizontal to vertical stresses obtained from *in situ* stress measurements (after Brown and Hoek, 1978).

2.4.2 Hydraulic fracture

This technique requires a section of a borehole to be sealed and isolated (by the use of *packers)* above and below the rock horizon in which the *least principal stress* is to be determined (Figure 2.10). One of two main types of results may be expected.

Firstly, if the vertical stress (S_z) is the least principal stress (S_3), a horizontal fracture will form in the wall of the borehole, at A in Figure 2.10, when the bore-hole fluid pressure p is

$$p > S_3 + T_0$$

where T_0 is the tensile strength of the rock. These conditions may frequently be satisfied in shallow boreholes.

However, here we are concerned with determining which of the horizontal stresses represents the *least principal stress*, S_3. These conditions are most frequently encountered in relatively deep bore-holes (i.e. greater than 2 km).

The theory relating to the second situation, in which the vertical stress is greater than at least one of the horizontal principal stresses, was first given by Hubbert and Willis (1957) and is also set out in Jaeger and Cook (1969). It can be shown, that the tangential stress S_t in the surface of the borehole corresponding to the fluid pressure (p) and S_1 and S_3 is given by:

$$S_t = (S_1 + S_3 - p) - 2(S_1 - S_3).cos2\theta$$

This expression varies from a maximum of $3S_1 - S_3 - p$, when $\theta = p/2$, to a minimum of $3S_3 - S_1 - p$, when $\theta = 0$. If

$$p > 3S_3 + T_0 - S_1$$

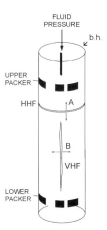

Figure 2.10 Simplified diagram of a borehole (b.h.) with a sealed section between upper and lower packers. Fluid pressure is generated between the packers until the borehole fails. Near the surface, the rock tends to develop a horizontal hydraulic fracture (HHF) – A. At depth, where the vertical stress is higher than the lateral rock pressure, a vertical hydraulic pressure fracture (VHF) usually forms – B.

tensile failure will take place along a radial plane normal to the axis of least principal stress (B in Figure 2.10). The value of T_0, of course, is obtained from tests conducted in the laboratory on specimens taken from the test horizon.

If the axis of the vertical principal stress is not exactly parallel to the borehole, the geometry of the borehole will determine the orientation of the fracture at the borehole, but the induced hydraulic fracture will curve into the plane, normal to the axis of the least principal stress, away from the borehole. The deviation between the borehole fracture and the true orientation of the hydraulic fracture away from the borehole will not usually be known.

2.4.3 Break-out

A further source of information regarding the direction of action of the greatest horizontal principal stress is obtained from the shape, in cross-section, of deep, unlined bore-holes. These, of course, are usually drilled by the oil and gas industry. From elasticity theory (Obert and Duvall, 1967; Jaeger and Cook, 1969), it can be shown that the stress concentrations around a hole subjected to biaxial compression, normal to the axis of the hole, is such that the maximum reduction in stress occurs in the hole where the axis of maximum principal stress meets the hole boundary. A simplistic representation of how and where tensile and compressive stresses, which develop about a small hole in a disk subjected to uniaxial compression, is shown in Figure 2.11a.

Deep drill holes in the crust contain drilling muds at relatively high pressure. It can be inferred from Figure 2.11b that, because of the stress concentration that occurs, the walls of the drill hole will fail in compression and spall into the drill hole, as indicated in the shaded area of Figure 2.11b. This spalling, or *break-out*, causes the hole to develop an elliptical cross-section, with the long axis coincident with the axis of minimum horizontal stress. From surveys of borehole break-out, reasonable estimates can be obtained of the regional trends of the axes of greatest and least horizontal stress (Cowgill, 1994).

2.4.4 Seismic data and analysis of geological structures

The orientation and relative intensity of the horizontal stresses can be inferred from seismic data and also from certain geological features and structures. For example, provided the dyke orientation is not controlled by joints or faults, it is reasonable to infer that the least principal stress acts normal to dykes, so that, when such features are vertical, it follows that the greatest horizontal stress acts in the plane of the intrusion. Similarly, the regional orientation of stresses can be inferred from studies of folds and faults. Earthquake data can also be used to infer orientation of stress axes. However, all such techniques, based on theory, require very careful application.

Fault movement and seismic data

When a fault is initiated in homogeneous and isotropic rock, it is reasonable to accept the theoretical concepts of brittle failure, and assume that the axis of intermediate principal stress lies in the plane of shear failure, and that the axis of maximum principal stress makes an angle with the fault plane which is determined by the physical properties of the rock. This angle is commonly close to 30°.

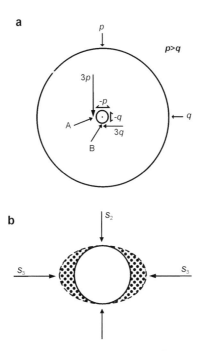

Figure 2.11 (a) Simplistic representation of the force distribution of a disk with a hole, subjected to biaxial compression, at *poles* and *equator* of the disk. There is a three-fold force concentration of stress at the hole boundary in the direction of the applied forces p and q and a tensile stress of $-p$ and $-q$ as indicated in the diagram. With $p>q$, the biaxial forces acting at the 'pole' and 'equator' of the small hole become significantly enhanced. (b) When these enhanced forces, per unit area, are translated into stresses, it can be inferred that the walls of the borehole may spall or 'break-out'.

Once shear failure takes place, the magnitude of the stresses changes. If we are dealing with a normal fault, initiated when S_1 is vertical, it can be shown that the erstwhile least principal stress (S_3), adjacent to the fault plane, increases in magnitude, so that it becomes the intermediate principal stress (S_2), while the erstwhile intermediate principal stress in the plane of the fault becomes (S_3). Once a fault is initiated, it develops to reach its maximum extent. This development, requiring a number of *slips* on the fault, may take place in a relatively short period, during which the orientation of the principal stresses may remain reasonably constant. Over a longer period, the axes of principal stress and their magnitudes, although causing fault movement, may undergo such rotation that all principal stress axes are oblique to the fault plane (Price and Cosgrove, 1990).

By postulating orientations of the axes of principal stress to a hypothetical plane, and by assuming one or more specific values for the coefficient of sliding friction and also the ratio of the magnitude of the principal stresses, it is possible to determine the conditions under which slip may take place on the hypothetical fault plane. Such an exercise is reported by Jaeger and Cook (1969), where the areas were determined in which the normal plane must fall if slip on the plane can take place, for the assumed conditions of friction and ratio of stress magnitudes (Figure 2.12).

From seismic data, one may infer the orientation of the axis of maximum principal stress, from *first arrivals*. However, frequently, the orientation of the fault plane is not known, nor are the orientations or relative magnitudes of the principal stresses. The fact that the seismic analyst is so frequently confronted with such 'unknowns' gave rise to the oft forgotten caution by McKenzie (1969), that the analyst can only expect to place the pole of the maximum principal stress in the *correct quadrant*!

Analyses of geological structures

These analyses embrace a number of techniques. The easiest of these is based on, for example, the reasonable assumption that *dykes*, sometimes in conjunction with *alignment of volcanic*

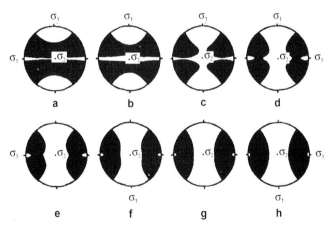

Figure 2.12 Slip conditions on a fracture in the crust are determined by the orientation and magnitude of the principal stresses and their orientation relative to the given fracture plane. The diagrams a–h indicate the zones of failure (black) for a variety of stress directions. The reader is directed to Jaeger and Cook (1969), and Price and Cosgrove (1990) for a more detailed treatment (after Jaeger and Cook, 1969).

vents, also provide good evidence for the orientation of the greatest horizontal stress. But these events are limited in occurrence and, moreover, the inferred stress data relate to the period when the dykes or vents were emplaced, and so may well be different from the current stress situation.

In recent studies of what has been termed *neo-tectonics*, the slip direction of recently active fault planes has become a favoured means of determining the orientation of the axis of maximum principal stress. This technique has been pioneered and developed by French structural geologists (e.g. Carey and Brunier, 1974; Carey, 1979; Angelier *et al.*, 1982). In these studies, several adjacent fault planes are usually exposed, and the slip direction can be inferred from striations and slickenside. It is assumed that the axis of maximum principal stress acts at 30° to the fault plane, rather than the 45° favoured by geophysical analysts of seismic data. However, the problems outlined above in the section on *Fault Movement and Seismic Data* still hold.

2.5 The world stress map (WSM)

This project is a cooperative effort to compile and interpret data on the orientation and relative magnitudes of *in situ* tectonic stress in the Earth's lithosphere. Zoback (1992b) reports that over 30 scientists from more than 18 different countries have been involved in this project. Doubtless, these numbers have increased since 1992. At that time, over 7300 data points had been compiled. The techniques used to establish the stress orientation and sometimes magnitude of the stress have been briefly outlined in previous paragraphs. The reader will, therefore, be aware that the inferred stress orientations obtained using different techniques are not of equal accuracy. Accordingly, WSM data are classified and grouped into five categories (A–E), of which only A–C are considered to be of sufficient accuracy to be worthy of analysis.

The proportion of the various categories of reliability (A–C) for the various techniques is indicated in Figure 2.13a, and the depth range over which the various techniques apply is shown in Figure 2.13b. Of the total number (7300) of data points, over 4400 are within the A–C categories.

2.5.1 Direction of horizontal stresses and absolute plate motions

As pointed out by Coblentz and Richardson (1995), early assessments of these data presented in the WSM relied upon visual assessment. Where data were especially plentiful, the spread of directions of the S_{Hmax} axes resulted in vague assessments of mean orientations (Figure 2.14). The stress patterns in several areas have been studied in some detail; for example, Eastern N America (Zoback, 1992b), Europe (Müller *et al.*, 1992; Grünthal and Stromeyer, 1992), East Africa (Bosworth *et al.*, 1992), China (Zuber, 1992) and India (Gowd *et al.*, 1992). However, here we concentrate on the relationship between the axes of maximum horizontal stress and the absolute motion of plates. Some simple statistical analyses have been carried out, and these do indeed provide a valuable, broad-scale insight into the trends, but are of limited use when considering areas, such as the Indo-Australian plate, where the orientations of S_{Hmax} are not consistent (Richardson, 1992).

In the study by Coblentz and Richardson, they applied elementary geostatistics to the WSM data. Their analysis was carried out in three steps.

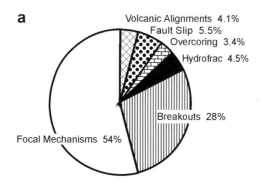

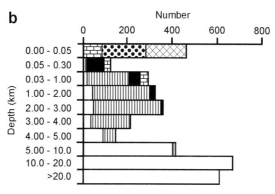

Figure 2.13 (a) Pie-diagram. (b) Depth-range of methods used.

(1) They first estimated the state of stress (compressional, strike-slip or extensional) within 5° × 5° areas of the Earth's surface, which they refer to as *bins*. By this time, the WSM data set consisted of over 9000 stress indicators. However, only Quality A–C (as defined by Zoback, 1992b), i.e. 4537, about half the total number, are considered good enough for reliable analysis.

(2) They applied the Rayleigh test, which is a standard statistical method of analysing directional data, to the orientation of each estimate of S_{Hmax} in each of the bins. This process tests the null hypothesis that these orientations are random (Mardia, 1972).

(3) The S_{Hmax} directions within each bin that passed the Rayleigh test were compared with absolute plate velocity azimuths and ridge torque directions to assess possible correlations.

From the map (Figure 2.15), it can be inferred that the data are heavily biased in favour of Europe and N America, with, as one would expect, scanty information from Antarctica, Africa and the Pacific.

Interested readers should refer to the original paper by Coblentz and Richardson (1995), regarding details of the mode of analysis. Here we only present the conclusions of their study. The bars in Figure 2.16 give the orientations of S_{Hmax} for 5° bins which pass the Rayleigh test. Open and solid bars indicate, at the 90 per cent and 97.5 per cent confidence levels respectively, that the orientations are *not* random. The significance of these bar orientations can be seen by making reference to Figure 2.17a, where the bars are related to the ridge-push torque poles, and Figure 2.17b where the bars are related to the absolute plate-velocity azimuths.

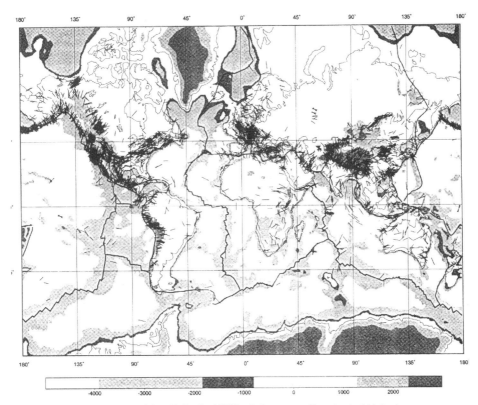

Figure 2.14 World stress map (after Zoback, 1992b. © American Geophysical Union).

Zoback (1992b) and Richardson (1992) have indicated that the large-scale trends of S_{Hmax} are consistent with ridge-push forces. Coblentz and Richardson endorse this conclusion, particularly for eastern N America, western S America, western Europe (where data are abundant) and also in central Asia and continental Australia. They also note that, in many areas, S_{Hmax} orientations are aligned with the absolute plate-velocity azimuths. This alignment is less than perfect in areas where plate motion is slow, such as in the African and Eurasian plates, which have relatively large uncertainties as regards their absolute velocities. Thus, the stress data tend to deviate from the ridge-push torque direction, or the plate-velocity azimuths in areas where the data are sparse, or where the absolute plate velocities are very slow, so that what is termed second-order tectonic effects may be important.

In addition, Zoback *et al.* (1986) note that only in a few restricted areas, such as in the uplifted, gently flexured, anticlinal zones of oceanic lithosphere, prior to the plate being subducted, do these axes of stress change, with the axis of least horizontal stress pointing in the direction of movement. We shall show in Chapter 3 that such a change-over of stress intensities can readily be explained. Consequently, although some of the inferred stress orientations may be less than perfectly accurate, the degree of agreement of so much of the stress axis data with ridge-push torque poles and plate-velocity azimuths, permits us to infer that the plates are dominantly being pushed.

However, one must not suppose that the stresses in a plate necessarily remain constant as to both direction and magnitude over long periods, especially if that plate is slow moving.

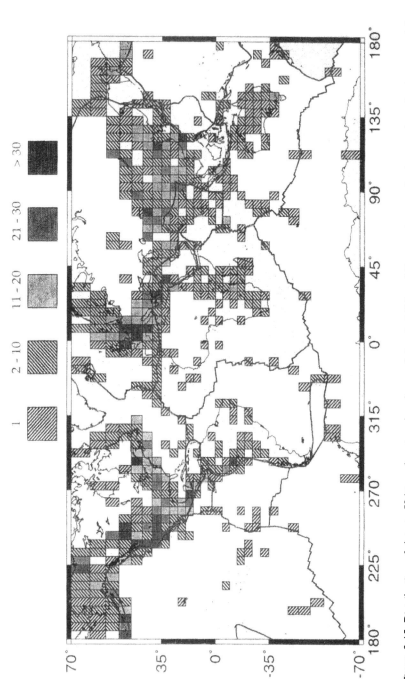

Number of indicators in bin

| | 1 | 2 - 10 | 11 - 20 | 21 - 30 | > 30 |

Figure 2.15 Distribution of data in 5° bins, showing abundance of data in USA and Europe, but paucity in oceanic and other areas (after Coblentz and Richardson, 1995. © American Geophysical Union).

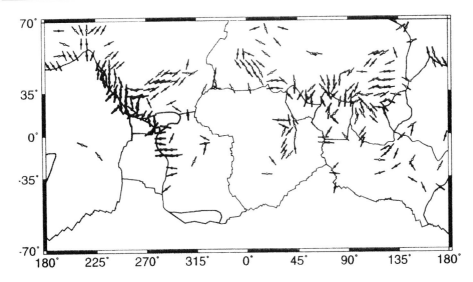

Figure 2.16 Bars, representing orientation of maximum horizontal stress, which have passed the Rayleigh test (after Coblentz and Richardson, 1995. © American Geophysical Union).

Indeed, Cowgill (1994) has shown that the stress orientations in the North Sea area, based on the break-out orientation of the boreholes, do, in fact, show markedly different orientations of the horizontal principal stresses for strata of markedly different ages.

2.6 Rock mechanics data and the strength of the lithosphere

Tests of rock-strength in uniaxial and triaxial compression have been conducted for most of the 20th century. A great deal of this work was directed to understanding the behaviour of rock types frequently found in, and for conditions likely to be encountered within, the continental crust. It was soon established that the main parameters influencing the strength of such rocks (i.e. the differential stress at which the rock failed, whether in a brittle, semi-brittle, or ductile manner (Figure 2.18)), were (1) confining pressure, (2) temperature, (3) strain-rate, and (4) fluid pressure within the rock. The strongest rock units in oceanic and continental lithosphere are the mafic rocks which exist beneath the crust. Except in special circumstances, it is widely accepted that these rocks are probably completely dry. We will, therefore, neglect the influence of the last of these parameters. The importance of fluids upon the 'strength' of rock is discussed at length in Fyfe *et al.* (1978) and Price and Cosgrove (1990).

(1) *Confining Pressure.* As indicated in Figure 2.19, from the classic experiments by Love (1944), it can be seen that the influence of confining pressure is to increase the magnitude of the differential stress that is required to cause failure, and also that different modes of failure are induced at increasingly higher confining pressures. These tests were conducted at room temperature, on cylindrical specimens of Cararra marble (which were encased by a thin copper jacket, to prevent the fluid, in which the confining pressure was built up, from entering the rock specimen).

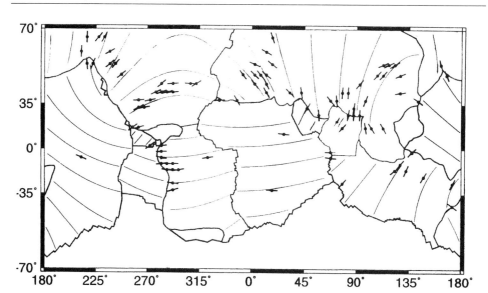

Figure 2.17a Ridge-push torque directions and the orientation of the stress bars that have passed the Rayleigh test (after Coblentz and Richardson, 1995. © American Geophysical Union).

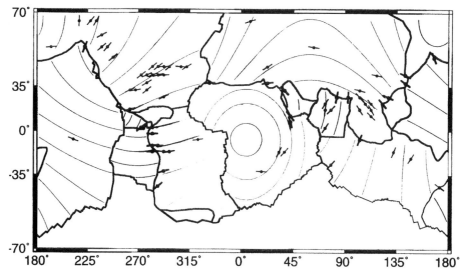

Figure 2.17b Absolute plate-velocity azimuths and trend of stress bars that have passed the Rayleigh test (after Coblentz and Richardson, 1995. © American Geophysical Union).

(2) **Temperature.** The simplest way of demonstrating the influence of temperature upon strength and mode of failure of a rock type, used by Griggs *et al.* (1960), was to conduct repeated experiments on a specific rock type, in which the duration of the experiment and the confining pressure were maintained constant. Only the temperature at which the individual experiments were conducted was changed. The results obtained for specimens of dunite are given in Figure 2.20a, while the manner in which the strength of a range of rocks and minerals varies with temperature is shown in Figure 2.20b.

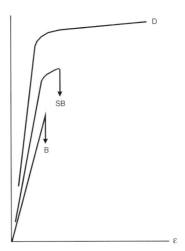

Figure 2.18 Stress-strain curves of a hypothetical cylindrical rock specimen subjected to biaxial compression at various confining pressures. At zero confining pressure, brittle failure (B) occurs; at moderately large confining pressure, failure is semi-brittle (SB); at high confining pressure, the rock speciment behaves in a ductile fashion (D).

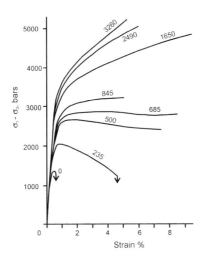

Figure 2.19 Strength and types of failure related to magnitude of confining pressure (after Love, 1994).

(3) **Strain-rate.** This type of experiment, which was mainly pioneered by Heard (1963), requires patience and extremely reliable apparatus. The test specimens were held at a specific confining pressure and temperature, while an electric motor, driving through a gear box, shortened, or permitted extension of, a jacketed, cylindrical specimen, at a constant rate, which usually ranged between 10^{-3} to $10^{-7.5}$ strain per second. (Nowadays, strain-rates are controlled and maintained constant by computer-controlled hydraulic systems.) Consequently, experiments lasted from a few hours duration, for fast strain-rates, to 70–80 days for the slower strain-rates.

The results derived from this type of experiment were plotted in the form shown in Figure 2.21, which enabled the constants to be evaluated for an equation of state of the form:

$$de/dt = constant.exp(-Q/RT)(S_1 - S_3)^n \tag{2.5}$$

where e is strain, Q is the apparent activation energy, R is the gas constant, T is the temperature in Kelvin, $(S_1 - S_3)$ is the differential stress and the exponent n is a material constant for some specific environmental conditions. It can be shown that this type of equation can be interpreted in terms of various mechanisms of crystalline deformation, so that the deformation map for a wide range of environmental conditions can be derived from a relatively limited spread of experimental data. Such a chart for olivine is shown in Figure 2.22.

2.7 Failure limits in oceanic lithosphere

Clearly, in order to estimate the strength of the lithosphere, it is necessary to establish the temperature gradient, so that one can relate the vertical stress and the temperature that obtains at any specific depth, thereby permitting the rheological behaviour of the lithosphere at that depth to be inferred. As noted earlier, the distribution of the temperatures based on such a cooling model is shown in Figure 2.5.

From rock mechanics data and for the specific strain-rate of $de/dt = 1.0 \times 10^{-14}\,s^{-1}$, Goertz and Evans (1979) also set out the relationship between temperature, depth and the differential stress that oceanic lithosphere could sustain, in compression and extension, as shown in Figure 2.23a and b respectively. In compression, the 10 kb isobar, appears within a few kilometres of the ridge, while in extension the 10 kb (1 GPa) differential stress zone comes into existence some 55 km from the ridge.

Based upon the experimental data and rheological model set out by Goertz and Brace (1972) and Goertz and Evans (1979) expressed the relationship between depth and the differential stress the lithosphere could support for a specific strain-rate in the form shown in Figure 2.24. It will be seen that the strength relationship is relatively simple, but asymmetric. In compression, the strength of the lithosphere increases linearly and rapidly down to a depth of 24 km, where it exhibits a peak value of about 17 kb (1.7 GPa). From depths of 24 to 50 km there is a rapid linear decrease in strength, so that, at 50 km, the strength is only 0.5 kb. Below 50 km, the diffusion processes of rock deformation dominate, so that at a depth of about 60 km the long-term strength is reduced almost to zero. The failure conditions in extension follow a similar pattern, but the differential stress attains a maximum value of –8.5 kb, and this peak value occurs at a depth of about 35 km. The base of the strong layer for the specified conditions is arbitrarily put at 50 km, where the differential stress which the litho-sphere can sustain, for the specified conditions, is 500 bar (50 MPa).

It will be noted that in compression, the strong layer, which we take as the 10 kb contour, and which extends to a depth of 50 km in Figure 2.25, is some 5 km deeper than the lower limit of the strong layer shown in Figure 2.24. This can be attributed to the strain-rate used. When Kirby calculated the stress contours, he chose a strain-rate two orders of magnitude faster than that assumed by Bodine *et al.* (1981).

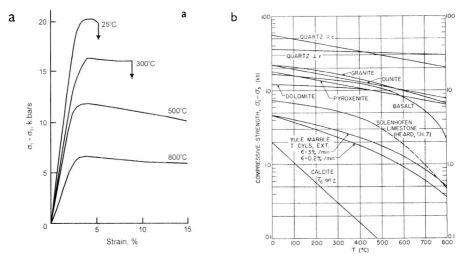

Figure 2.20 (a) Effect of temperature on strength (Griggs et al., 1960). (b) Variations of strength with temperature for various rock types (Griggs et al., 1960).

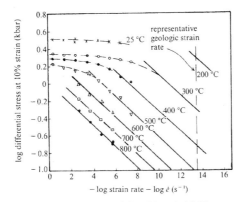

Figure 2.21 Variation of strength with strain-rate (after Heard, 1963).

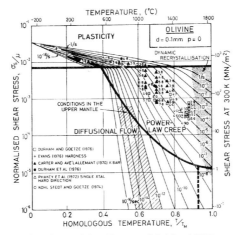

Figure 2.22 Deformation map for olivine (after Frost and Ashby, 1982).

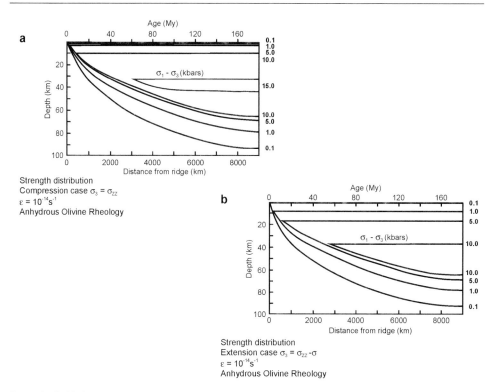

Figure 2.23 (a) Distribution of isobars of strength in compression (after Kirby, 1980). (b) Distribution of isobars of strength in extension (after Kirby, 1980).

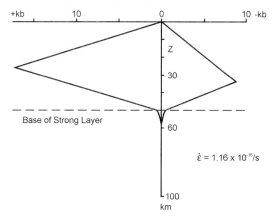

Figure 2.24 Estimates of variations of strength of oceanic lithosphere in compression (+kb) and extension (–kb) (after Goertz and Evans, 1979).

2.7.1 Stresses in oceanic lithosphere because of Gravitational Loading

Analyses which have been made of the flexure of oceanic lithosphere are usually of two types. The first of these relates to the broad anticlinal upwarping of the lithosphere adjacent to trenches. We shall defer discussion of this type of feature to Chapter 3. The second group of studies relates to the mainly downward deflection of the lithosphere brought about by volcanic chains. The form of such a chain in the vicinity of Hawaii, has received special attention. This follows from the abundant data relating to gravimetric anomalies and bathymetry, at such a large distances from the spreading ridge that the thickness of the lithosphere is about 80–100 km; in addition, the area is considered to be remote from any extraneous influence.

The aspects of elasticity theory that enable one to calculate the deflection of an elastic layer subjected to a vertical line load, are presented by Turcotte and Schubert (1982). The first step of such an analysis is to evaluate the distribution of compressive and tensile strains and stresses that develop in a simple flexured beam (Figure 2.26a) subjected to *bending moments*, M (Figure 2.26b). The loading by an island or seamount is represented as a point load, which causes deflection of a continuous floating beam. Knowing the magnitude of the load, and the thickness of the beam (i.e. the strong layer in oceanic lithosphere) and its elastic moduli, then, by solving a fourth order differential equation, the elastic deflection can be obtained. The inferred (half) profile of the (exaggerated) deflection of the sea-floor obtained by Turcotte and Schubert (Figure 2.26c) is in reasonable agreement with the observations.

The thickness of the beam in this diagram is, in part, determined by the value of the elastic modulus used. In the linear, two-dimensional type of analysis used here, the value of Young's modulus is assumed to have a constant value (say) $E = 5 \times 10^5$ bar. However, it is really more apposite that one uses *bending-plate* theory and cite the *Flexural Rigidity*, D, where:

$$D = E.m^2.h^3/12(m^2 - 1) \tag{2.6}$$

E and m are as previously defined, and h is the thickness of the plate, in the engineering sense, or, here, refers to the thickness of the *strong layer*.

When dealing with engineering problems, it is reasonable to assume that a specific metal will have a constant value of E, but one cannot reasonably make this assumption for the upper oceanic lithosphere. It is therefore pertinent to refer to data relating to the elastic moduli of the mantle as listed in Preliminary Reference Earth Model (PREM) or another seismically inferred data set of mantle elasticity.

Seismology gives values for the bulk modulus (K), the shear modulus (G) and Poisson's ratio (which can be readily transposed into Poisson's number m). If one assumes that the mantle is homogeneous and isotropic, one can calculate the value of E from the known values of G and m, using the relationship:

$$E = 2.G(m - 1)/m \tag{2.7}$$

The values of E and m for a few specific depths are given in Table 2.1.

Table 2.1 Changes in values of E and m, with depth Z

Z (km)	E (bar)	m
0	4.0×10^5	3.56
12	5.1×10^5	3.92
22	5.1×10^5	3.92

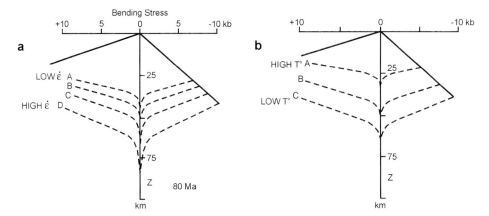

Figure 2.25 (a) This diagram, redrawn after Bodine *et al.*, indicates the influence of strain rate on the thickness reduction of a 'strong layer' with reduction of strain-rate. Curve D represents the effective thickness of the layer when subjected to high strain-rates ($10^{-2}\,s^{-1}$) such as those which may result from seismic events. Curves C, B and A represent the reduction in thickness of the strong layer at successively lower strain-rates of $10^{-15}\,s^{-1}$, $10^{-18}\,s^{-1}$ and $10^{-21}\,s^{-1}$ respectively. (b) This diagram, also redrawn after Bodine *et al.* (1981) indicates that the influence of temperature on the thickness of the strong layer is even more profound. The temperatures cited for C, B and A are respectively 60°C, 80°C and 130°C.

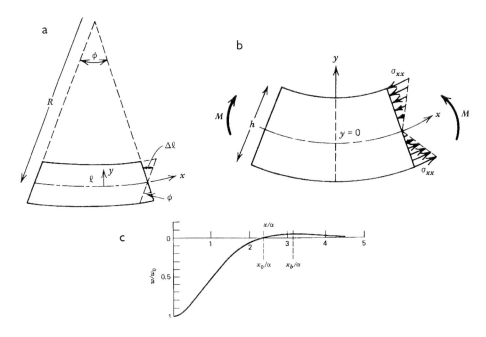

Figure 2.26 (a) Strains in a bending beam as the result of bending moments *M*. (b) Variations in compressive and tensile stresses in a beam caused to flex as the result of bending moment *M*. (c) Half profile of deflection of a floating beam in response to a point load (after Turcotte and Schubert, 1982).

It will be seen that the values for E and m are identical at depths of 12 and 22 km, though the value of m is somewhat smaller than the constant value $m = 4.0$ assumed earlier in this chapter and also by Watts *et al.* (1980).

The cited depths of 12 and 22 km are located within the depth for the strong layer. However, the estimated depth and thickness of such a strong layer have been based on linear elastic behaviour, with a single value of Young's modulus, E. In the light of the values of E cited in Table 2.1, errors in the thickness of the strong layer incurred by assuming a single value for E throughout the layer are not likely to be large.

Vening-Meinesz (1941), who conducted one of the early submarine gravity measurements, demonstrated that the load of the Hawaiian Islands was not compatible with the Airy model, but was regionally supported. Subsequently, Walcott (1970), Watts and Cochran (1974) and Watts (1978), using more recent gravity and bathometric data, showed that the compensation of the islands can be explained in terms of a simple elastic plate mechanism.

Watts *et al.* (1980) note that the objections to the early model of flexure of an elastic beam, were based on data obtained in the 1950s and 1960s, when the collection of such data, they suggest, was hampered in the sea areas by large shot spacing and navigation that lacked the accuracy of present-day methods. Accordingly, the supporters of this theory reiterate the applicability of the model based on flexure of a thick elastic layer in the oceanic lithosphere. Modelling carried out by Bodine *et al.* (1981) involved determining the rate of displacement of a highly viscous substratum, on which the elastic beam is supported. A simplified section of the island of Oahu, Hawaii, whose weight causes deflection of the lithosphere on which it rests, together with the areas in which the load causes failure, is shown in Figure 2.27a. The form of the calculated deflection of the lithosphere is shown in Figure 2.27b for a range of strain-rates, from which one may infer that the depth of the depression caused by a constant load would bring about a continuing, but decreasing, rate of deepening of the trough with time.

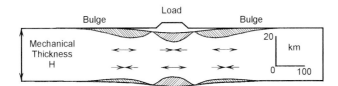

Figure 2.27a Deflection of oceanic lithosphere as the result of loading by the Hawaiian Ridge.

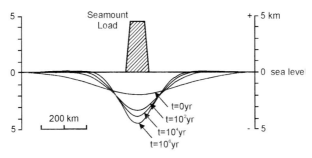

Figure 2.27b Showing an increased degree of flexure of the lithosphere at progressively slower strain-rates (after Bodine *et al.*, 1981).

When viewed in more detail (Watts *et al.*, 1980), the section of the ridge near Oahu is about 150 km wide and has a bathymetric profile as shown in Figure 2.28a. The vertical exaggeration of this profile, of 40 : 1, enables one to see that there is an initial dip of the sea-floor which is flanked by an upward bulge, on either side of the island ridge.

The variations in magnitude of stresses in the flexure caused by the weight of Oahu are shown in Figure 2.28b. The largest compressive and tensile stresses, respectively, are +6.8 kbar, which occurs beneath the island load, and –2.8 kb which occurs in the flanking regions.

To obviate this problem of the disparity between the upper and lower stresses, Watts *et al.* suggest that a better approach is to use the yield strength envelope derived by Goertz and Evans (1979) to estimate the bending stresses which develop in the plate, in terms of estimated applied *bending moment* coupled with the limitations set by the Goertz and Evans failure envelope (Figure 2.29). The thickness of the intact strong layer (able to support a differential stress of 7.0 kb) can be inferred from this diagram to be 34 km.

However, as many of the cited authors emphasise, it must also be borne in mind that the inferred thickness relates specifically to the ambient temperature gradient, mainly determined by the rate of plate motion and the distance from the spreading-ridge, coupled with the strain-rate at which deformation develops (see Figure 2.23).

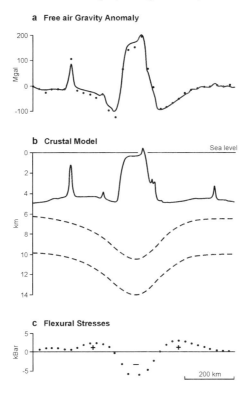

Figure 2.28 (a) Free-air gravity anomaly across Hawaiian Ridge near Oahu (solid line) and computed gravity anomaly (dots). (b) Comparison of crustal flexure based on seismic reflection data to a calculated flexure curve (dashed line). (c) Bending stresses at the top of the plate associated with the calculated flexure curve in (b).

Note that these authors are using the physicist's nomenclature with the plus sign denoting tension and minus for compression (from Watts *et al.*, 1980).

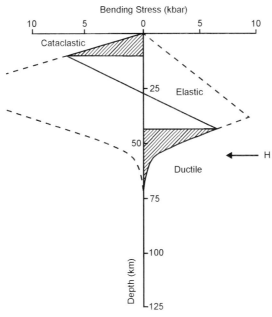

Figure 2.29 Bending stress as a function of depth for oceanic lithosphere (age 80 Ma) (modified from Goertz and Evans, 1979). The dashed lines indicate the limits of strength shown in Figure 2.24. Shown within the limits of the dashed lines are the failure conditions that can occur by the bending of the oceanic plate. The shaded areas show the limits of maximum bending stress that can be supported before brittle failure occurs. The strain-rate limit is taken to be 1.16×10^{-16} s^{-1}.

We would add a further caveat, namely that it is also necessary to include the effect of the ambient horizontal stress normal to the direction of the chain. In Figure 2.30a, we take the horizontal stress, which develops as the result of gravitational loading at a specific depth, to be one third that of the vertical stress. For simplicity and convenience, in Figure 2.30b, we have ignored the complication added by the additional vertical stress incurred by the volcanic chain or island. It will be seen from Figure 2.30b that the highest magnitude stress of equal positive and negative value which can be fitted into the modified failure envelope is 8 kb, and the thickness of lithosphere which can sustain such a stress is 22 km. For a corresponding stress situation, where the stress limits are +5 kb and –5 kb, the thickness of the strong layer will be 34 km. Moreover, we shall argue in Chapter 3 that the horizontal principal stress acting normal to the direction of absolute plate motion is less than the horizontal principal stress that acts in the direction of absolute plate motion. Hence, one can infer that the thickness of the strong layer may be even smaller that the dimensions cited above.

Hence, to summarise; the consensus among the model analysts, cited above, is that oceanic plates contain a strong layer, which stretches from a few tens of kilometres to the furthest limit of that plate, from the spreading-ridge. From the results presented by these various cited authors, it is reasonable to conclude that the differential stress which this strong layer can sustain, for periods of at least 100 Ma, is likely to exceed several kilobars, and in areas distant from the ridge, the strong layer may support a differential stress approaching, or even exceeding, 10 kb. However, as regards the thickness of the strong layer, relying as this

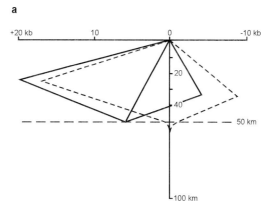

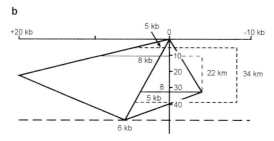

Figure 2.30 (a) The dashed lines show the failure conditions represented in Figures 2.24 and 2.29 and the full lines represent the failure envelope when the horizontal stress, which results from gravitational loading, is taken into account. (b) It will be seen from this diagram that the maximum symmetrical stress that satisfies the failure envelope is 8 kb and this can only be sustained over a thickness of lithosphere of 22 km. A symmetrical stress of 5 kb, however, can be sustained in a thickness of lithosphere of 34 km.

does on strain-rate, the elastic moduli and the ambient horizontal stress, we can be far less certain. If the differential stresses are in compression, as we may infer from Figure 2.30b, it is likely that the strong layer may not be more than about 20–25 km thick, for the very slow strain-rates of 1.16×10^{-16} sec^{-1}, rather than a thickness of about 34 km, which would be inferred from the thickness estimates given in Figure 2.23. This implies that the thicknesses shown in Figure 2.24 also over-estimate the probable thickness of the strong layer.

It should also be noted that Figure 2.30 assumes that the elastic limit is not exceeded, except by fracture, down to a depth of 50 km. However, the differential stress which the rock mass can sustain at this depth is as little as 500 bar, when it will begin to deform mainly in a ductile manner. The probable horizontal stress as a result of gravitational loading is indicated by the dashed line in Figure 2.30b. This aspect will be considered in more detail in the following chapter.

2.8 Magnitude of interplate stresses

The lithosphere of the central area of the Indian Ocean is an area of high seismicity (Wiens and Stein, 1983, 1984; Bergman and Solomon, 1984, 1985) and these seismic data have been used to estimate the magnitude of the differential stress that gave rise to earthquakes in the area. Cloetingh and Wortel (1985, 1986) calculated that the values of stresses required are very high, i.e. in excess of several hundred megapascals (i.e. greater than 3 kb). These estimates of differential stress were approaching an order of magnitude higher than those previously obtained by Richardson *et al.* (1979) and were questioned by Richardson (1987, 1989).

Two or three decades ago, it was not unusual for geophysicists, who based their estimates on 'seismic moments', to cite the stress-drop, and (hence the strength of faulted rock) to be in the order of 10–100 bars. Such strengths are, of course, possible in weak sedimentary rock, and in somewhat stronger rocks in which the fluid pressure was high. However, from rock mechanics tests, which we have already cited, dry, unweathered, igneous rock subjected to high confining pressures in laboratory tests, can be capable of sustaining quite large differential stresses. From slow, constant strain-rate experiments conducted at temperatures and confining pressures that are consistent with a depth of cover of 20 km, many igneous rocks are capable of sustaining a differential stress of as much as 10–20 kb. Hence, a stress-drop of 3–5 kbar, associated with a seismic event can be considered as intrinsically reasonable, and even modest.

Gravity highs exist in the Central Indian Ocean which are attributed to basement undulations, which result from buckling of the lithosphere. McAdoo and Sandwell (1985) and Zuber (1987) calculated the compressive stress necessary to buckle the lithosphere to the specified wavelength and amplitude of the observed flexures, and reached the conclusion that the required magnitude of compressive stress must approach 5 kb, so supporting the values of stress inferred by Cloetingh and Wortel (1985, 1986).

In the light of this debate with Richardson (1989), Govers *et al.* (1992) reassessed the problem. In this reappraisal, whenever it was necessary, they 'made assumptions which (1) are considered to be realistic and (2) give low-end differential stress estimates'. The results of their calculations are shown in Figure 2.31 where, it will be seen, the highest differential stress is just over 400 MPa (4 kb). The thick bar in each of the columns represents the values previously calculated by Cloetingh and Wortel. In the majority of instances, the recalculation increases the estimated differential stress required to bring about the seismic event.

It is interesting to note that the magnitude of the stresses inferred from this study increases progressively from the south to the north. Also, we suggest that because reliable records of earthquakes have only been available for a few decades, it is most unlikely that the recent seismic events are the greatest that have ever been experienced in the central Indian Ocean. Consequently, it seems likely too that, in the light of the efforts of Govers *et al.* to minimise the estimated magnitude of the differential stress, the maximum differential stress that can be supported by the Indian Plate may well be significantly in excess of 5 kb.

The Indian Plate is providing the energy that is driving India into Tibet, with the attendant deformation to the Asian continent which is currently unequalled anywhere else in the world. Consequently, a compressive stress in the Indian Plate of at least 5 kb is inherently probable. In contrast, Govers *et al.* also studied seismicity in the Pacific Plate, where the seismic activity is lower, and only in one location does the estimated differential stress reach 4 kb. Mainly, the estimated differential stress is less than 2.5 kb (Figure 2.32). (These latter data are also dealt with in Chapter 3, where we consider the mechanisms driving plates.)

2.9 Conclusions

By using simple models to represent a plate moving in a specific direction for long periods, it may be inferred that such a plate must be pushed, pulled, or moves as the result of both

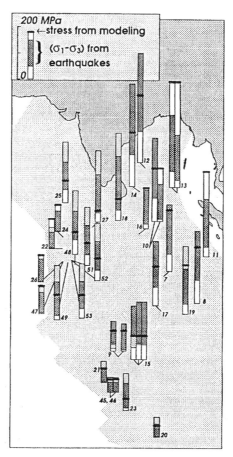

Figure 2.31 Inferred magnitude of stresses in Indian Plate (after Govers *et al.*, 1992. © American Geophysical Union).

pushing and pulling. Whether the dominant force, or pressure, results from pushing or pulling, it should give rise to a dominant strain and, therefore, a dominant stress pattern within the plate. If the dominant force is one generated by some pulling mechanism, the least horizontal stress (S_3) should generally be aligned parallel to the direction of absolute plate motion. However, if plate movement is dominated by some pushing mechanism, it is the axis of greatest horizontal stress that will be parallel to the direction of absolute plate motion.

The orientation of the maximum horizontal stresses in the Earth's crust, based on a whole array of methods and techniques, shows that the majority of the inferred axes of maximum horizontal stress, in any but the slowest moving, or almost stationary, plate, is in reasonably close alignment with the direction of absolute plate motion. Hence, we conclude that Push is the dominant mechanism.

It has long been inferred that, because oceanic plates are able to support the load of solitary, or chains of, volcanoes for tens of millions of years, then, of necessity, a strong and elastic layer must exist within oceanic lithosphere. Based upon accumulated rock mechanics data and an assessment of the variations of thickness and temperature distribution from the spreading-ridges to the abyssal areas of oceanic plates, it is clear that the strong layer comes

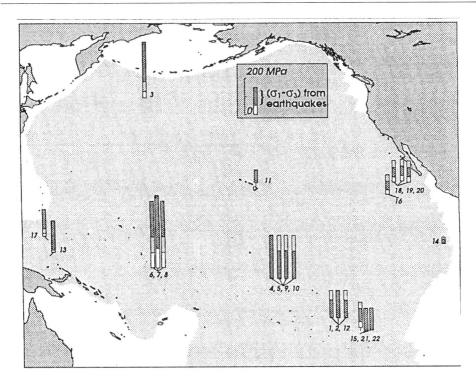

Figure 2.32 Inferred magnitude of stresses in the Pacific Plate (after Govers *et al.*, 1992. © American Geophysical Union).

into existence at only a little distance from the spreading-ridge. From conservative estimates of its strength, it is almost certainly capable of supporting horizontal compressive pressures of up to at least 8–15 kb.

The thickness of the strong layer is, however, much less certain. Depending, as it does, on strain-rate and the distribution of the value of elastic moduli in the upper 50 km of the lithosphere, the vertical extent of this strong layer may be somewhat less than the estimates to be found in the literature. Nevertheless, the existence of such a strong layer is not in doubt. Finally, analyses of seismic data and of flexures in the Indian Ocean plate demonstrate that differential stresses with a magnitude of at least 4 kb, and probably in excess of this magnitude, exist in the lithosphere. In the light of these various observations and inferences, it is now apposite to ascertain what combination of mechanistic elements can be put forward to explain plate motions, satisfactorily. In the following chapter we shall consider the possible pull and push mechanisms, and the magnitude of stress these various mechanisms may be capable of generating.

Note

1 The conventions used to indicate the relative values of principal stresses can cause confusion. The earliest work on evaluating stresses in real material was conducted by physicists and engineers. The first work was conducted on the stretching of wire, attached to a hook in a beam or ceiling, by putting a weight in a pan attached to the lower end of the wire. As weights were added, it was noted that the wire increased in length, i.e. the 'stretching was positive'. The

stretching induced a tensile stress, so tensile stresses were also considered to be positive. When, many years later, geologists began to become interested in stresses in the Earth's crust and mantle, they argued that, except near the Earth's surface, the total stresses were always compressive and that it was therefore convenient to designate all compressive stresses as positive.

Chapter 3

Assessment of mechanisms that cause plate movements

3.1 Introduction

In this chapter we shall discuss the main driving and resistive mechanisms which are of primary importance in defining plate motions. It is generally considered that there are two main such driving forces, namely (1) the so-called *ridge-push mechanism* and (2) that mechanism which gives rise to tensile forces as the result of the oceanic lithosphere becoming subducted into the asthenosphere. We use the term *slab-pull* to explain this tensile force, where 'slab' is an abbreviated form of 'subducting slab'. (The trench is a surface feature brought about by subduction and has, itself, no significant mechanistic role in subduction.)

We argue that the combined mechanisms of slab-pull (once the resistive effect of the subducting slab forcing its way through the asthenosphere, is taken into account) plus the ridge-push are insufficient to overcome the various elements resisting plate motion. Consequently, we resurrect and expand upon a mechanism which has been largely neglected for three decades. This is followed by an outline of the types of structures, flexures and fractures, that are known to develop in oceanic plates and we note how such features have developed and fit in with the conclusions of this and the previous chapter.

The discussion, to this point, is concerned with a plate that is *up and running*. To attain this state, one must explain the initiation and development of the spreading-ridge and the trench where subduction of the oceanic lithosphere is formed. Only the second of these topics is discussed in this chapter; the formation of a spreading-ridge we leave until later.

The reader will realise that the conceptual modelling that we carry out is extremely simple and approximate. Nevertheless, these models contain the essential physics of the problem and permit the stresses, strains and strain-rates to be quantified. We do not attempt a comprehensive review of all the pertinent elements of plate motion. For example, we do not consider the magnitude of the stresses involved in mountain building. However, we note that these stresses can be inferred from earthquake data and evaluation of the stresses required to produce specific geological structures (Sibson, 1975), from which one can infer that the maximum differential stress is probably in the range 4–6 kb.

Locally, when faults are generated, the rates of movement are fast; but only for brief transient periods. The dominant rates of plate movements are ponderously slow, usually in the range of 1–10 cm a^{-1}. The models discussed in this chapter are completely unable to explain the abrupt changes of rate and direction of plate motion which are, from time to time, exhibited in the geological record.

3.2 Mechanisms driving and resisting plate motions

The conventional mechanisms that may be expected to determine plate dynamics and kine-matics, which have been given brief mention in Chapters 1 and 2, can be grouped under three headings, namely:

(1) Basal effects deriving from convection cells in the mantle.
(2) Resistive, or retarding effects.
(3) Driving mechanisms associated with ridge-push and slab-pull forces.

3.2.1 The convection cell mechanism

This concept initially attracted a great deal of support (Griggs, 1939; Holmes, 1965). It was argued that the heat generated in the interior of the Earth (largely as the result of radioactive decomposition of minerals) must escape through the mantle by convection and by conduction at the lithospheric boundary, and that this accounted for the high temperature gradient of about 25°C near the surface of the Earth. The density anomalies in the mantle, which are necessary to drive currents at a rate comparable with that of continents, may be easily met, in that the required temperature difference between the up-going and down-going currents, at a specific depth, need only be small (see Chapter 1).

In this early period, it was thought that plate motion was dominated by convectional movements in the asthenosphere beneath the plates, in much the same manner in which surface 'skin' in a cooking pot is driven into ridges by the mobile convectional movements of the more fluid content of the pot. This concept was supported by a persuasive paper (and film) of a model in which David Griggs (1939) 'built a mountain chain' by causing opposing currents to develop in a viscous fluid, dragging a lighter, plastic crust to form a *geosyncline*. In this phase, the crust became compressed and thickened. The crust was then uplifted to form a mountain chain, once the motivating currents were stopped (Figure 3.1).

However, it was soon realised that small plates would require to be driven by a convection cell in the asthenosphere of a comparable plan-area. For a very much larger body, such as the Pacific plate, a correspondingly large plan-area convection cell would be required, for it is difficult to envisage how a relatively large number of small convection cells could combine to provide a resultant force capable of driving such a large mass as the Pacific plate at a relatively high velocity.

Furthermore, hotspots, some of which are presumably rooted in the mantle beneath the plates, are almost stationary, as regards lateral movement, while the plates move at significantly higher velocities. Hence, it was rendered even more difficult to envisage how plates can be driven by mantle convection, while hotspots are relatively unaffected. This difficulty was further enhanced (Turcotte and Schubert, 1982; Bott, 1993), once it was realised that oceanic plates were underlain by the low-viscosity zone (LVZ) that might be 50–200 km thick and in which the coefficient of viscosity is at least an order of magnitude less than that of the mantle in general. Consequently, the possible coupling between a mantle convection cell and the overlying plate was seen to be exceedingly improbable. In motoring terms, there was a 'slipping clutch' between the engine and the drive shaft.

Anderson *et al.* (1992) expressed what is, we believe, the generally held view that plates slide on the LVZ because it presents a relatively small, unit area, resistance to plate movement. The *plume hypothesis*, mentioned above and also in Chapter 1 (Morgan, 1971) which refers

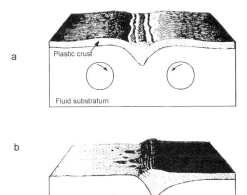

Figure 3.1 A mechanical scale-model designed by Griggs to simulate the action of convection currents in the mantle by using rotating drums. (a) This figure shows how down-folding of the crust has developed in response to the inward drag of the currents in the mantle. (b) Shows crustal development when only one drum is rotated. The 'continent' to the right is thickened and shortened as the result of lateral transfer of crust from the left. It will be noted that the mountain chain is transposed towards the non-rotating drum. It is held in that position because the right-hand wall of the tank prevents lateral motion of the crust. If the wall were distant from the drums, so that the crust could freely move to the right, it is probable that the mountain chain would not have developed.

to upwelling columns in the mantle, has been used to explain the initiation of spreading-ridges, hotspots and a wide range of igneous provinces (all of which will receive comment in later chapters). The plume mechanism was originally considered to be responsible for the generation and continued development of a spreading-ridge. However, there are certain obvious problems in applying this concept, namely that all spreading-ridges migrate and, moreover, they frequently *jump*, i.e. they show abrupt displacement along transform faults (Figure 3.2). The reasons for such abrupt displacements is not well understood. However, less rapid differential movement of a spreading-ridge can be induced by the behaviour of strong crustal elements in the oceanic lithosphere. For example, the Nazca/Sala-y-Gomez Ridge abuts the S American plate (Figure 3.3). This ridge is too thick, wide and strong to permit its easy subduction beneath the S American plate. Consequently, it acts as a battering-ram which locally closes the trench, and also prevents current activity of previously active volcanic cones in the Andes, in the latitude of the ridge. However, S America is advancing westward and is, therefore, also pushing the 'battering-ram' westward, thereby, causing differential displacement of that ridge. It is not easy to see how a surface feature, such as the Nazca/Sala-y-Gomez Ridge can disrupt a deep-seated wall-like plume.

3.2.2 Driving and retarding mechanisms

Several attempts have been made to establish the importance of various elements involved in the main mechanisms which could reasonably be thought to give rise to plate motions. One of the more important early syntheses is illustrated in Figure 3.4. The forces acting on such a plate can be classified into two categories, namely (a) *basal* and (b) *edge* forces.

a

b

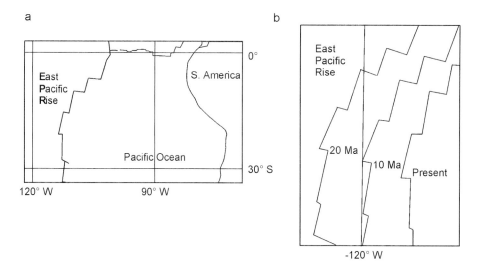

Figure 3.2 (a) Map of the present-day East Pacific Rise (EPR), is an Atlas 3.3 representation. (b) Map showing the position of the ridge at 0, 10 and 20 Ma, indicating the migration of the ridge and 'jumps' in the ridge which represent displacement of the spreading-ridge by transform faults.

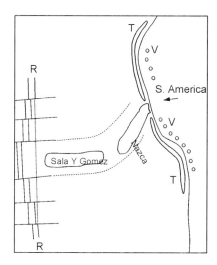

Figure 3.3 The Nazca and Sala-y-Gomez Ridge impinges on the S American continent. It has locally obliterated the trench and has caused differential westward migration of the East Pacific Rise. R = ridge; T = trench; V = volcano.

The basal forces, which may be either a driving or a retarding force, depending upon the differential motion, are represented by a series of symbols in Figure 3.4. For ease of presentation we shall represent in the text these basal force with the symbol F_b and mainly restrict the application of these forces, acting on the base of the plate to result from the viscous coupling of oceanic lithosphere with the asthenosphere.

The continental and oceanic lithospheres may behave differently, as the LVZ is less well defined, or even non-existent, below continents. Moreover, the continental lithosphere may also have a relatively deep 'keel'. If such a keel exists beneath a continental lithosphere, and plate motion is independent of mantle flow direction, then it is reasonable to infer that the basal force will resist movement in plates containing a large proportion of continental material, more than in plates dominated by oceanic lithosphere. Hence, one would expect plates with a high percentage of continental crust to move relatively slowly. For current major plate geometries this expectation is generally supported (Table 3.1). From the data in this table it is reasonable to conclude that drag on the base of the continental lithosphere is a major retarding force on plate motion.

There are a number of forces which act upon the edge of plates. Studies of these edge-forces have been made by several authors./ Forsyth and Uyeda (1978) set out and commented upon the various forces involved in plate motion (on which Figure 3.4 is based). Runcorn (1980) presented a survey of the causes and magnitudes of various contributory forces that can contribute to plate movements. Richardson (1992) analised the torques acting on plates. Other authors to which the reader is directed for a broad view of forces involved in plate movements include Bott (1993), Wilson (1993) and Ziegler (1993).

Without doubt, most early papers favoured the importance of slab-pull. Indeed, Forsyth and Uyeda concluded that slab-pull was an order of magnitude more important than any other force. Their conclusion would appear to be confirmed by the degree of correlation between absolute velocity and length of trench boundary of plate as a percentage of the total circumference of the plate, as can be seen for the ten most important plates cited in Table 3.2.

Table 3.1 Plate absolute velocity.

Plate	Absolute velocity (cm y^{-1})	Continent (per cent)
Eurasian	0.7	74
North America	1.1	60
South America	1.3	49
Antarctica	1.7	25
African	2.1	39
Arabian	4.2	90
Indian	6.1	25
Philippine	6.4	0
Nazca	7.6	0
Pacific	8.0	0

Table 3.2

Plate	Absolute velocity (cm y^{-1})	Circumference (km)	Effective trench (per cent)	
Eurasia	0.7	42100	0	(0.0)
North America	1.1	38800	1000	(2.6)
South America	1.3	30500	300	(1.0)
Antarctica	1.7	35600	0	(0.0)
African	2.1	41800	900	(2.1)
Arabian	4.2	9800	0	(0.0)
Indian	6.1	42000	8300	(19.8)
Philippine	6.4	10300	3000	(29.1)
Nazca	7.6	18700	5200	(27.8)
Pacific	8.0	49900	11300	(22.6)

As regards the retarding (drag) forces, these primarily act on the base and on the vertical boundary walls of the plate and on transform faults. A further resistive force is presented by the colliding plates (F_{CO}). In addition, there are resistive elements which relate to the bending of the lithosphere as it is subducted, as well as the resistance of the asthenosphere to the descending slab.

The basal shear is related to the area of the plate and is dominantly resistive. Moreover, because this resistive element is considered to be related to the plate moving over the LVZ, it will be a viscous restraint, and, therefore, will also be related to the velocity of plate motion relative to the asthenosphere. Unfortunately, the magnitude of this basal resistance is not well constrained, but it can be inferred from maps that the Pacific plate has an area of the order of 150,000,000 km², so that in this instance at least, the total basal resistance to movement must be considerable; yet it is the fastest moving plate.

Beneath the oceanic plates, the layer of viscous restraint is likely to be contained within the LVZ, which is also regarded as a low-viscosity layer. Below the continental lithosphere, the LVZ appears to occur only sporadically, if at all. Also, it is known that some continental lithospheric units contain quite deep roots. Both these features would result in a relatively high resistance of continental lithosphere to basal slip.

The shear resistance at the boundary and transform faults (F_{TF} in Figure 3.4) will be related to the area of these features, given by its length multiplied by the average vertical height. The resistance to brittle, strike-slip shear can be attributed to simple frictional resistance. For a leaky transform fault, the restraint will be viscous (so that this resistance will be proportional to the velocity of differential plate motion across the transform fault).

Let us now consider the main elements described by Forsyth and Uyeda, and estimate the probable magnitude of individual elements which constitute driving, or retarding, forces or forces and stresses.

3.2.3 Spreading-ridges

It is generally accepted that the spreading-ridges contribute a driving element to plate motion. This conclusion is based on the lack of symmetry of the lateral pressures of fluids above and below the oceanic lithosphere (Figure 3.5). The fluid above the lithosphere is, of course, the

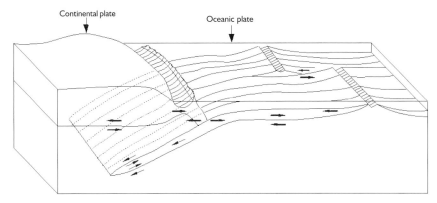

Figure 3.4 Driving and retarding forces, at external and internal surfaces, which contribute or detract from plate movement (after Forsyth and Uyeda, 1978).

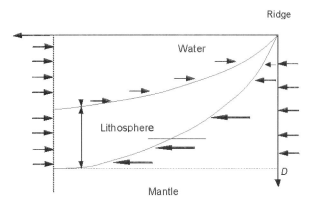

Figure 3.5 Ridge-push mechanism (after Turcotte and Schubert).

ocean, which exerts a lateral pressure on the ridge which is commonly 3–4 km higher than the abyssal plain. The other fluid is that of the asthenosphere which exerts a corresponding lateral pressure on the base of the lithosphere directed away from the ridge.

If we consider mature oceanic lithosphere to be about 150 Ma in age, then the greatest thickness of the lithosphere will be about 100 km. As can be inferred from Figure 3.5, because sea water has a density only a little over 1.0 g cm^{-3} and has a total depth of only 4 km, on which the pressure acts, the lateral pressure will be relatively small. However, the density of the asthenospheric mantle (about 3.3 g cm^{-3}) is higher than water and, therefore, generates a correspondingly higher pressure than water, for every kilometre of depth. Moreover, this mantle pressure operates on a surface that extends from close to the elevation of the ridge itself, down to a depth of about 100 km, so that this pressure acts on a surface with a vertical extent of about 25–30 times the elevation of the sea-floor from abyssal plain to spreading-ridge: and, of course, it acts in the direction necessary to bring about ridge-spreading.

The derivation of the equations necessary to quantify the ridge-push pressure is given by Turcotte and Schubert (1982), who conclude that the total ridge push acting on a 100 Ma plate, per unit length of ridge, is about 4×10^{12} N m^{-1}, giving rise to a stress of 4×10^{7} Pa (400 bar). Other authors (Lister, 1992; Parsons and Richter, 1980; Meijer and Wortel, 1992; and Bott, 1993) estimate the force to somewhat smaller values, which fall in the range 2–3×10^{12} N m^{-1}, giving rise to a stress of 200 to 300 bar.

It should be noted, however, that the water acts against a strong ocean floor, whereas the much larger asthenospheric pressure acts on the lower oceanic lithosphere, which is very weak. Hence, the ridge-push may be significantly dissipated by deformation of the lower lithosphere. Only if this deformation gives rise to viscous drag on the base of the elastic layer, will this description of 'ridge-push' generate a significant horizontal stress in that strong layer.

3.2.4 Subducting lithosphere

One can readily show that a subducting, or down-going slab of lithosphere must, potentially, give rise to a very significant tensile stress on the horizontal, oceanic lithosphere. This is brought about by the fact that cold lithosphere is more dense (3.3 g cm^{-3}) than hot astheno-sphere (3.1 g cm^{-3}) of almost identical composition (Figure 3.6).

Turcotte and Schubert (1982) took into account the olivine-spinel phase change and showed the tensile force generated by the slab to be about 5×10^{13} N mm^{-1}. If this force (F) is transmitted to the elastic lithosphere, with a strong layer thickness of 50 km, it would generate a tensile stress of -1 GPa (-10 kb). It can therefore be inferred that, when the slab is inclined at about 45°, the slab-pull is likely to be 25–30 times greater than the previously calculated ridge-push of about 0.2–0.4 kb. However, we have so far ignored the important effects of the resistance presented by the asthenosphere to slab penetration (r).

This factor is, of course, dominant in reducing the calculated *potential* value of the tensile pull of the down-going slab. The estimates of the stress derived or cited above do not make any reference to the shear resistance at the upper and lower boundaries of the slab, or to the resistance to penetration by the front end of the slab. The values of these resistive forces are not readily determined, for they are based on a range of boundary conditions which are not easily demonstrated or quantified.

We accept the 'judgement of Solomon' approach taken by Turcotte and Schubert who suggest that, 'Trench (slab) pull must be mostly off-set by large resistive forces encountered by the descending lithosphere as it penetrates the mantle. The nett force at the trench is probably comparable with that of the ridge-push'. One could infer from this statement that the forces of trench-pull and 'ridge-push' would give rise to a differential stress of not more than 10^8 Pa (1.0 kb). However, we have indicated that the 'ridge-push' mechanism may have little effect upon the elastic layer of the oceanic lithosphere. Hence, at the trench, we are theoretically left with a small 'slab-pull'.

3.2.5 Retardation at the lithospheric/asthenospheric interface

The quantification of this mechanism has been very largely neglected. Richardson *et al.* (1979) and Richardson (1992) have suggested that the restraint to basal movement is 10^{-2} MPa (0.1 bar) or less per unit area. These estimates have been derived to fit the limitations set by the ridge-push/trench-pull mechanisms, and are, we suggest, not appropriate values.

The total resistance of basal slip will, of course, be related to the whole area of the plate and, because the restraint provided by the LVZ is a viscous one, it will also be related to the velocity of plate motion. If the plate is large and the motion relatively rapid, the total resistive force presented by this basal restraint will not be trivial! We shall discuss this topic

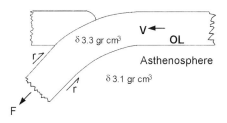

Figure 3.6 Slab-pull mechanism. V = velocity of movement of oceanic lithosphere, d = 3.3 g cm^{-3} and 3.1 g cm^{-3} for density of oceanic lithosphere and asthenosphere respectively, r = the retardation force on the upper and lower surface of the down-going slab and F = the tensile force generated by the difference in density of the oceanic lithosphere and the asthenosphere.

later in this chapter and present what we consider to be more realistic values for the basal restraint to plate motion than those cited above.

3.2.6 Restraints caused by boundary walls and internal transform faults

These restraints cannot be quantified with accuracy. The number of boundary walls in a simple model are, of course, two, with the plate bounded by a spreading-ridge at one end and by a subduction trench at the other. Real plates, as will be seen from Figure 2.2, are much more irregular and complex. Also, real plates contain a number of internal vertical, strike-slip, transform faults which off-set the spreading-ridge boundary to a plate. These fractures usually trend normal to the pole of rotation of the plate. How these strike-slip faults develop is not yet clearly understood. We consider the most feasible explanation is presented by Gudmundsson (1995), to which we direct the interested reader.

The structure of transform faults is known to be far from uniform (the reader may wish to refer to various papers on this topic *Oceanic Fracture Zones, Journ. Geol. Soc. London*, **143**, 5, 1986). However, Garfunkel (ibid) argues that the great majority of such ridge-to-ridge transforms are *leaky*. At such 'leaking' faults (which, from time to time, extrude lava at the surface) the resistance to strike-slip motion will be relatively small.

The length of such features varies from place to place. For example, the Clipperton and Clarion Transforms, in the Pacific, extend for about 5000 km, with several others in the E Pacific having a length of 2000–3000 km. However, transform faults in the Atlantic and other oceans are usually less than 1500 km, but the portion of the fault that moves is largely, or wholly, restricted to the distance between the off-set, spreading-ridge. Consequently, one can infer that transform faults develop in areas where the oceanic lithosphere is relatively thin. Hence, we suggest that they will constitute a relatively small percentage of the total restraint to movement of a major plate.

Plate boundary faults may be non-leaking and, of course, deeper and also of greater lateral extent than transform faults. In a large, mature oceanic plate, the average depth of the boundary surfaces may be about 60 km, and their total length reaches 50,000 km, so that the area of such features is about 3×10^6 km^2, with perhaps only half this area cutting the elastic region of the oceanic lithosphere. However, the area of the Pacific plate is of the order of 100×10^6 km^2, so that the boundary and transform fault areas are likely to constitute only about 2 per cent of the total restraining areas of a major plate.

Hence, when one compares the total area of the major boundary plates combined with that of the transform faults, with that of the basal area of the plates, it is likely that the resistance to movement produced by the boundary faults will be only a small fraction of that produced by the basal slip plane on a large plate. The ratio of the boundary-fault/retardation may increase somewhat for small plates.

However, if small sections of the boundary faults, on opposite sides of the plate, converge slightly, this would increase the retardation effect. Such converging boundaries would locally build up large stresses. If the stresses become too large, then, of course, the converging part of the boundary will be 'sheared-off', hence, there is likely to be a 'feed-back mechanism' which tends to set a limit on this 'convergence' effect (cf. the San Andreas fault complex).

Clearly, precise evaluation of the various retarding effects on boundary and internal strike-slip faults is constrained by our lack of knowledge regarding these features. However, it is probably prudent not to underestimate the contribution that these boundary and internal strike-slip faults make to plate movements.

3.2.7 Critique

From the above discussions, it would follow that the total combined push-pull stresses that can be accounted for by using the simple models described above are not more than about 1.0 kb (and may be as little as a few hundred bar). It is not reasonable to expect this magnitude of stress to overcome all the elements that resist plate motion, and also account for the deformation that results from continent/continent collision and the attendant mountain building, such as the Himalayas. As we have already noted, it has been inferred that deformation which develops as the result of such tectonic processes must sometimes attain a magnitude of at least 4–6 kb (Sibson, 1975). In this assessment we have initially given slab-pull equal weighting with ridge-push. However, the S American plate is not influenced by slab-pull, yet is still able to move westward at a respectable plate velocity. Consequently, we may infer that if this plate lacks slab-pull and yet moves and also sustains a mountain chain, then the dominant element in sustaining plate motion must be push and, moreover, it requires a push many times larger than can be accounted for by the model of ridge-push outlined above. A mechanism capable of generating a push-dominated, differential stress with a magnitude of several, perhaps ten, kilobars must be missing from our assessment.

3.3 Iceland, roll-over structures and gravity-glide

Iceland sits astride the N Atlantic spreading-ridge. Western Iceland belongs to the N American plate while E Iceland is part of the Eurasian plate. It must at once be admitted that this island is not completely typical of the submarine spreading-ridge. Nevertheless, because it has a significant, sub-areal outcrop, it provides important data regarding ridge-spreading. This island, which is considered to be situated above a plume, dominantly consists of lava flows, the oldest of which, found at the surface, date back to 14.5 Ma and, as will be seen from Figure 3.7, the 'Atlantic ridge' continues through the island, albeit along an offset (transform) track.

It was during a traverse of the recently opened ring road, in 1982, with Agust Gudmundsson, that we saw evidence that convinced us of the mode and mechanism by which plates moved and generated high-magnitude compressive stresses directed away from the spreading-ridge. Lava flows in Iceland are intermittently, but, on the geological time-scale, frequently generated along the active volcanic zone shown in Figure 3.7. In this central area, the lava flows dip gently away to east or west, from the fissures that permit their development. However, as one travels eastward from the volcanic zone, and encounters older flows, one notes that the dip of the lavas becomes reversed and progressively increases, until, at the east coast, they dip westward by as much as 10°. A similar, but mirror image, change is to be seen in a traverse westward from the central zone. A schematic E–W section through the island is shown in Figure 3.8a.

We reached the conclusion that one could most readily explain the inward dip of the lava flows on either side of the volcanic zone as part of two *roll-over* structures, back to back. Roll-over structures are well known, and are often found in the soft sediments of a major delta, which dip sea-ward at about 1–3°. Instability occurs along particularly favoured bedding surfaces, or narrow weak zones, so that the upper units slide down-slope towards the bottom of the delta, where they may form folds and/or thrusts. Up-slope, the glide-mass becomes detached by movement along *listric faults* to form the attendant anticlinal roll-over structures (Figure 3.8b). On what plane could the Icelandic lavas have slipped? We suggested that, in

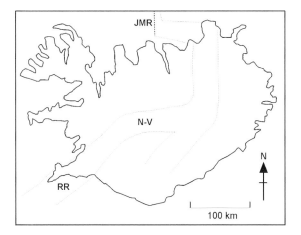

Figure 3.7 Sketch-map of Iceland. RR = Reykjanes Ridge, JMR = Jan Mayen Ridge, N-V = Neo-Volcanic Zone.

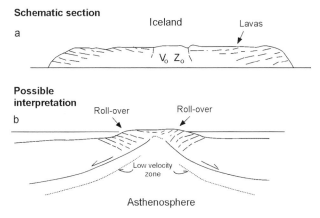

Figure 3.8 (a) E–W section through Iceland. (b) Roll-over structure and section through lithospheres of Iceland.

the case of Iceland, slip took place on the LVZ at the base of the oceanic lithosphere. Beneath the volcanic zone in central Iceland the asthenosphere approaches within a few kilometres of the surface, but deepens relatively rapidly away from either side of the volcanic zone (Figure 3.8b).

The development of such a roll-over structure requires the stretc.hing and shearing deformation of the uppermost rocks in the structure. It may be suggested that such deformation would not easily be accommodated in basalt, a rock known to be strong and, therefore, not easily deformed. However, such a suggestion does not take into account the cooling fractures which develop in lava flows. These lava flows are often extruded onto earlier lava flows, the upper surface of which is usually slightly weathered. The extruded lava of the fresh flow cools rapidly to develop mainly vertical, open fractures which intersect, or abut, to form, in plan, a polygonal pattern. Such fractures will, of course, easily open further to accommodate stretching.

In such a decidedly wet environment as is afforded by Iceland's climate (and as would be present in any submarine environment) the open fractures soon permit the surfaces of the polygon to become weathered and smeared with clay minerals. Each column of basalt is quite strong, but they tilt relatively easily along the fracture surfaces, thereby also permitting shear strain.

The reader may wish to demonstrate this latter mechanism by taking a dozen or so new pencils with a hexagonal section. Bind the pencils together lightly with one or two thin elastic bands. Lightly press the points into a rubber or soft plastic pad, already in place on a table, so that the pencils are vertical. It will be found that the pencils can then be tilted, with little difficulty, through an angle of a few degrees, in any direction, merely by manipulating the top of the column of pencils as one would a 'joy-stick'.

3.3.1 Gravity-glide of oceanic lithosphere

Gravity-glide (Figure 3.9a) as a possible mechanism of generating a horizontal compressive stress was first proposed in a brief note by Hales (1969). He suggested that this force resulted from oceanic lithosphere sliding down an inclined and 'lubricated' lithosphere/asthenosphere surface.

This concept has either been completely ignored, or has received a 'bad press'. For example, Wilson (1993) omits this mechanism when discussing the constraints and controversies regarding plate motions. Runcorn (1980), on the other hand, does discuss the mechanism briefly and admits that it is capable of generating a significant compressive stress in the down-slope direction. However, surprisingly, he rejects the mechanism as being an unlikely one, without in any way justifying this conclusion.

Jacoby (1980) rejects Hales' argument, on the assumption that the stresses in the lithosphere are near hydrostatic, as shown in the boundary conditions represented in Figure 3.9b, and that in such a condition the upper block, resting on a low velocity layer, would show no tendency to slip. It will be clear to the reader that what Jacoby has modelled is a 'bowl of water', i.e. an entirely fluid system, rather than one in which part of the system (the strong layer of the lithosphere) is a rigid elastic body. However, we are concerned with an 'up and running' oceanic plate which is being subducted (Figure 3.9c). Consequently, the boundary conditions assumed by Jacoby are inappropriate to the real situation, so that his conclusions are not valid as regards subducting plates.

Peel (1982) resurrected Hales' concept and applied it to a more realistic plate model (Figure 3.9d), albeit of modest dimensions, and concluded that the horizontal stress that could be generated was about 2.0 kb. Price *et al.* (1988) took a comparable shape for oceanic lithosphere, with the downward-curved, lower boundary, as well as the lower limit of the strong elastic layer, which we have represented in Figure 3.9d to Figure 3.11a. These diagrams have a vertical exaggeration which is, for long plates, in excess of 100 : 1. Therefore, except for positions close to the ridge, the angle of dip (q) of this sub-surface is very small and, consequently, only very small errors are introduced if the oceanic lithosphere is reduced to a section with a simple triangular prism (Figure 3.10b).

One can infer that the weight (W) of the prism, of unit thickness, is given by:

$$W = d(L \times z_{max})/2 \tag{3.1}$$

where d is the density of the lithosphere, L is the length of the section and z_{max} is the maximum depth.

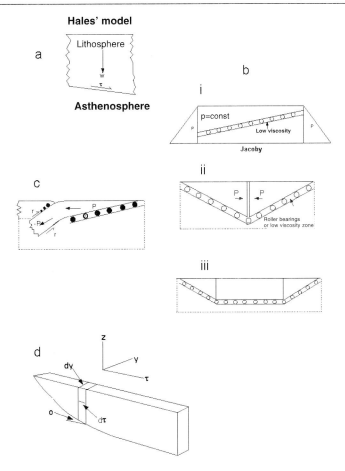

Figure 3.9 (a) Hales' model. (b) (i–iii) Jacoby model showing how the lithosphere can be reduced to a 'bowl of water'. (c) Real situation. (d) Section through oceanic lithosphere used by Peel (1982) and Price *et al.* (1988). P = horizontal compression stress; τ = shear stress; r = restive stress.

The weight W, free to slide on a frictionless underlying layer, could give rise to a horizontal force F_h, where:

$$F_h = W.cos\theta.sin\theta \qquad (3.2a)$$

However, when θ is very small, this reduces to:

$$F_h = W.sin\theta \qquad (3.2b)$$

The horizontal stress S_h that can be generated per unit width of the prism (normal to the plane of the paper) is given by

$$S_h = F_h/z_{max}$$
$$= W.sin\theta/z_{max} \qquad (3.3)$$

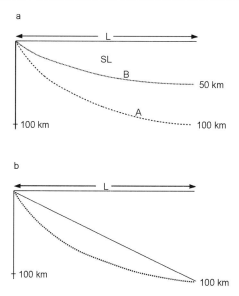

Figure 3.10 (a) Section through oceanic lithosphere (A) and the elastic zone (B). (b) Comparison of curved and triangular sections.

In Figure 3.11a we take the total length of the plate to be 12,500 km, the maximum depth to be 100 km and the density to be 3.3 g cm^{-3}. With these dimensions and density, and the use of Equations 3.1 to 3.3, it follows that the maximum, average, *potential* horizontal pressure that can theoretically develop at the end of the wedge is 1.65 GPa (16.5 kb).

Average horizontal potential pressures, at distances of 2,500 to 12,500 km, show how these potential average horizontal pressures build up from zero at the ridge to a maximum at the most distant parts of the plate away from the spreading-ridge (curve A, Figure 3.11b). The difference between the 'curved' basal line in Figure 3.11a and the straight lines, seems far from trivial. However, it should be borne in mind that the vertical exaggeration of this diagram is 125 : 1.

A second feature of Figure 3.11b is that, provided slower-moving plates reach a thickness of 100 km, the average, potential, horizontal stress (i.e. that which develops at the most distant section of the plate from the spreading-ridge) remains constant at 1.65 GPa (16.5 kb). This relationship follows from the fact that the mass of a short plate is smaller than for a long plate. However, the straight-line dip, from the ridge to the maximum depth of 100 km, is greater for the short plate. It is emphasised, however, that the cited, potential stresses with a magnitude as high as 16.5 kb will never be realised at the strain-rates involved in plate tectonics. Firstly, because this stress level is approximately twice the average 'failure' stress that can be sustained by the 'strong layer' in the oceanic lithosphere and, secondly, because a further proportion of the 'potential stress' is used in overcoming the resistance to movement of the oceanic lithosphere offered by the basal and boundary-wall retardation.

As we have seen in the previous chapter, the distribution of 'strength' in oceanic lithosphere is determined by a number of parameters which are not readily quantified. For example, as regards strain-rates, the transition from the 'flat-lying' oceanic lithosphere, through an anticlinal flexure to a relatively 'straight' down-going slab, results in a quite significant change in sign (i.e. from compression to extension and possibly back to compression) as well as in the rate of strain.

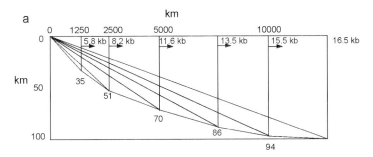

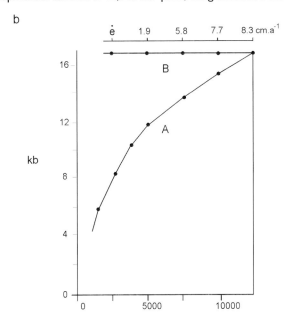

Figure 3.11a Development of stresses in 12,500 km plate, using successive triangular sections.

Figure 3.11b Curve A shows distribution of potential horizontal stress from spreading-ridge to most distant part of plate at 12,500 km. Line B indicates that the maximum potential horizontal stress is a constant for all mature ocean plates, i.e. plates that attain a depth of 100 km.

In the 'flat-lying' portion of the oceanic lithosphere, variations in strain-rates may be induced as the result of vertical loading caused by basaltic extrusions. The importance of the flexure brought about by such loading will be related to the periphery of the intrusion, so that even in the W Pacific, where large and small oceanic flood basalts abound, anomalous strain-rates will be experienced in only a small percentage of the total area of the oceanic plate.

As regards mature oceanic lithospheres (i.e. those that have attained a depth of about 100 km) the maximum potential stress is constant at about 16.5 kb, so it can be inferred that the strain-rate will be related to the length of the 'horizontal' oceanic lithosphere, and hence to the average rate of absolute plate motion.

As may be inferred from Figure 3.11a, the rate at which the potential horizontal stress develops, decreases with distance from the spreading-ridge. However, for all parts of flat-lying, mature oceanic lithospheres, the range of strain-rates is unlikely to exceed an order of magnitude. Hence, as may be inferred from Bodine *et al.* (1981), the change of magnitude of the potential, maximum horizontal stress owing to this factor is not likely to be large.

As oceanic lithosphere develops and a given point travels further away from the spreading-ridge, it is subjected to an increasing magnitude of horizontal compressive stress. This will result in an elastic shortening in the strong zone of the oceanic lithosphere of 2–4×10^{-3}. If this shortening is attained in 150 Ma, the average strain-rate will be of the order 10^{-18} s^{-1}. Hence, only the uppermost sections of the oceanic lithosphere will be able to sustain a differential stress which will, at deeper levels, be sufficiently large to give rise to brittle failure.

Based on the arguments expressed in Chapter 2 (Figure 2.30), it will be seen, in Figure 3.12, that at a depth of about 25 km in the 'strong' 0–50 km band of the oceanic lithospheric unit, the rock is sufficiently strong to support a horizontal compression somewhat greater than 16.5 kb. The failure conditions established in the previous chapter are represented by the triangular area b.x.c. Any stress condition that falls outside this triangle exceeds the failure conditions and cannot be sustained. Only stress conditions which fall within this triangle are valid. This triangular area defines the depth limits over which the average horizontal stress is 8.25×10^8 Pa (8.25 kb). However, if one takes into consideration the relative weakness of the rocks in the parts of this strong zone (i.e. from 35–50 km) indicated by the dashed curve, the average sustainable stress will be probably somewhat less than 8 kb. In particular, this level of stress of about 8 kb is further reduced, as the result of basal restraint of movement of the oceanic lithosphere over the LVZ, and also of the frictional/viscous restraint at the plate boundaries.

However, we have neglected the contribution of stress that can be attributed to the slab-pull and ridge-push mechanisms. As we have noted, the magnitudes of these mechanisms cannot be quantified with accuracy. Nevertheless, if we attribute a stress of 1 kb to these mechanisms, it follows that the average sustainable stress in the strong elastic layer of oceanic lithosphere will almost certainly be close to, or even a little in excess of, 8.0 kb. We submit,

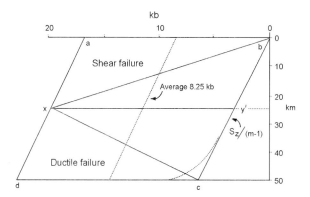

Figure 3.12 The potential average horizontal stress is shown by the parallelogram abcd, which represents a constant potential compressive stress of 1.65 GPa (16.5 kb) over the 50 km thick elastic layer. The triangle xbc represents the failure strength conditions for the strong layer. By comparing the area of the triangle with that of the rectangle, it follows that the average realisable horizontal stress at the end of the 12,500 km section of lithospheric section most distant from the spreading-ridge is 8.25 kb. However, the probable horizontal stress which develops at the lower levels (indicated by the dashed line) reduces the average horizontal stress in the strong layer to about 8.0 kb.

therefore, that it is this average stress in the oceanic lithosphere, of about 8 kb, extending from the surface to a depth of about 50 km, that is the driving element that sustains plate motions and even the generation of mountain chains (see p. 14).

The average sustainable stress at a vertical interface between an oceanic and a continental lithospheric unit will generate a force over a sectional slice of the elastic layer 50 km high and of 1 m in thickness, normal to the plate motion of

$$8.0\ GPa \times (50 \times 10^3\ m^2) = 4.0 \times 10^{13}\ N$$

This is the force, per metre width of a plate, normal to its direction of plate motion that must drive a plate at a specific velocity by overcoming the sum total of all resistive forces, whether they be owing to viscous basal drag, viscous or frictional/viscous drag on vertical internal or boundary faults or, when appropriate, mountain building.

3.3.2 Restraint to slip of the low viscosity zone

The very large *potential* horizontal stresses in the oceanic plate, which should result from the gravity-glide mechanism are not fully realised because of the development of fractures in the strong layer and also viscous drag at the base of the lithosphere, as it moves over the *low viscosity zone*, together with the frictional or viscous restraints of the *vertical*, boundary, or internal, shear zones. (Here we initially assume that the resistive elements opposing sinking of the slab are balanced by the forces that arise because of the density difference between slab and asthenosphere.)

Unfortunately, the dimensions and viscosity of the LVZ are not known with precision. Turcotte and Schubert (1982) report that interpretations of postglacial rebound permit one to infer the presence of a layer of the order of 100 km in thickness. However, from seismic data, they consider that the *low velocity zone* has a thickness of about 200 km. This LVZ is characterised by low seismic velocities and the seismic waves, especially the shear waves, are attenuated. These authors point out that the thickness of the LVZ is not direct evidence of the existence of a low-viscosity zone of comparable thickness. However, they argue, 'the physical circumstances responsible for the reduction in seismic wave speeds and the attenuation of the waves (high temperature, small amounts of melting) also favour the formation of a Low Viscosity Zone'.

The assessments of the LVZ from glacial rebound have an in-built tacit assumption, that the degree of uplift is the result of unloading of a horizontal, strong layer *that is not subject to horizontal compression.* However, we have argued that horizontal stresses of high magnitude generally exist in plates. As we have seen, such stresses would affect the estimates of the thickness of the strong layer, so it is possible that the LVZ is perhaps even less than 100 km thick.

We are concerned that we do not underestimate the effects of viscous drag at the base of the oceanic lithosphere, so it is initially assumed that the average thickness of the LVZ is only 50 km and that the average viscosity within this zone is 4×10^{19} Pa s^{-1} (Turcotte and Schubert, 1982). Also, it should be noted that the viscosity in this zone may be somewhat smaller than the cited figure, so that this, too, builds in a 'safety factor' which ensures that the effects of gravity-glide are not overemphasised.

It is first shown how the average, nett forces resisting slip can be calculated, so that the nett sustainable horizontal stress at specific distances from the spreading-ridge can be established. In addition, the average, horizontal stress which develops in LVZs that are 75 km and 100 km thick respectively are also presented. This permits one to draw three curves showing how the magnitude of this nett stress varies with distance from the spreading-ridge. By referring to estimated stress magnitudes presented in Figure 2.32, an assessment is made regarding which of the three curves is likely to represent the correct thickness for the LVZ.

The restraints to plate motion caused by the external boundaries and the internal vertical strike-slip faults are also poorly constrained. We have noted that the circumference of the Pacific plate is about 50,000 km. Let us assume that the height of the boundary ranges from zero to about 100 km, with an average height of 60 km. Hence, the area of the vertical boundary will be about 2 per cent of the basal area of the plate. Let us also assume, that the average resistance to shear of the boundaries is ten times greater that the average basal resistance; accordingly, we assume that the vertical fractures, both bounding and internal to a plate, present a degree of restraint which is equal to 20 per cent of the viscous restraint at the base of the plate. (Not all the boundary stresses need be strain-rate dependent. However, for convenience, it is further assumed that all restrictive forces are related to viscous deformation.)

Before one can quantify the restraint to gravity-glide that such a 50 km thick LVZ, with the specified average viscosity, will have upon a prismatic block of the dimensions shown in Figure 3.11a, two further simplifying assumptions will be made. The first of these is that the viscous layer behaves as a *Newtonian fluid*, with a constant *viscosity* (m) so that the shear-stress (t) at any location in the flow is given by

$$t = \mu.du/dt \tag{3.4}$$

where *du/dt* is the rate of development of shear-strain.

The second, related, assumption is that deformation of the LVZ is the result of simple *Couette flow*, so that a hypothetical straight line, in the shearing zone, rotates, but remains straight (Figure 3.13). Let us now apply these equations to estimate the restraint to gravity-glide of certain hypothetical models.

In this model, we shall take the plate length in the direction of motion to be 12,500 km, where the oceanic lithosphere attains a thickness of 100 km and the average plate velocity is 8.3 cm a^{-1}. Thus, we are modelling a section that is roughly comparable with the Pacific

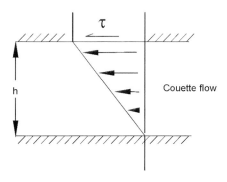

Figure 3.13 Couette (i.e. linear) flow of viscous material between two parallel plates at a fixed distance apart (*h*).

plate from the Easter Island area to the Mariana Trench. Also, it is assumed that the LVZ, with an average vertical extent of 50 km and an average viscosity of 4×10^{19} Pa s, underlies the model oceanic lithosphere.

If the oceanic plate is experiencing a displacement of the lithosphere at a value of 8.3 cm a^{-1}, then the retarding shear-stress at the lithospheric/LVZ boundary can be derived as follows. The reader will recall that there are 3.15×10^7 s in a year.

Therefore, the rate of shear-strain, per second, for a 50 km thick LVZ is given by

$$du/dt = (8.3/3.15 \times 10^7)/(50 \times 10^5)$$
$$= 5.27 \times 10^{-14} \text{ u s}^{-1}$$

The shear stress (τ) is given by:

$$\tau = (4 \times 10^{19}) \times (5.2 \times 10^{-14})$$
$$= 2.1 \times 10^6 \text{ Pa}$$

The traction force (f_t) generated per unit square metre of the glide surface is given by

$$f_t = 2.1 \times 10^6 \text{ N}$$

If we take the plate to extend 12,500 km in the direction of absolute motion, the total force f_t per metre width of the front of the plate will be

$$F_t = (2.1 \times 10^6) \times (1.25 \times 10^7)$$
$$= 2.63 \times 10^{13} \text{ N}$$

Over an area 50 km high by 1.0 m wide, this resistive force will be equivalent to an average resistive stress of

$$S_R = (2.63 \times 10^{13})/(5 \times 10^4)$$
$$= 5.26 \times 10^8 \text{ Pa } (= 5.26 \text{ kb})$$

We have also assumed that the boundary and transform faults add a further 20 per cent of resistance to glide. Hence, the total resistive stress (S_T) on the vertical plane within a plate 12,500 km from the spreading-ridge (which is 50 km high and 1 m wide) is 6.6×10^8 Pa. We note from Figure 3.11a that, at a distance of 12,500 km from the ridge, the average stress that can be attributed to the gravity-glide mechanism is 8.0×10^8 Pa. Consequently, the *nett* horizontal stress (S_N) that results from gravity-glide minus the total viscous restraint is given by

$$S_N = (8.0 - 6.6).10^8 \text{ } Pa = 1.4 \times 10^8 \text{ Pa}$$

This value of average, *nett* horizontal stress, at the end of a 12,500 km plate, which is specifically related to the plate moving with a velocity of 8.3 cm a^{-1}, is shown as the end of curve **A** in Figure 3.14.

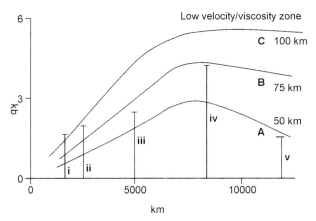

Figure 3.14 Distribution of nett horizontal stress which can develop in a 12,500 km long oceanic lithosphere. This section is an approximate model of the Pacific plate. The vertical lines i to v represent the estimated magnitude of differential stress, based on seismic data (see Figure 2.32) at specific distances along a line approximately parallel to the direction of absolute plate motion, from the vicinity of Easter Island to the Marianas.

The potential horizontal stress within the wedge changes in magnitude in a smooth curve (Figure 3.11b). The reduction in the nett horizontal stress for specific distances from the spreading-ridge can be estimated by using diagrams of the form shown in Figure 3.12. Such changes in magnitude are a consequence of the reduction of the effective mass of the plate, as the spreading-ridge is approached, and also because of the decrease in thickness of the elastic layer.

Because it has been assumed that the low-velocity zone is constant in thickness, the basal viscous restraint will increase linearly from the spreading-ridge to the maximum extent of the plate. Consequently, the nett horizontal stresses at specific distances from the spreading-ridge are as listed in Table 3.3.

If one assumes the LVZ to have a thickness of 75 km or 100 km, the total stress curves are as shown by B and C respectively in Figure 3.14. The values of the available nett horizontal stress at these distances are plotted together with the values of stress inferred by Govers *et al.* (1992) from seismic data in the Pacific (Figure 2.32). The sites of these earthquakes lie adjacent to a line from the vicinity of Easter Island to the Mariana Trench, which is also approximately parallel to the direction of plate motion.

There are only five data points (I–V), in this section, regarding intraplate stresses, based on interpretation of seismic evidence. Stress bar I supports, and IV lends some support to, curve C. Stress bars I, II and IV are in reasonable agreement with curve B, while stress bars

Table 3.3

Distance from ridge (km)	Average sustainable glide stress (10^8 Pa)	Nett resistive stress (10^8 Pa)	Total available stress (10^8 Pa s)
12500	8.0	6.6	1.4
10000	7.6	5.2	2.4
7500	6.8	3.9	2.9
5000	5.7	2.6	3.1
2500	4.0	1.3	2.7

I–IV all exceed curve A. Thus, of the three curves, we suggest that the stresses derived from seismic data are best satisfied by curve B, with curve C as the next best fit. Curve A can reasonably be ruled out because stress bars I–IV exceed this curve, and it is statistically unlikely that four out of five seismic events would give estimated differential stress magnitudes in excess of the mean values of curve A.

This conclusion is, of course, predicated, in part, on the assumption that the seismic stresses have been accurately assessed. If it is assumed that the LVZ is 100 km thick, the forces that retard plate motions are relatively small. However, as noted earlier, we do not wish to risk underestimating the magnitude of these retarding effects, for we wish to ensure that the gravity-glide mechanism is fairly, or even harshly tested.

It can also be noted that, if one takes curve B, the average stress at 12,500 km is compressive and is about 3.8×10^8 Pa. If we liken this section to that of the Pacific plate, at 12,500 km from the ridge, this point would be close to the Mariana Trench. The lithosphere in this area is upwardly flexed, in order that it can pass from the horizontal to an inclined subducting slab. We have noted (Figure 2.26c) that the anticlinal rise to the Pacific side of the trench has been attributed to slab-pull. We cast doubt on this suggestion, and consider, in the light of the conclusions regarding the nett horizontal stresses indicated in Figure 3.14, that the upward flexure can be attributed to horizontal compression and is, in all probability, an anticlinal buckle. Indeed, it is probable that the compression of the lithosphere is transmitted for some distance into the down-going slab, so that the extension phase caused by the difference in density of the cold slab and the surrounding hot asthenosphere is not likely to come into effect until perhaps several tens of kilometres of subduction has been brought about.

We set up a simple section which simulated a gravity-glide model of the Pacific plate, and estimated the stresses that developed in the strong layer of the lithosphere, in response to the body-weight and the various boundary forces driving and retarding the model oceanic plate. It was gratifying to find that the stresses in the Pacific plate, inferred from seismic data, fitted the range of estimated stresses obtained from the model, and enabled us to infer that the average LVZ thickness was probably about 75 km. The correlation between the stresses inferred from our model and the magnitude of the stresses inferred from seismic data can be looked at in two ways. Firstly, the seismic data can be thought of as confirming the model. Alternatively, the model can be thought of as an explanation of why the stress-drops inferred from the seismic data vary systematically across the Pacific.

3.4 Intraplate structures in oceanic lithospheres

Let us now consider briefly some of the various structures that occur in oceanic plates between the trench and the spreading-ridge. From the intensity of seismic activity that occurs in down-going slabs, one can infer that a multiplicity of fractures also develop as the lithosphere is flexed, to become subducted, and then straightened as it penetrates deeper into the mantle. We shall deal with the development of some of these fractures later in this chapter. Here, we shall be concerned only with structures in the horizontal oceanic plate which can be seen at the surface, or inferred from seismic traverses and earthquake data.

3.4.1 Anticlinal flexing of the lithosphere adjacent to the trench

Such flexures are often seen to occur immediately adjacent to the trench. Indeed, such flexing helps define the trench. A representative structure of this group of features is that associated with the Mariana Trench (Figure 3.15a), and has been studied by a number of authors (e.g. Hanks, 1971; Watts and Talwani, 1974; Watts *et al.*, 1980). Turcotte and Schubert (1982) also present a theoretical basis for the development of such structures. The boundary conditions they use to derive the flexure profile are as shown in Figure 3.15b. Their derivation of the *universal lithospheric deflection profile* is worthy of study by the reader.

In several of the analyses cited above, the approach was similar to that of Turcotte and Schubert. However, we suggest that an element is usually missing from such analyses, namely that of the downward drag of the mass of subducting oceanic lithosphere. This will provide a significant *pull* along the central line of a section through the lithosphere, which will tend to decrease the amplitude of the anticlinal flexure.

However, the flexure exists, so an explanation is necessary. Either the influence of the pull of the down-going slab is negligible, or there is an alternative solution. It has been noted that there is evidence which indicates that oceanic lithosphere is capable of supporting very significant compressive, horizontal, differential stresses. Moreover, in the Indian Ocean, large-wavelength, low-amplitude flexures occur, which, it has been inferred, are compatible with compressive stresses acting in the direction of plate motion of about 0.5 GPa.

It has also been shown (Figure 3.14) that a model which approximately represents the Pacific plate is likely to exhibit an average horizontal stress, 12,500 km away from the spreading-ridge and, therefore, adjacent to the Mariana Trench, of as much as 0.4 GPa. Consequently, at this point, we can at least suggest that *buckling* contributes to the development of the Mariana flexure and similar features.

Whether the flexures are the result of bending or buckling, the outer fibres of the structure, which may be exposed at the surface, are likely to be about 25 km above the neutral surface. Consequently, such areas, at the surface, are likely to have the axis of least horizontal stress parallel to the direction of absolute plate motion, rather than the more general situation, for unflexured lithosphere, in which the greatest horizontal stress is parallel with that motion.

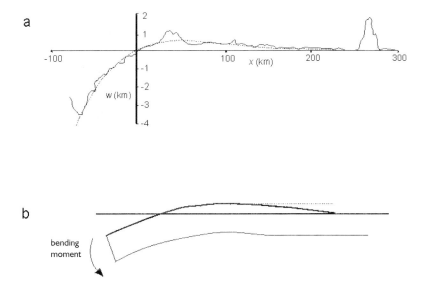

Figure 3.15 (a) Section through the Mariana Trench. (b) Boundary conditions on which the universal flexure profile is based.

3.4.2 Intra-oceanic lithospheric fractures

Some seismic reflectors in the oceanic crust of the western N Atlantic (Figure 3.16) show reflectors with apparent dips towards and away from the ridge. Some of the structures are approximately linear in section, while a few are curved. The linear fractures may dip in either direction. However, the well-defined curved fractures (Line 711, at CDP Numbers 3400–3800) and other smaller features (CDP Numbers 4400, 4800, 7600 and 9000) are all concave, away from the ridge.

The traverses shown in Figure 3.16 are approximately normal to the ridge. Also, it can be inferred from the gravity-glide model that the direction of glide in the N Atlantic will be at high angle to the spreading-ridge. Consequently, the axis of greatest horizontal compression will also be at high angles to the ridge. The majority of the planar features shown in these sections dip at about 30°. In the light of the gravity-glide model put forward here, we suggest that these reflectors are probably thrust-faults which developed to relieve the potentially large compressive stresses that develop, at shallow depths, away from the spreading-ridge.

It has been noted (Figure 3.8) that the oceanic lithosphere moves away from the ridge along relatively steeply dipping, curved surfaces which are concave upward and away from the ridge. These are glide surfaces which are comparable with some normal listric faults in the continental crust. However, because the glide surfaces in the oceanic lithosphere near the ridge must be at high temperature and moderate confining pressure, we suggest that melting and/or granulation will occur along these surfaces, which will result in a thin glide-zone, comparable with the *Lochseitenkalk* beneath the Glarus Nappe, in the Swiss Alps. Such zones could well become seismic reflectors.

These surfaces and zones are, of course initiated close to the spreading-ridge. However, as new crust is formed at the ridge, the glide zones will be abandoned, one after the other, and become preserved, from time to time, in the crust of the oceanic plate far from the ridge.

The suggested modes of development of the recorded reflectors as the result of thrusting, well away from the ridge, and by normal, listric faulting near the ridge are of course conjectural. However, they are completely compatible with the mode of deformation and development of oceanic plates, both as regards geometry and the inferred stresses which develop in these plates. It will also be noted, that these various faults occur at relatively shallow depths. At depths of greater than about 14–15 km (Figure 3.16), there are no obvious signs of faulting.

From the analyses presented earlier in this chapter, we concluded that the *elastic layer* of oceanic plates, distant from the spreading-ridge, has the potential to develop an average horizontal differential stress of up to 0.8 GPa (8 kb). It has also been shown that, if the thickness of the low viscosity zone is taken as 75 km, then for a 12,500 km long plate, the potential horizontal stress is reduced to about 0.4 GPa.

For slower moving, shorter plates (with, say, a length of 3400 km) it can be inferred that the nett average stress is almost double that in a long fast moving plate, so that, it can be inferred, much of the upper elastic layer must have failed. Why then do we not see evidence of such faulting in Figure 3.16?

We suggest that there are two main reasons why traces of faulting cease to be observed below a specific depth. The first of these relates to the type of fault that will develop at different depths. Thus, in Figure 3.17, we show the failure envelope for material which forms the oceanic lithospheres. Let us assume that the uniaxial strength is 3.2×10^8 Pa, so that the cohesive strength (C_o) is 0.8×10^8 Pa, and that the angle of friction (θ) is 25°. These data are in reasonable agreement with the failure conditions cited by Kirby (see Chapter 2). The

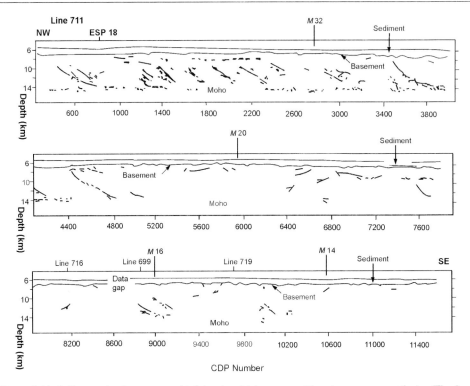

Figure 3.16 Reflectors in the western N Atlantic which are considered to represent faults. The low angle linear faults we interpret as thrusts and the more steeply dipping features which are concave in the direction of plate movement we interpret as *fossil* ridge-spreading faults (after Morris *et al.*, 1993).

upper failure envelope refers to unfractured rock, while the lower envelope represents failure conditions along fractures which make an angle with the maximum principal stress of $(45° -\theta/2)$. (Readers not familiar with this method of representing failure conditions are referred to Price, 1966; Jaeger and Cook, 1968; Price and Cosgrove, 1990.) We also assume that the failure envelope is linear, and that the oceanic lithosphere behaves as a linear elastic material down to a depth of at least 18 km.

Two sets of semicircles are represented. The first set is 'anchored' at 0.2 GPa and the second set at 0.6 GPa (6 kb). For both these sets, the larger semicircle represents the condition to cause shear failure in hitherto unfractured rock, while the smaller semicircle represents the conditions required to cause further movement on the fracture, set up by the larger stress semicircle.

The stress point at 0.2 GPa (2 kb) represents the vertical stress that is generated at a depth of 6 km, and the stress point at 0.6 GPa represents the vertical stress that is generated at about 18.0 km. The stress semicircle that will initiate failure for these two stress points is then drawn to touch the upper failure envelope, thereby giving the two maximum values of S_1 of 7.3 and 1.7 GPa (7.3 and 17.0 kb). Assuming that the strain is zero in the direction normal to plate motion, and using Equations 2.2 to 2.5, it is possible to calculate the magnitude of the remaining principal stress. The magnitudes of the two sets of stresses for the two cases cited in Figure 3.17 are given in Table 3.4.

Table 3.4

Depth	*I* 6.0 km		*II* 18.0 km	
S_z	0.204 MPa	(S_3)	0.60 MPa	(S_2)
S_y	0.770 MPa	(S_1)	1.72 MPa	(S_1)
S_x	0.244 MPa	(S_2)	0.58 MPa	(S_3)

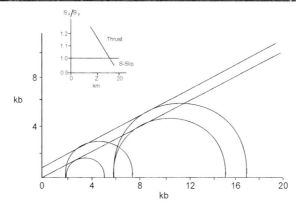

Figure 3.17 Stress circles and failure envelopes.

The important thing to notice as regards these two cases is that for I, the axes of least and intermediate principal stresses are respectively vertical and horizontal, while for II, the situation is reversed. Case I represents the failure conditions which give rise to *thrusts*, with inclined fault planes which dip at about 22.5°, while Case II gives rise to *strike-slip, vertical faults.*

Thrust planes are relatively easily inferred from seismic traverses, while vertical, strike-slip faults are much more difficult to detect. This explains, therefore, why we only see thrust faults down to a specific depth in Figure 3.17. At what depth does the change-over in the style of faulting take place? This we can infer from the inset diagram in Figure 3.17. The ratios of S_x/S_z in Cases I and II are 0.97 and 1.196 respectively. Change-over in the mode of faulting occurs when $S_x = S_z$, i.e. when $S_x/S_z = 1.0$. It will be seen from the inset diagram in Figure 3.17, that this change-over takes place at a depth of about 15 km. In light of the approximations, which must arise when one substitutes idealised concepts for the behaviour of real materials, the degree of agreement between observation and model is gratifying.

An important point to note is that the development of faults does not render the lithosphere infinitely weak. For the two depths represented in Figure 3.17 (inset), of 6.0 and 17.0 km, provided the maximum horizontal principal stresses at these levels do not attain values of 0.59 and 1.69 GPa respectively, it can be inferred from Figure 3.17 that the lithosphere will behave as an unfractured body.

3.4.3 Elastic shortening of oceanic lithosphere

It can be inferred from Figure 3.18a, that the horizontal stresses in a plate will increase in magnitude from the spreading-ridge. For a 7500 km long plate (length L), the average value of the nett compressive stress (S_h) in the direction of movement will be dependent upon the plate velocity. Let us assume that the average nett horizontal differential stress is 3×10^8 Pa (3 kb). This will give rise to elastic shortening of the plate by a total distance l. The value of l is given by the relationship:

$$l = L.S_h/E \qquad (3.5)$$

where E is Young's modulus, which we have earlier taken to be about 50 GPa (5×10^5 bar). Substituting these various values for L, S_h and E into this expression, $l = 45$ km. Thus, there is a significant amount of elastic strain and, therefore, also of strain-energy stored in such a relatively long plate.

3.4.4 Stress in a continental 'inclusion'

Consider now a modified plate in which a sliver of continental lithosphere is included (Figure 3.18b). Let us assume that this sliver, from front to back, has a length (L') of 500 km. The continental sliver will be subjected to a magnitude of compressive stress which has been dependent upon its position in the plate. If it is situated near the spreading-ridge, the stress will be relatively low, while if it is situated near the trough it will be modestly high. Let us assume that the sliver is sited about 2500 km from the spreading-ridge, and it is subjected to a nett average compressive stress of 2 kb. Then, from Equation 3.5, this will result in a shortening in the mantle beneath the continental crust of 2.0 km, or 0.4 per cent strain. This strain will be transmitted to the continental crust above, where it will cause a differential stress to develop in near-surface, sedimentary rocks. If we assume that these rocks have an average value of Young's modulus of $E = 10^5$ bar, a horizontal differential stress of 400 bar will result. The maximum horizontal stress will, on average, be parallel to the direction of absolute plate motion. Hence, the various models presented above are compatible with the observed relationship, that the axis of maximum horizontal stress will be compressive and approximately parallel to the direction of absolute plate motion.

3.5 Development of a subduction zone

It can be inferred from the patterns of striping formed by the alternating magnetic anomalies, produced by spreading-ridges, that more such ridges existed some 80 Ma ago than are currently in existence (McKenzie, 1977). Indeed, there is still a greater length of spreading-ridge in

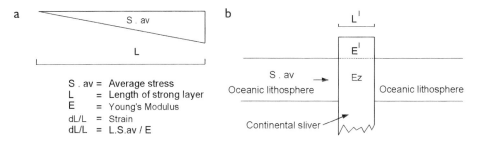

Figure 3.18 (a) Elastic shortening in an oceanic plate. (b) Elastic shortening in an oceanic sliver included in an oceanic plate.

existence than the current length of troughs. Moreover, McKenzie further notes that, whereas most ridges are relatively young, i.e. many are less than 80 Ma old, the major trench systems of the Indian and Pacific Oceans have probably been in existence for over 200 Ma. Hence, he concludes, ridges are easily started, while trenches are difficult to initiate.

Turcotte (1983) suggests that 'although the subduction process is reasonably well understood, the processes that lead to the formation of new ocean trenches are obscure'. Somewhat earlier, Turcotte *et al.* (1977) considered that a tensional environment could lead to a foundering of oceanic lithosphere. However, McKenzie (1977) proposed the opposite view, namely that subduction occurred in a compressional environment. We concur with this latter view. Indeed, we consider McKenzie's argument to be essentially correct, except that he found it necessary to stipulate that pre-existing thrust faults cut through oceanic lithosphere, although he did not demonstrate how such a through-cutting thrust came into existence. For such conditions, he estimated that the necessary horizontal stress was 800 bar (8×10^7 Pa). From arguments presented in this and the previous chapter, it is clear that this differential stress is about an order of magnitude too low.

Fyfe and Leonardos (1977) argue that subduction requires the oceanic lithosphere to be flexured, and that the necessary degree of flexuring can only be induced by a thickness of sediment of about 10 km. However, Cloetingh and Woertel (1985), using finite element analysis, studied the behaviour of oceanic lithosphere that is deflected downward, as the result of differential loading caused by sediments. They concluded that the magnitude of such loads could not realistically be expected to give rise to failure of oceanic lithosphere. However, these latter authors neglected the possibility of oceanic plates being concomitantly acted upon by horizontal compression of considerable magnitude.

Off-shore basins are plentiful and some of them exceed 5 km in depth. Consequently, the depth of basin which can give rise to subduction is likely to be 'exceptional'. The suggestion made by Fyfe and Leonardos, that the 'sedimentary piles which can give rise to subduction must be in the 10 km range', is therefore likely to be correct.

3.5.1 Off-shore basins and oceanic flexure studies

Off-shore basins have been widely investigated, so are therefore well documented. A section through such an example of sediments accumulated off the N Carolina coast of the USA is shown in Figure 3.19a, where it will be seen that the sediments have accumulated a 'half' basin to a depth of at least 3.0 km, and that the 'half' width of the basin may be about 200 km.

A stratigraphic cross-section of the North Sea Graben is shown in Figure 3.19b. It will be noted that this basin contains pre-Tertiary sediments. Also, it has almost certainly developed above a major graben. Hence, we may infer that this basin was initiated prior to the opening of the N Atlantic. It has been argued that the development of these basins may have occurred in a stretching phase, which ultimately gave rise to the fracturing of the continental mass and the opening of the N Atlantic.

Watts *et al.* (1982) carried out an analysis of these and other basins, in an attempt to correlate stratigraphic curves for various sedimentary boundaries with the time required for the various depths of sediments to accumulate. They used both simple elastic and visco-elastic models in their analysis, and decided that the elastic analysis provided the best fit to stratigraphic and gravitational anomaly data. We would agree with their conclusions regarding the suitability of the elastic model.

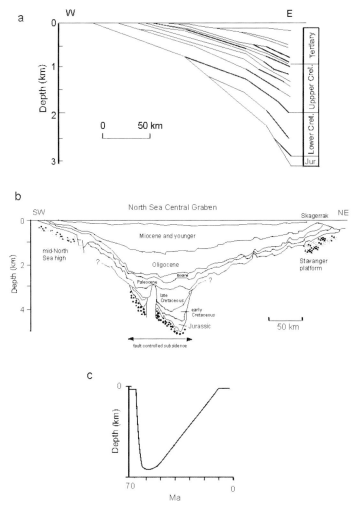

Figure 3.19 (a) Section of off-shore 'half' basin, off the US Atlantic coast (after Sleep and Snell, 1976). (b) Central North Sea Graben (after Ziegler and Louwerens, 1977; Ziegler, 1977). (c) Schematic representation of 'basin inversion' of a North Sea Graben.

The initiation of the N Atlantic coincided with a period of basin inversion in the North Sea (Figure 3.19c) which started approximately 65 Ma ago. We would interpret the basin inversion as being the result of lateral compression of several kilobars, brought about by gravity-glide from the spreading-ridge of the developing oceanic lithosphere.

Mature oceanic lithosphere is covered by a basaltic crust with an average thickness of 7 km. Continental lithosphere, however, is covered by a relatively low-density crust with a thickness of, perhaps, several tens of kilometres. The sub-crustal lithosphere in oceanic and continental plates will be similar in composition, but because of the different depth of 'burial' they will generally be at different temperatures, so that their strengths at comparable depths are likely to be quite different. Also, there is, of course, a lateral transition zone between adjacent lithospheres, especially where sediments derived from adjacent continental areas spread out to form a continental shelf over the oceanic lithosphere. Thus, there must generally be a transition zone between adjacent lithospheres.

This is a difficult situation to model in simple terms. Therefore, we assume here that the junction between oceanic and continental lithospheres occurs along an approximately vertical surface. This pressure will prevent any differential vertical movement of one lithosphere relative to the other by frictional restraint (Figure 3.20a). Moreover, if there is a mismatch of strong layers in the oceanic and continental lithospheres, the initial vertical plane will be deformed and indented (Figure 3.20a). This indentation would further inhibit vertical differential movement of the two lithospheres along their junction. Hence, there would be no tendency for the oceanic lithosphere to founder because of its higher density relative to that of the asthenosphere.

Let us now consider what happens when the continental shelf begins to develop. Once a super-continent is divided, its drainage patterns are disrupted and often shortened, so that rates of sedimentation tend to increase. This enables sediments to be more rapidly deposited, thereby building up a continental shelf of thick sediments in a relatively short period. Such a shelf is represented in Figure 3.20a.

The weight of sediments will, of course, tend to depress the oceanic lithosphere. But the oceanic and continental lithospheres are locked at their junction, so that the sediments cause deflection of the strong layer of the oceanic lithosphere, adjacent to the junction as a modified cantilever. However, the cantilever does not have a free end, for it is a continuous unit reaching back to the spreading-ridge. Consequently, the strong layer is obliged to form a major tilted syncline. Because the thickest sections of off-shore sediments will develop and migrate outward from the interface between continental and oceanic lithosphere, the vertical loading will give rise to an asymmetrically tilted syncline in the oceanic lithosphere.

The arrow, shown at the left end of the section, is there to remind the reader that this strong layer is being continually subjected to a significant horizontal stress, which will contribute to the flexuring of the strong layer.

One can estimate the general distribution of fibre stresses which result from the flexing of the oceanic lithosphere. These are also shown in Figure 3.20a, for the brittle portion of the elastic layer, together with the orientation of the normal faults or thrusts, which may be engendered as the result of these fibre stresses.

Downward flexing of the oceanic plate near the junction between ocean and continent as the result of the development of an off-shore basin, is also represented in Figure 3.20a. It will be seen that the downward curvature of the oceanic lithosphere, between (i) and (ii), will cause compression below and extension above the neutral surface of the strong oceanic layer. However, between the inflexion points (ii) and (iii), there is an upward curvature of this layer which will give rise to the opposite distribution, with compression above and extension below the neutral surface. Between (iii) and (iv) the strain regimes are again reversed, so that, beyond (iii), the anticlinal flexure will give extension above, and compression below, the neutral surface.

If failure conditions are reached in all the deformed zones above the downward and upwardly curved neutral surface, the orientations of the conjugate normal or thrust faults are as shown in Figure 3.20a. Below the strong layer the mode of failure is ductile. Such ductile failure is unlikely to generate a single shear zone capable of cutting through the strong layer. Above the strong layer, extension occurs in the flanks of the syncline which would give rise to reduction of the horizontal stresses. Such stress reduction, alone, could possibly give rise to normal faults, with the orientations shown in Figure 3.20a. However,

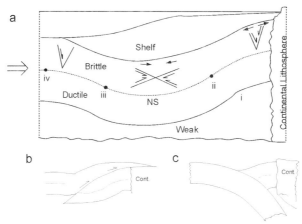

Figure 3.20 (a) Section through a hypothetical junction between oceanic and continental lithospheres, with off-shore sediments giving rise to flexuring of the oceanic lithosphere, which, in turn, determines the potential location of normal and thrust faults. (b) Failure leading to thrusting of oceanic lithosphere over the continent. (c) Failure leading to subduction.

these fibre stresses are developing in a zone of strong regional compressive stress generated by the oceanic lithosphere, so they may not, or at least are unlikely to, give rise to a general failure of the elastic layer. This argument also holds for the area below the neutral surface in the centre of the syncline. Hence, the most likely modes of failure of the strong layer of the oceanic lithosphere are along one or other of the thrust planes with the orientation of the solid lines in Figure 3.20a.

If up-thrust failure takes place on a major thrust plane, beneath the sedimentary pile, which dips away from the continent, and develops as a through-going fracture, *obduction* of *ophiolites* could result (Figure 3.20b). Alternatively, if the thrust plane which dips toward the continent fails and dominates deformation, breakthrough of the lower levels of the oceanic lithosphere, by ductile shearing, could result in the initiation of subduction of the oceanic plate (Figure 3.20b).

White *et al.* (1987) show seaward-dipping reflectors which have developed in oceanic lithosphere, beneath a thin sedimentary cover, in the Hatton Bank Margin (Figure 3.21). In this figure, the reflectors dip at about 45°. However, it can be inferred from the bounding scales of this diagram, that it has been drawn with a vertical exaggeration of 2.4 : 1. Hence, it can be inferred that the real dip of the reflectors is about 23°. It is, therefore, reasonable to infer that these reflectors mark the position of a series of seaward-dipping thrust planes, which could possibly develop into one or more major overthrusts, potentially giving rise to the emplacement (obduction) of ophiolites onto the continental lithosphere.

The mode of development whereby ophiolites are pushed ashore requires these dense rocks to be lifted up tens of kilometres above their original level, and also transported laterally over sediments or indurated, older continental rocks. Thus, as a very considerable amount of work has to be done, such over-thrusting is likely to be relatively curtailed.

As regards subduction, however, once a through-going landward-dipping thrust or shear zone develops, the breakthrough of the thrust results in the release of elastic strains, stored in the flat-lying areas of the lithosphere, which gives an initial 'push' to the chisel-shaped oceanic lithosphere. Thereafter, the work that must be done to develop subduction is much less than that required to cause the oceanic lithosphere to be up-thrust. Also, because the

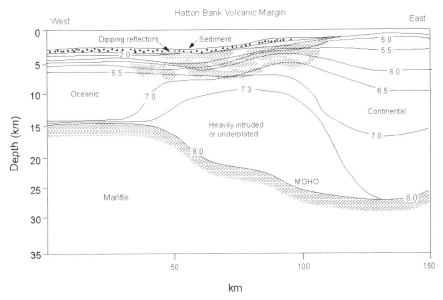

Figure 3.21 Junction between oceanic and continental lithospheres (after White, 1992).

oceanic lithosphere is, eventually, more dense than the asthenosphere, the former will show an automatic tendency to sink, so that subduction can continue until the whole of the oceanic lithosphere is consumed (Figure 3.20c).

To quantify the fibre stresses which develop in the oceanic lithosphere prior to thrusting, we must not only estimate the amount of deflection, but also define the thickness of the strong layer. Once the fibre stresses, for a particular deflection, have been ascertained, this quantity can be added to the estimated nett compressive stress caused by gravity-glide of the oceanic lithosphere on the LVZ. If a combination of the nett compressive stress, plus the largest positive value of the fibre stresses, exceeds the strength of the layer, then a thrust could develop, or, more likely, an existing thrust (which developed in mid-ocean, see Figure 3.16) would experience regeneration (i.e. re-shear). Such thrusting would weaken the outer fibres, and so enable a reduction in the radius of curvature to occur. Eventually, by continually reducing this radius, the fracture could develop, and eventually give rise to through-going rupture of the whole of the oceanic lithosphere and subsequent subduction.

As regards the average thickness of the strong layer, Watts *et al.*. (1982) have compiled a survey of 25 oceanic flexures associated with the ridges, islands, seamounts and trenches. We are here concerned with the initiation of a subduction zone, so we have selected only the six examples which relate to trenches. They estimated the average thickness of the strong layer for the trenches to be close to 30 km: a figure we shall take as a starting point.

For convenience, in the model presented in Figure 3.22a, the radii of the different segments of the flexure have been made equal. However, as we have noted, in a real situation the curvature beneath the thickest sediments is likely to induce a somewhat reduced radius of curvature. Consequently, one would expect this site to be preferred for the initiation of thrusting and subsequent subduction.

The combination of complex vertical loading with a buckling stress does not lend itself to a simple, rigorous analysis. Consequently, we shall follow the method used by Price *et al.*

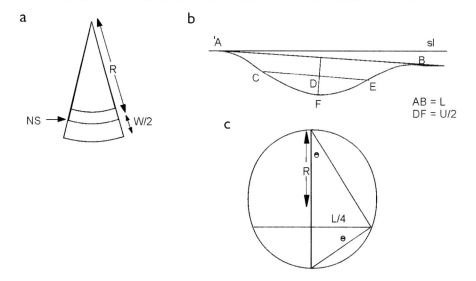

Figure 3.22 (a) Flexures of the strong layer in oceanic lithosphere induced by loading of a sedimentary basin. (b) Fibre-stresses induced by flexure of the strong layer leading to likely failure conditions. (c) Geometrical construction used to calculate radius of curvature (all after Price *et al.*, 1988).

(1988) who assumed that the flexured structure could be represented by a series of circular arcs. The model they adopted is shown in Figure 3.22b. Let us now carry out some calculations based on this simplistic model to establish the magnitude of the fibre stresses that may develop as the result of flexing.

The magnitude of the *fibre stress* (S_f) induced by bending, varies as the distance of the *fibre* from the neutral surface (Figure 3.22a). At the upper and lower limit of the strong layer (thickness w) the fibre stresses are given by

$$+/- S_f = w.E/2R \tag{3.6}$$

where E is Young's Modulus, w is the thickness of the strong layer (which we take to be 30 km) and R is the radius of curvature of the synclinal or anticlinal flexure.

The situation envisaged by Price *et al.* is shown in Figure 3.22b in which sediments on, or adjacent to, the continental margin (the shaded area) cause a downwarp curve ACFEB, which is made up of a series of arcs of circles of equal radii. The distance $AB = L$. (Point B is on the ocean floor and A is near sea-level, hence the tilt of the diagram.) Here we are specifically concerned with the section CE (length $L/2$) and the vertical displacement $DF = U/2$. To calculate the radius of curvature (R) of the central part of the basin, they used the construction in Figure 3.22c, from which it follows that

$$tan\theta = 2U/L = L/4\{2R - (U/2)\}$$

so that

$$R = (L^2 - 4U^2)/16U$$

But

$$S_j = wE/2R$$

and

$$S_j = 8wEU/(L^2 + 4U^2) \qquad\qquad (3.7)$$

In the paper by Price *et al.*, the modulus of elasticity was given a value which we now consider to be too high. This invalidates their conclusion that the magnitude of stress required to cause the breakthrough of the strong layer in the oceanic lithosphere can occur when the basin has the modest depth of about 4 km.

We are initially concerned with the compressive fibre stresses set up in the synclinal part of the strong layer to depths of about 30 km, and take the average value of Young's modulus (E) to be 5×10^5 bar.

In Figure 3.22b, in which we set $L = 400$ km, $U = 6$ to 12 km and $w = 30$ km, the maximum compressive fibre stresses set at the upper limit of the strong layer are as listed in Table 3.5.

The magnitudes of these compressive fibre stresses at the deeper levels are obviously significant. However, we emphasise that they are *maxima* and only *tend* to develop at the outermost fibre of the strong layer, for as soon as the fibre stresses increase the maximum horizontal stress to a value that just exceeds the resistance to shear on a thrust plane, the fibre stresses are reduced by causing an increment of shear on an existing fracture. Only by very considerable local deepening of the sedimentary basin (so that the radius of curvature of the strong layer is significantly reduced from the original 400 km) could the fibre stresses be continually enhanced, nearer and nearer to the neutral surface, until a major, through-cutting thrust could develop. Such a drastic reduction in the radius of curvature of the flexure presents obvious geometric and analytical difficulties.

3.5.2 Sedimentation and changes of thermal and vertical stress gradients

Fyfe and Leonardos (1977), it will be noted, considered that sediments in a basin must be at least 10 km thick, in order that the conditions that will give rise to subduction can be met. These piles of sediments may be accumulated over a period of the order of 100 Ma. Consequently, thermal effects will have the time necessary to approach equilibrium conditions. Adjustment of the vertical and horizontal stresses in the sediments, as the result of the accumulation of sediments, will, of course, be progressive.

The geothermal gradient in oceanic lithosphere is not constant throughout. Near the spreading-ridge the gradient is several hundred °C per km of depth. This high gradient

Table 3.5

Differential depth of basin (km)	Maximum fibre stress (kb or 10^8 Pa)
6	4.5
8	6.0
10	7.5
12	9.0

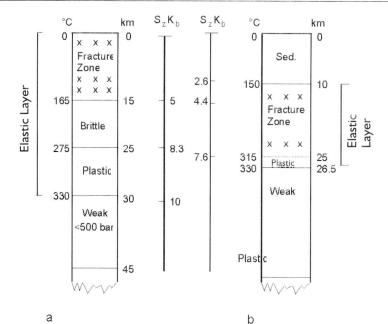

Figure 3.23 (a) Temperature and pressure gradients in upper section of oceanic lithosphere, uncovered by sediments derived from a nearby continent. (b) As (a), when the oceanic lithosphere is covered by 10 km of sediments.

decreases rapidly away from the ridge until, in a mature oceanic lithosphere with a thickness approaching 100 km, the thermal gradient is close to 13°C km⁻¹. The increases in temperature with depth within a mature section of oceanic lithosphere is shown in Figure 3.23a. The upper 25 km, which initially represented the original *elastic layer*, shows the depth to which the rock is essentially brittle. We have argued and adduced evidence that the uppermost 15 km of this elastic layer will be cut by thrusts, while the lower level may be cut by strike-slip faults. From a depth of 25–45 km, the behaviour of the rock material will change from brittle and become progressively more *plastic* (i.e. ductile) and weaker.

Let us now consider this section of oceanic lithosphere to be covered by a 10 km thick sheet of sediments (Figure 3.23b). The geothermal gradient in such a pile of sediments will, in part, depend upon the type of sediments it contains. For example, a 1.0 km thick layer of clays and muds may exhibit a temperature difference between the bottom and top of as much as 100°C. Here we shall be conservative and assume that the sediments have attained thermal equilibrium with an average gradient, throughout the 10 km of sediments, of an extremely modest 15°C km⁻¹. Hence, the temperature at the interface between a 10 km thick pile of sediments and the oceanic lithosphere will be 150°C. In our model, this results in the specific temperature levels in the oceanic lithosphere being transposed upwards. The elastic layer is significantly reduced in thickness from 30 km to 16.5 km. With such a reduction in thickness of the elastic layer, the maximum compressive fibre stresses, for the thickness of basin sediments of 10 km and 12 km, are respectively 4.1 kb and 5.0 kb.

In addition, it is necessary to note that the upper, brittle portion of the strong zone is transposed to contain the zone in which it is likely that thrusts have already developed, and that the 15 km thick section, which comprises the 'brittle' and 'plastic' zones in Figure 3.23a, are reduced to a mere 1.65 km in Figure 3.23b. Moreover, the vertical (confining)

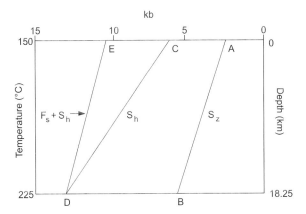

Figure 3.24 Failure conditions, described in text, with fibre stresses superimposed. F_s = fibre stress; S_z = vertical stress; S_h = horizontal stress.

pressure (Figure 3.23b), at any specific depth, is significantly lower that shown in Figure 3.23a, so that re-shearing along existing thrusts will be enhanced.

Let us now see how the various changes induced by the sedimentary blanket over the oceanic lithosphere will influence the latter's behaviour.

The magnitude of the least principal stress in the fractured zone down to the neutral surface (NS) is given by line AB in Figure 3.24. The re-shear conditions on existing thrusts, represented by line CD, will be sustained by the gravity-glide mechanism set out earlier in this chapter. The loading of the oceanic lithosphere by a thick layer of sediments can cause downward bending of the oceanic lithosphere, which gives rise to compressive fibre stresses. If the thickness of the sedimentary pile is 10 km, the maximum *notional* fibre stress is 4 kb (line DE). These notional fibre stresses in the upper layer of the oceanic lithosphere will be dissipated by causing, or enhancing, re-shear, thereby extending existing thrust-fractures.

At these high levels in the oceanic lithosphere, the re-sheared thrusts can disrupt the sediment–lithospheric interface, thereby permitting water from wet sediments to penetrate downward into the erstwhile dry basalts. The development of clay surfaces on the thrust zones will greatly reduce the differential stress required to induce further shear movement. Consequently, the uppermost regions of the oceanic lithosphere will become significantly weaker, thereby reducing the effective thickness of the downward-flexing beam formed by the 16.5 km thick strong-zone, from depths of 10–26.5 km.

The thinner the bending beam, for a given bending moment, the greater the degree of flexure. Consequently, the outer fibre-stresses will tend to cause the onset of an accelerating process, in which the neutral surface migrates deeper into the oceanic lithosphere. One or more thrust planes develop and extend, by re-shearing, until the strong layer is completely cut through by one or more such fractures.

The initial resistance of pushing a wedge of relatively cold oceanic lithosphere into hot asthenosphere will be considerably smaller than the head-to-head confrontation of the near irresistible force of the oceanic lithosphere and the near-immovable continental unit. This will result in a significant reduction of the horizontal stresses in the oceanic plate. This, in turn, releases stored strain-energy, which permits elastic recovery, which will extend the length of the oceanic lithosphere by an amount which could be several tens of kilometres;

thereby, for a short period, enhancing the rate of development of the subduction zone. Thus, if we take the elastic recovery to be 30–40 km, the subduction process will have been initiated, coupled with the early development of the 'down-slab' pull. This process will have taken place with no apparent change in length of the oceanic lithosphere, between spreading-ridge and subduction trench, as determined at the surface, so that there will be no noticeable change in the rate of plate motion.

The model we have discussed above is complex and, therefore, difficult to quantify precisely. However, we believe we have shown that the gravitational gliding stresses can certainly fracture the upper zone of the lithosphere, but may not be able to generate sufficient energy to cause a through-going fracture to develop in the oceanic lithosphere. However, where continental and oceanic lithospheres abut, a thick blanket of sediments will cause a downward flexure to develop in the oceanic lithosphere. If the layer of sediments approaches a thickness of about 10 km, the weight of the sediments brings about a change in geometry of the oceanic lithosphere, coupled with a change in geothermal gradient, such that the changes in the environment so weakens the strong layer of the oceanic lithosphere that failure of the whole lithosphere takes place and a subduction zone is initiated. These conditions are met only infrequently.

3.6 Discussion and conclusions

In this chapter, the mechanisms which are widely accepted as being important in giving rise to plate motions have been briefly considered. It is reasonable to accept that the *slab-pull* provided by the subducting lithosphere is potentially capable of generating tensile stresses of very large magnitude. However, the extent to which this potential is decreased by the resistance of the surrounding asthenosphere is undoubtedly also large, but difficult to quantify with precision. Indeed, this resistance is thought to be so high, that the slab-pull may only be able to engender a real tensile stress of, at most, 50 MPa (500 bar) at the top of the slab.

The widely accepted mechanism of *ridge-push* is also a valid one. However, much of this mechanism is directed at the base of the oceanic lithosphere, where it is very weak. Consequently, it is difficult to infer just how much of this ridge push is actually transmitted to the strong, elastic layer.

It is reasonable to conclude, therefore, that whereas the generally accepted mechanisms of slab-pull, possibly coupled with some portion of ridge-push are valid, the magnitude of the stresses associated with these mechanisms is small (almost certainly less than 1.0 kb) and completely incapable of generating significant plate movement and will certainly not be able to generate mountain chains.

When the *gravity-glide* mechanism is included, the situation is transformed, for this latter mechanism has the potential to provide as much as 15–20 times more stress than the combined stresses for other cited mechanisms. Moreover, it can be inferred that the gravity-glide mechanism dictates that the axis of maximum horizontal stress must usually approximate closely to the direction of absolute plate motion.

We have shown that a simple gravity-glide mechanism is able to provide potential compressive stresses in oceanic plates, which comfortably exceed the level of stress which the strong layer is able to sustain. Following failure of the weaker parts of the elastic layer, the horizontal force generated at the leading edge of the oceanic lithosphere is still capable of transmitting an *average* compressive stress of about 8×10^8 Pa (8 kb). This potential compressive stress at a vertical section of an oceanic plate most distant from the spreading-ridge is not completely realised, because of viscous constraints at the base of the plate (as well as those provided by boundary and internal vertical faults). For a very long, relatively fast-moving oceanic plate (cf. the Pacific plate), these resistive elements are extremely important so that, at a distance of 12,500 km from the ridge, the average potential stress of 8 kb is probably reduced to about 2–4 kb.

Consequently, it is concluded, the gravity-glide mechanism, even without assistance from other possible mechanisms, is capable of generating many of the known features of an *up-and-running plate*.

In order that a plate can attain such a state, two features must be created. They are (1) the spreading-ridge and (2) a companion trench to accommodate a subducting slab. In this chapter we have indicated how such trenches may develop. The conditions necessary to generate subduction have been attributed to a combination of vertical loading, generated by a thick sedimentary pile, of about 10 km in thickness, which causes downward flexure of the oceanic lithosphere adjacent to a continental lithosphere, combined with a horizontal stress, generated in the oceanic lithosphere by the gravity-glide mechanism. Large horizontal compressive stress is most easily generated in a relatively short, slow-moving oceanic unit. These combined conditions are only infrequently met. Finally, it should be noted that there exist a number of relatively small, arcuate, intra-oceanic, subduction zones, several of which have developed in the last 100 Ma. These intriguing features cannot be satisfactorily explained by the models discussed in this chapter. The problem of the development of such arcuate, intra-plate subduction zones is discussed in Chapter 7.

Chapter 4

Hotspots, plumes and lithospheric thinning

Large igneous provinces and the splitting of continents

4.1 Introduction

The quantity and rate of continued production and eruption of basalt at the Earth's surface, comprise one of the most impressive characteristics of Earth processes. This production of basalt is associated with three main classes of feature which are: (1) *spreading-ridges*, (2) *large igneous provinces* (LIPs) and (3) *hotspots*.

Of these sources, the most prolific and *continuous* formation of basalt is associated with the development of spreading-ridges. Thus, if we take the total, global length of such ridges to be about 60,000 km, the average thickness of oceanic crust to be 7 km, and the average spreading-rate to be 5 km Ma^{-1}, then the total volume of basalt produced at the ridges is in excess of 2,000,000 km^3 every million years.

The second category of eruption is related to the development of continental flood basalts (CFBs) and oceanic plateau basalts (OPBs). Such outbreaks of vigorous igneous activity are well recognised and, as will be seen from Figure 4.1, such flood basalts may cover wide areas. Here we use the adjective *vigorous* to indicate the emplacement of large volumes of basalt of the order of 10^{6-7} km^3 Ma^{-1}. These events can therefore exceed the average rate of basalt produced at the ridges each year. However, the emplacement of large igneous provinces may be largely accomplished in less than a million years, whereas the eruption at the spreading-ridges is a continual process. As noted above, the volume of material produced to form CFBs is usually in the range 10^{6-7} km^3, while the corresponding volume that gives rise to the larger oceanic flood basalt (OFB) is almost an order of magnitude larger, at up to 6×10^7 km^3.

The third category of eruption is associated with more than a hundred hotspots. The volcanic activity associated with a hotspot ranges from a single volcano to extensive submarine chains containing a number of volcanic islands and seamounts (e.g. the Hawaiian-Emperor Chain). One can infer from the variations in width and elevation of such chains that the volcanic activity usually appears to have been non-uniform throughout its active life, which may extend for as much as 90 Ma. It is, therefore, difficult to be specific regarding the average output of basalt erupted from a single hotspot. However, in general, it appears to range between 2000 and 10,000 km^3 Ma^{-1}, though this upper limit may well be exceeded during a period of intense activity. From the estimates presented above, it can be inferred that it is unlikely that a single mechanism can be invoked to explain how these volumes are erupted at such disparate rates. Following the data and classifications set out by Coffin and Eldeholm (1992), we shall first list the various major igneous bodies (Table 4.1) and indicate the disposition of these features around the world (Figure 4.1). These bodies are briefly classified and described in general terms as large igneous provinces (LIPs), whether they be the result of hotspots, continental flood basalts or oceanic plateau basalts.

Table 4.1a Large igneous provinces emplaced over the past 250 Ma.

LIP	Abbrev. (Figure 4.1)	Type	Spatial or temporal association with continental break-up?
Aden Traps	ADEN	CFB	YES
Alpha Ridge	ALPH	*SR/OP	?
Austral Seamounts	AUST	SMT	NO
Bermuda Rise	BERM	OP	NO
Broken Ridge	BROK	SR	NO
Cape Verde Rise	CAPE	OP	NO
Caribbean Flood Basalts	CARI	CFB/OBFB	?
Caroline Seamounts	CARO	SMT	NO
Ceara Rise	CEAR	OP	NO
Chagos-Laccadive Ridge	CHAG	+SR	NO
Columbia River Basalts	COLR	FB	NO
Conrad Rise	CONR	OP	NO
Crozet Plateau	CROZ	OP	NO
Cuvier Plateau	CUVI	OP	YES
Deccan Traps	DECC	CFB	YES
Del Cano Rise	DELC	OP	?
Eauripik Rise	EAUR	OP	NO
East Mariana Basin	EMAR	OBFB	NO
Etendeka	ETEN	CFB	YES
Ethiopian Flood Basalts	ETHI	CFB	YES
Galapagos	GALA	SML	NO
Hawaiian-Emperor Seamounts	HAWA	SMT	NO
Hess Rise	HESS	OP	NO
Iceland/Faeroe-Greenland Ridge	ICEL	+OP/SR	NO
Karoo	KARO	CFB	YES
Kerguelen Plateau	KERG	OP	NO
Line Islands	LINE	SMT	NO
Lord Howe Rise Seamounts	LORD	SMT	NO
Louisville Ridge	LOUI	SMT	NO
Madagascar Ridge	MADA	+SR	?
Madeira Rise	MADE	OP	NO
Magellan Rise	MAGR	OP	NO
Magellan Seamounts	MAGS	SMT	NO
Manihiki Plateau	MANI	OP	NO
Marcus Wake Seamounts	MARC	SMT	NO
Marquesas Islands	MARQ	SMT	NO
Mashall Gilbert Seamounts	MARS	SMT	NO
Mascarene Plateau	MASC	+OP	YES
Maud Rise	MAUD	OP	NO
Mid-Pacific Mountains	MIDP	SMT	NO
Naturaliste Plateau	NATU	OP	YES
Nauru Basin	NAUR	OBFB	NO
New England Seamounts	NEWE	SMT	NO
Ninetyeast Ridge	NINE	SR	NO
North Atlantic Volcanic Province	NAVP	CFB	YES
Ontong Java Plateau	ONTO	OP	NO
Osborn Knoll	OSBO	CP	NO
Parana Flood Volcanism	PARA	CFB	YES
Pigafetta Basin	PIGA	OBFB	NO
Rajmahal Traps	RAJM	CFB	?
Rio Grande Rise	RIOG	+OP	NO
Roo Rise	ROO	OP	NO
Shatsky Rise	SHAT	OP	NO
Siberian Traps	SIBE	CFB	NO
Sierra Leone Rise	SIER	OP	NO
Tahiti	TAHI	SMT	NO
Tasmantid Seamounts	TASM	SMT	NO
Tuamotu Archipelago	TAUM	SMT	NO
Wallaby Plateau	WALL	OP	NO
Walvis Ridge	WALV	+SR	NO

Reproduced from Coffin and Eldholm (1992), by permission of the Geological Society Publishing House.

Notes: CFB, continental flood basalts; OBFB, ocean basin flood basalts; OP, oceanic plateau; SMT, seamount; SR, submarine ridge. * referred to in the literature as both a submarine ridge and an oceanic plateau. + oceanic plateaux or submarine ridges which can be tied to LIPs originating during break-ip, but for which volcanism post-dates break-up.

Table 4.1b Dimensions, age ranges, and emplacement rates of five large igneous provinces.

Large igneous province	Area (10^6 km^2)	Volume (10^6 km^3)	Age range (Ma)	Emplacement rate (km^3 yr^{-1})	Spherical diameter[a] (km)
Ontong Java					
Ontong Java Plateau[b]	1.86	49.0–61.3[c]	117.7–118.2[d]; 121–124[e]	16.3–98.0; 20.4–112.6[f]	
Ontong Java Plateau[g]	1.86	26.9–39.3[c]	117.7–118.2[d]; 121–124[e]	9.0–53.8; 13.1–78.4[i]	
Ontong Java Plateau[h]	1.86	8.4	117.7–118.2[d]; 121–124[e]	2.6	
Nauru Basin	1.75	0.856	117.7–118.2[d]; 121–124[e]		
Manahiki Plateau	0.77	8.8–13.6[c]	117.7–118.2[d]; 121–124[e]	2.8–17.0; 4.5–27.2[f]	
Pigafetta and East Mariana basins	0.50	0.25	114.6–126.1		
Total	4.88	36.4–76.0[c]	117.7–118.2[d]; 121–124[e]	12.1–72.8; 25.3–152.0[f]	614–1427
Kerguelen (Cretaceous)					
Kerguelen Plateau	1.54	9.9–15.4[c]	109.5–114	2.3–3.5	
Kerguelen Plateau[h]	1.54	7.7	109.5–114	1.7	
Elan Bank	0.24	1.2–1.9[c]	109.5–114	0.3–0.4	
Broken Ridge	0.51	4.1–6.9[c]	88.0–89.2	0.9–1.5	
Total	2.30	15.2–24.1[c]	109.5–114	3.4–5.4	459–973
North Atlantic volcanic province	>1.3	6.63	54.5–57.5	2.2	348–663
North Atlantic volcanic province[h]	1.3	1.8	54.5–57.5	0.6–2.4[j]	
Deccan					
Deccan traps	0.5–1.5	8.2	64.5–65.5; 65–69		
Deccan traps[h]	0.5–1.5	>1.5	64.5–65.5; 65–69	0.3–2.4[j]	
Seychelles	0.25	?	64.5–65.5; 65–69		
Total	0.75–1.75	8.2	64.5–65.5; 65–69	2.1–8.2	374–679
Columbia River basalts	0.1637	1.3	6–17.5; 15.7–17.2	0.1–0.9	202–368

Source: Reproduced from Coffin and Eldholm (1992), by permission of the Geological Society Publishing House.

Notes
a At 15–30 per cent partial melting.
b Assumes crustal thickness from seismic refraction experiments.
c Minimum and maximum volumes assume off- and on-ridge emplacement, resp.
d Timescale of Harland et al. (1982).
e Timescale of Harland et al. (1990).
f First range, minimum volume (pre-existing crust, maximum and minimum age range;

second renge, maximum volume (no pre-existing crust), maximum and minimum age range.
g Assumes crustal thickness from satellite altimetry data.
h Extrusive component.
i Assuming 0.5-m.y. eruption period for two-thirds of the basalt.
j For 80 per cent of the lavas.

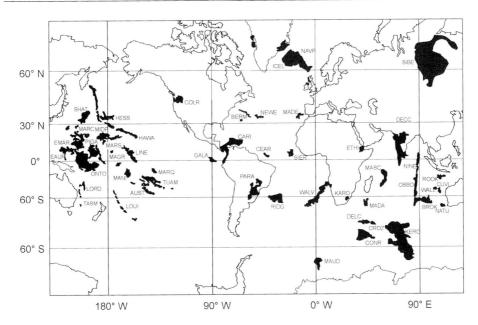

Figure 4.1 Map of large igneous provinces (after Coffin and Eldholm, 1992).

Of the eruptive features listed and described by Coffin and Eldeholm, we shall first consider those associated with *hotspots*. We present this group of features first, because their characteristics are reasonably well defined, and the mechanism by which these relatively small to medium-sized bodies are emplaced is thought to be reasonably well understood in terms of *plumes* of relatively hot mantle material, which may originate at the core/mantle boundary, or possibly in the vicinity of the transition zone. The amount of melt that can be produced by such plumes has been assessed by mathematical modelling. The volume of melt that such plumes can deliver at the base of the lithosphere is, however, somewhat limited and is only sufficient to give rise to single islands, seamounts, or chains of such bodies.

In the second part of this chapter we deal with the emplacement of continental flood basalts and oceanic plateau basalts (CFBs and OPBs). The mechanisms which are held to be responsible for the emplacement of these large basaltic provinces are contentious, and hotly debated. Initially, the concept of an extremely active plume, i.e. a *super-plume*, was invoked. Alternatively, some other possibly additional mechanism was thought to be required. This second mechanism, it has been suggested, relates to lithospheric thinning, brought about by stretching. However, a high degree of stretching is required to give rise to the volume of melt-rock at the base of the lithosphere that is necessary to generate major CFBs and OFBs. We argue that the degree of stretching required far exceeds what can feasibly be expected.

It has been further suggested in the literature that plumes, coupled with stretching of continental lithosphere, can give rise to the splitting of super-continents. This leads us to discuss how stretching of continental lithosphere may be induced by strike-slip faulting, and how some rift valleys form. We indicate how a relatively small degree of stretching may give rise to some, but possibly not all, rift valleys. Moreover, we point out that the generation of such faults does not give rise to a general tensile stress in the continental crust.

4.2 Large igneous provinces

The obvious question to ask is what constitutes a *large igneous province* (LIP)? Is it possible to choose a lower limit of size, defined by area or volume, or time limits for the duration of emplacement of such bodies? Unfortunately, such pertinent data are largely lacking for many igneous provinces. As a result of the inability to define precisely which igneous province may be described as large, when setting out a list of LIPs, Coffin and Eldholm (1992) were reduced to citing those bodies which, by consensus, are held to be 'large'. These authors compiled a list of about sixty such bodies (see Table 4.1). In this table they classify the type of LIP of each named site, and indicate which province is associated, in time or space, with the opening of an ocean. They also provide extensive references for the interested reader.

The igneous provinces are divided into the following categories:

- continental flood basalts (CFB)
- oceanic plateau basalts (OPB)
- oceanic basin flood basalts (OBFB)
- seamounts and submarine ridges (SMT) and (SR)

Let us now briefly consider, in turn, the characteristics of these various classes of igneous bodies, the location of individual examples of which are shown in Figure 4.1.

4.2.1 Continental flood basalts (CFBs)

Continental flood basalts are, of course, the most easily accessible and, consequently, the most closely studied of all igneous provinces. They are found to be mainly derived from tholeiitic magma, which has erupted on continents over a relatively short time-span. This period commonly lasts for 1–2 Ma, though some appear to exhibit a decreased activity for at least a further 10 Ma. The period over which the main eruptive activity usually lasts is, however, open to some debate. For example, repeated attempts have been made to determine accurately the time-span over which the Deccan Traps was erupted. The most consistent results come from the palaeomagnetic studies of samples obtained from the current outcrops of the basalt flows, regarding which, Gallet *et al.* (1989) conclude that they occurred in a period of less than 3.0 Ma and were quite probably emplaced in less than 1.0 Ma. Baksi (1990) endorsed this conclusion and suggested that more than 80 per cent of the erupted rocks were emplaced in about 1.0 Ma; Courtillot (1990) went further and suggested that the bulk of such basaltic material was erupted within 10,000 years.

Gallet *et al.* (1989) state that the mean date of emplacement was 65 Ma (i.e. centred around chron 27R). For a variety of reasons, the Ar/Ar and K/Ar dating techniques produce a wider range of ages (from 60–70Ma) though the mean values are the same. The range of dates using these techniques is, in part, attributable to the intrinsic uncertainties in the age-determining method (usually about +/– 1.0 Ma), while the history of weathering or contamination of the specimen may give rise to further uncertainty. For example, when discussing the ages of the Deccan Trap and the Seychelles rocks, Devey and Stevenson (1992) quoted dates, with errors, which range from 82 +/– 16 Ma to 63.7 +/– 1.1 Ma. Clearly, more recent analytical methods have the greater accuracy. The petrology and geochemistry of these basalts and their mode of surface emplacement, which is mainly by sub-horizontal lava flows, is reasonably well established. However, little is known as regards their crustal structure and even less is known regarding their deep structure, if any, in the mantle. The origin of the components which make up the surface flows, whether they came from the lithospheric mantle or from the upper or lower mantle of the asthenosphere and what proportions of derived material may be attributed to these different depths, is still open to debate.

4.2.2 Oceanic plateau basalts (OPBs)

Oceanic plateau basalts are broad features with, as their name implies, a relatively flat top. They have an elevation above the surrounding sea-floor sometimes in excess of 2 km. These plateaus are sometimes supported by anomalously thick crust. Obviously, their age must be younger than the surrounding sea-floor. However, the age difference may sometimes be surprisingly small.

The volume of erupted rock that forms OPBs extends over more than two orders of magnitude. The largest of these structures, namely the Ontong-Java, in particular, is associated with vast volumes (up to 57×10^6 km^3) of erupted material. However, this feature is not connected in any way with continental break-up. Indeed, OPBs are often quite distanced from major continents.

4.2.3 Ocean basin flood basalts (OBFBs)

These provinces form extensive areas of extrusive basalt which are only marginally younger than the ocean floor upon which they rest. They are different from most OPBs in that they do not form well-defined plateaus, but, as their name implies, form in oceanic basins. Coffin and Eldholm (1992) have referred only to three examples (all in reasonably close geographical association with the Ontong-Java OPB) which they consider to be definitely under this heading, and one other (the Aves Ridge, in the Caribbean) which may also be so classified. This form of LIP has attracted the least amount of study and is the least well understood.

4.2.4 Seamount groups and submarine ridges (SMTs and SRs)

Seamounts are basaltic volcanoes which rise from the ocean-floor. They form submarine peaks or, if they have been eroded by marine action, they are flat-topped (*guyots*). These mounts may occur singly or in groups, or even in linear chains, and are not usually related, either by time or space, to continental break-up.

Submarine ridges, which are of wholly marine origin, are elongated features with relatively steep sides, and commonly exhibit marked topographical variations. They may be created either on or off the axis of spreading. Those that are related to the axis of spreading may span the ocean, cf. the Greenland–Icelandic–Faroes feature in the N Atlantic and the complex ridge that spreads in opposite directions from Tristan de Cunha. However, some submarine ridges (e.g. the Hawaiian-Emperor chain) are remote from any spreading-ridge.

4.3 Hotspots and their mechanism of emplacement

There are over 100 'recognised hotspots', the location and types of which will be discussed later. Hotspots occur beneath both oceans and continents. In the latter situation, the hotspots tend to give rise to domal uplift and areas of high heat-flow associated with igneous activity and/or geothermal activity (e.g. Yellowstone Park, USA). In a marine environment, the hotspots can give rise to individual islands, or seamounts (usually accompanied by an area of domal uplift) as well as chains of such extrusive bodies.

Wilson (1963) suggested that such chains of volcanic islands formed as the lithosphere passed over a *relatively stationary hotspot*. Perhaps the most famous of these is the Hawaiian-Emperor chain (Figure 4.2), which extends for 6000 km, with a kink approximately half way along, and took approximately 70–90 Ma to develop. Moreover, as will be seen in this figure, there are other island and seamount chains in the Pacific which show similar trends to the Hawaiian-Emperor chain. Other hotspots and ridges, such as Walvis/Tristan/Rio Grande and continental hotspots in the Sahara, are shown in Figure 4.3.

It is now recognised that hotspots are not 'immutably fixed for all time'. Burke *et al.* (1978) and Molnar and Atwater (1973) established that there has been relative movement between hotspots, within the same plate, over the last 120 Ma, of at least 180 km (i.e. a drift of about 0.15 mm a^{-1}), while Molnar and Stock (1987) have determined relative motions between some hotspots which attain a rate of 0.20 mm a^{-1}. Thus, to assume hotspots remain absolutely stationary over this period can result in a small error. Nevertheless, the hotspots do provide a most useful frame of reference.

4.3.1 Surface manifestations

We have noted that plumes, in continents, give rise to domal uplift. Erosion and sedimentation tend to affect the topography of these subaerial features. Away from spreading-ridges, the topography of the ocean floor is usually relatively simple compared with some continental areas; hence the best examples of the surface manifestations of a hotspot are to be found in the oceans.

A number of locations where relatively large quantities of eruptive rock occur on or near rifted margins and which exhibit circular or elliptical domes, are listed in Table 4.2. Two

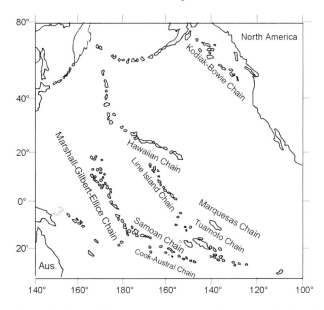

Figure 4.2 Linear chains of volcanic islands, seamounts and ridges in the Pacific which are related to hotspots. The younger islands in the chains are always to the east.

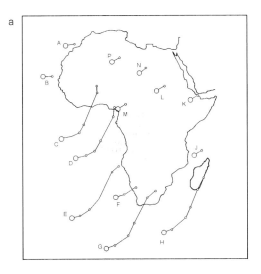

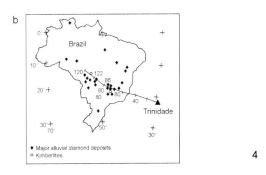

4

Figure 4.3 Hotspots and ridge tracks in Africa, the Atlantic and S America.

Table 4.2

Swell	Crustal age (Ma)	Residual depth anomaly at centre (m)	Width of swell (km)
Circular or Elliptical			
Bermuda	110	1000	800
Cape Verde	125	1900	1550
Cook-Austral	46	650	1250
Crozet	67	800	>1000
Marquesas	45	800	1500
Reunion	67	1050	–
Society	70	1100	–
Ridges			
Hawaii	90	1200	1200
Iceland/Greenland	65	2000	2000

examples of volcanic island chains which have developed on relatively deep ocean floor are also cited. From these data, it is inferred that the effect of sub-ocean plumes is to cause uplift of the existing floor mostly by about 1–2 km. The residual depth anomaly (RDA) (that is the difference in height between the highest part of the dome and that of the outer limit of the structure) and the width of the swell for most of the cited examples are listed in Table 4.2.

The record, cited here, stretches back for only 125 Ma. However, the range of ratio of widths is relatively small at 1 : 2.5, while the ratio of depths is 1 : 3.0. In this book we are particularly concerned with events which took place in the last 250 Ma. Hence, if we assume that the average dimensions of the surface feature of a plume is 2000 km diameter, with an RDA of 2 km, we are not likely to underestimate the magnitude of the effects of plumes.

4.3.2 Plumes

Following the insight of Tuzo Wilson, Morgan (1972) suggested that lithospheric hotspots are the result of deep mantle convection and the interaction of the up-welling material (the plume) with the lithosphere. The magma is thought to derive from pressure-release melting and differentiation in the up-welling plume. Olson *et al.* (1988) suggested that such plumes originate at the core/mantle interface, at which there is a significant thermal boundary layer, with the core material being considerably hotter than the mantle. Moreover, the interface may not be perfectly smooth, but could exhibit topographic relief of the order of 5 km. The individual perturbations on this interface are held to be responsible for a more rapid flow of heat from the core into the lowest portions of the mantle, thereby giving rise to the initiation of a plume of up-rising hot mantle material. Alternatively, plumes may originate at shallower depths within the mantle, such as at the base of the transition zone.

Plumes have been simulated by computer modelling, as indicated in Figure 4.4. Saunders *et al.* (1992) indicate how a plume head is an intrinsic part of the evolution of an upward migrating *stem*. It will be seen that, in section, such plumes are assumed to have the form of a long 'stem' with a 'mushroom' head. The plume is axi-symmetric, so that in three dimensions it has, indeed, a real mushroom-like geometry.

It can be seen (Figure 4.4) that the hottest material is about 200–300°C above the ambient temperature outside the stem of the plume, and is restricted to the core of the stem, where the hottest material is thought to have a radius of about 40 km. The upward flow in this central part of the stem, because of its relative high temperature and, hence, lower density, is more rapid than in the outer zone of the stem. The flow is deflected, as it approaches the asthenosphere/lithosphere interface, to form the mushroom head, where the flow-rate becomes slower.

In the vicinity of the LVZ, at the base of the oceanic lithosphere, the pressure and temperature (PT)conditions conducive to partial melting are met. Consequently, as the hot material in the central part of the stem approaches the LVZ decompression, it experiences melting and also, because of its excess temperature relative to the surrounding mantle material, it may cause further melting of the *in situ* material below the LVZ. (The reader with a particular interest in this subject should refer to the papers in Jackson (1998). The melt in the stem of the plume can become segregated and this more energetic component may penetrate the base of the mushroom head of the plume and could possibly cut through the lithosphere, thereby giving rise to volcanic activity at the surface. A subsequent decrease in the flow rate in the stem would then cause the volcanism to abate. This mechanism would supply the

a

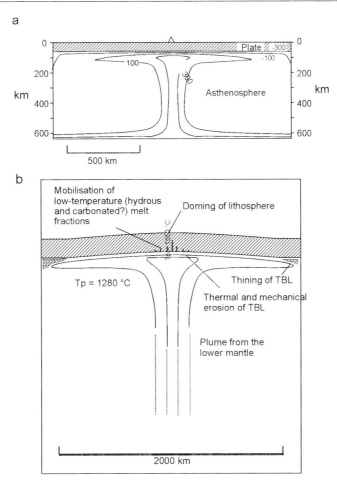

b

Figure 4.4 Computer simulation of a plume originating at the core/mantle interface, with the plume head, generated when the plume encounters the asthenospheric/lithospheric (after Saunders et al., 1992).

eruptive material and variations in flow rate in the plume and explain the intermittent volcanic activity exhibited by island and seamount chains. Griffith and Campbell (1991) modelled plume head development and have indicated that the surface uplift and development induced by the rising plume head is likely to exceed 30 Ma. (This is a point to which we shall return.)

More recent numerical modelling by Farnetani and Richards (1995) indicates that only the very hottest material undergoes melting. If this modelling is correct, it is necessary to infer that melts only come from source material (enriched in incompatible elements) spreading near the top of the plume. This approach requires special pleading.

Hard evidence relating to the geometry of the lower levels of a plume has yet to be adduced. Nataf and VanDecar (1993) conducted a seismological study of the Bowie hotspot, west of the Canadian coast and probed to a depth of 700 km, to infer the existence of a 150 km diameter column which, they suggest, is 300°C hotter than the surrounding mantle. However, the paper's title, 'Seismological detection of a mantle plume?' with its terminating question mark, indicates the degree of confidence the authors have in their findings. Indeed, they state that 'the detection of the plume cannot be considered definite. However, we believe that our analysis ... is the best that can be done with current data and techniques'. Anderson *et al.* (1992) have gone further, and consider that plumes are never likely to be detected using currently known geophysical techniques, with their inherent resolution problems.

4.3.3 Modelling

As indicated above, the morphology of plumes and the manner in which they give rise to the various geophysical signatures have been mainly based on mathematical modelling. Ribe and Christensen (1994) indicate that models can be divided into two main categories: *thermal* and *dynamical.*

Thermal models concentrate on the thermal structure of the lithosphere, but ignore, or treat in a simplified manner, the upward flow in the asthenosphere. Such models generally relate the effects of the assumed up-welling plume to the thermal boundary conditions applied to the base of the lithosphere. The earliest models were put forward by Detrick and Crough (1978) and Crough (1978), who concluded that the Hawaiian Island chain could be explained by the reheating and thinning of the lithosphere. More recently, plume-induced reheating has been studied by Spohr and Schubert (1982) and Nakiboglu and Lambeck (1985). Sleep (1987) proposed a thermal model which also incorporated simple flow kinematics and concluded that extensive lithospheric heating should occur 'downstream' in the Hawaiian Island chain.

Dynamical models are inherently superior to thermal models. However, until recently, with the advent of more powerful and faster computers, dynamical models have been restricted to rather simple two-dimensional or axi-symmetrical models.

Courtney and White (1986) proposed an axi-symmetric model for the Cape Verde Rise and compared their results with the known data for that area. This feature is a large E–W trending bathymetric dome, 500 km west of Senegal. It ascends to a little over 2000 m (from an old (125 Ma), slow moving (1.2 cm a^{-1}), ocean floor) over a horizontal distance of about 600 km. The anomalously high heat-flow, geoid and bathometric data were used by these authors to constrain various theoretical models of hotspot mechanisms. It was established that models which depended solely on lithospheric reheating, whether involving transient or sustained thinning of the lithosphere, fail to model the broad heat-flow anomaly, and the long duration of uplift and volcanism obtaining at the Cape Verde Rise. Axi-symmetric convection models provided a better fit to the observed data. This permitted the authors to suggest that the swell is primarily a consequence of dynamic uplift, generated by an ascending thermal plume in the underlying mantle. The good fit between the data and the axi-symmetric model can largely be attributed to the fact that the ocean floor beneath the rise is very slow moving. One would not expect this model to give an equally good fit to data obtained from the relatively fast-moving plate supporting the Hawaiian Island chain.

This latter problem was tackled by Ribe and Christensen (1994) who aimed at a full three-dimensional model that overcomes the limitations of earlier axi-symmetric thermal and dynamical models. The analysis is based on the geometry indicated in Figure 4.5a. The details of this analysis are beyond the scope of this book, so we shall merely indicate the results they produced, and the degree to which the theoretical results fit topographic and other data. The section of the plume is shown in Figure 4.5b, the topography and geoid

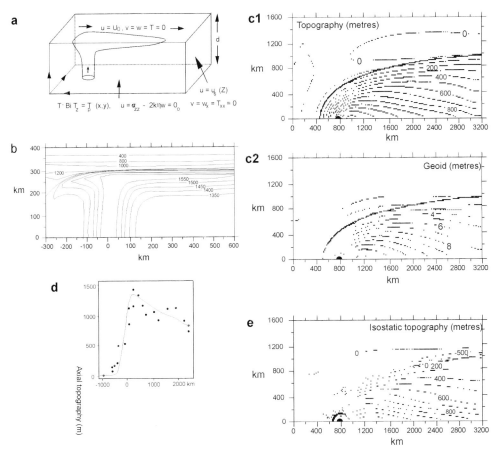

Figure 4.5(a–e) Three-dimensional dynamic computer model of a plume interacting with a relatively fast-moving lithospheric plate. The model demonstrates how the heating effect of the plume spreads downstream. The model data are compared with that obtained from the Hawaii area. See text for details (after Ribe and Christensen, 1994).

anomalies are shown in Figure 4.5c and Figure 4.5d, while Figure 4.5e is a comparison of the known topography with that produced from the model. The paper contains other details, but Figure 4.5 is sufficient to convince most readers that Ribe and Christensen have demonstrated, beyond reasonable doubt, that the Hawaiian Island chain and other similar chains are derived from a plume with an asymmetrical head, which tails away in the direction of plate motion.

4.3.4 Types of hotspots

It will be apparent from the description of the few geological examples, already cited, that there appear to be different types of hotspots. Some are thought to be almost circular in plan, while others give rise to well-defined trains, some of which are remarkably linear.

Marzocchi and Mulargia (1993) used a statistical, pattern-recognition approach, to identify the characteristic features of hotspot volcanism, which identifies associations as well as single parameters. The parameters used in their analysis include: absolute plate velocity of the site, minimum distance between neighbouring hotspots, minimum distance from the ridge, minimum distance from a trench and geoid anomalies.

These authors identified three types of hotspots.

Hotspots of the first type, shown in Figure 4.6 as black circles, are not obviously related to plate boundaries. This type is found to occur in sites characterised by positive values of geoid anomalies with harmonic degree between 2 and 10, and give rise to chains of volcanic islands and seamounts (e.g. Hawaiian-Emperor chain) which are linked to long-wavelength, positive geoid anomalies and also occur in continental areas. These anomalies, it is suggested, are sublithospheric in origin, with dynamics which are not influenced by surface tectonics.

The second type (indicated in Figure 4.6 as black circles cut by a vertical bar) consists of clusters of volcanoes, with a distance between volcanoes in the cluster of less than approximately 900 km and which develop in slow-moving plates (i.e. those plates with an absolute velocity of less than 2.5 cm a^{-1}). These, they suggest, are of a more superficial origin.

A third type (denoted by black circles cut by a horizontal bar) is represented by isolated volcanoes, which are more than 900 km distant from any other hotspot, and which are anomalously close to a mid-ocean ridge.

(There are also other (unmarked) sites, that are not recognised as belonging to any of the three types noted above.)

It would appear, therefore, that there are *hotspots* and there are 'hotspots'! We suggest that only the first group, identified by Marzocchi and Mulargia is worthy of the appellation *hotspots*, which originate deep within the Earth. Perhaps the others should more accurately be termed *pseudo-*, or *superficial hotspots*.

Thus, although Marzocchi and Mulargia suggest that it is only the hotspots of group 1 that should be attributed to mantle plumes, we shall see that the mantle plume concept has been widely accepted as giving rise to eruptive features of far greater magnitude than the volcanoes and seamounts which form chains.

One can safely conclude, therefore, that *plumes* exist and that they are of sufficient size to give rise to individual volcanoes or volcanic island/seamount chains. However, the depth at which the plumes originate is conjectural, and it is not known with certainty how the plumes are initiated. It is commonly assumed that large eruptions of basaltic material onto the surface of continents and also upon the ocean floor have been initiated by major-, super- or mega-plumes. Hence, it is necessary to infer the size of the plume from the volume of erupted material. Before we address this problem, let us first outline the evidence relating to the development of CFBs and OPBs and consider how these features are thought to have developed.

4.4 Continental flood basalts and the break-up of continents

The name, location and approximate age of the continental flood basalts (CFBs), which have developed within the past 250 Ma are listed in Table 4.3. Of the continental flood basalts listed, the largest, the Siberian flood basalt, and the smallest, the Columbia River flood basalt, and the two Antarctic events are not obviously associated with continental break-up.

A hybrid continental/oceanic eruptive unit is formed by the Greenland/Iceland/Faroes Ridge and other eruptions in Greenland and Norway, Scotland and England. This important continental/marine igneous province (Figure 4.7) is not now a CFB, but could have been initiated as one. This event not only produced a large volume of eruptive rocks, but is also attributed, by advocates of the plume mechanism, to have initiated the opening of the

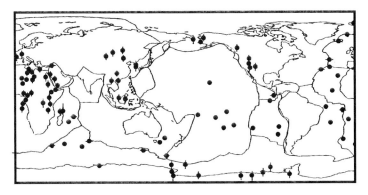

Figure 4.6 Showing the location of various type classes of hotspots (after Marzocchi and Mulargia, 1993). The three types of symbols (see text for clarification) enable the positions of the three classes of hotspots to be identified. Those locations, mainly from continental areas which do not fit the computer patterns have been omitted.

Table 4.3 Location and age of continental flood basalt

Name	Location	Age (Ma)
Siberian	N Russia	248
Karoo	S Africa	193–178
Antarctica I		176
Antarctica II		170
Paraña	S America	135
Deccan Traps	India	65
Aden and Ethiopian	Africa–Arabia	39
Columbia River	USA	17

North Atlantic. White and McKenzie (1989) site the plume beneath Kangerlussuaq, E Greenland. The extent of the plume-head and inferred limits of updoming are determined by the position of extrusive volcanic rock (shaded black in Figure 4.7) and the extent of early Tertiary igneous activity in the region (diagonal hatching).

4.4.1 Plumes and CFBs

White and McKenzie have suggested that, except for the Antarctic CFBs, which they did not discuss, the CFBs listed in Table 4.3 are the result of plumes. They applied the concepts of plumes coupled with lithospheric thinning, in an attempt to explain the development of specific continental flood basalts. They also suggest that the splitting of a super-continent can result from the topographical elevation of the continental lithosphere attained as the result of the plume. Of the CFBs discussed by them, it is the application of plume development to the Deccan Traps which is most well known and which we shall consider first. The Deccan Traps is, perhaps, the best studied example of continental flood basalts to have occurred in the last 250 Ma. The eruption occurred approximately 65 Ma ago. The basalts attain a maximum thickness of 3.5 km and extend over an area equal to one sixth of the sub-continent. The total amount of extruded material is somewhat in excess of 10^6 km^3 and was emplaced in a period of, possibly, less than 10^6 years.

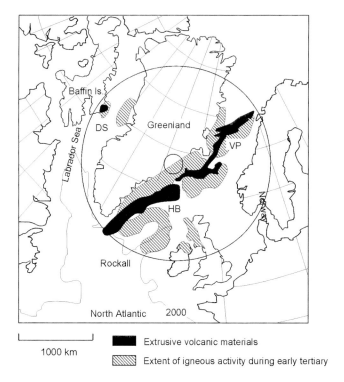

Figure 4.7 Greenland/Iceland/Faroes Ridge.

This event, White and McKenzie suggest, can be attributed to the development of a major plume below India. Indeed, they further claim that this plume was so important, that not only did it give rise to the Deccan Traps but also to a contiguous massive out-pouring of basalts which currently form the Seychelles and Mascarene Banks. (We will comment upon this conclusion in Chapter 6.)

These authors did not consider the Antarctic bodies in their paper, but suggested that the other CFBs listed above could be attributed to the arrival beneath the lithosphere of a powerful plume. They noted that, with the exception of the Siberian and Columbia River flood basalts, the other bodies, they consider, were all associated with continental break-up.

Just how a hypothetical plume gave rise to the opening of the N Atlantic is not very clear. White and McKenzie indicate that the elevation of the central part of the dome, which results from the assumed upwelling plume, relative to the limits of the plume-head, permits a gravitational glide potential, which can cause rupture in a favoured orientation which could eventually give rise to ridge spreading. We have already indicated, in Chapter 3, that the magnitude of the stresses engendered by such sliding is insufficient to give rise to the break-up of a continent.

However, when dealing with the North Atlantic Igneous Province, Holm *et al.* (1992) attest that the inferred plume caused no thermal doming of the lithosphere. Consequently, rifting in E Greenland was not generated by a gravitational slide off a plume-initiated thermal bulge. Indeed, Skogseid *et al.* (1992) argue that the spatial correlation between tectonic activity, magmatism and subsidence permits one to infer that rifting occurred before the assumed plume reached the lithosphere. Possibly with these points in mind, Gill *et al.* (1992) mention that the plume which is held to have given rise to the Greenland–Icelandic features was a plume with a 'ridge-like configuration'. There is the further problem that the influence of the hypothetical plume appears to become apparent so suddenly, that it precludes a prelude of volcanic activity. Thus, the plume concept appears to be at variance with the geological evidence adduced by several geologists. However, Iceland, which now sits astride the N Atlantic spreading-ridge, marks the site of a hotspot that has been in existence for at least 60 Ma, so that the situation is not clear-cut. Let us therefore turn to the tomographic evidence.

4.4.2 Tomographic evidence

If a continent passes over a hotspot it will be heated to an abnormally high temperature in the lower regions of the continental lithosphere. Consequently, it is reasonable to suppose that a large and relatively recent thermal event must leave a readily detectable *heat-signature*. However, as regards the Deccan Traps, Anderson *et al.* (1992) state that no such signature exists. This conclusion is based on their study of high-resolution upper-mantle tomographic models in terms of plate tectonics, hotspot and plume theories. They state that if this body of plateau basalts were the result of a major plume, one would expect to detect a large area of thin lithosphere and very slow seismic velocities under most of India. However, the tomography shows fast seismic velocities under the Indian subcontinent, down to a depth that includes the full thickness of the lithosphere of India, which indicates a thick, cold lithosphere and a cold asthenosphere. Therefore, they state, there is no evidence for plume head or lithospheric thinning under this major flood basalt.

Let us now extend the remarks by Anderson *et al.* (1992) to other continental flood basalts that have been attributed by White and McKenzie to be the result of plumes.

As regards the Paraña flood basalts (135 Ma), Anderson *et al.* comment that 'the time since the conjectured plume head ... event ... is too short for the lower lithosphere or (upper) asthenosphere to cool off substantially ... The regions of S America and Africa which should have been affected by the plume head do not differ from the surrounding continental areas. The seismic evidence is therefore not favourable to the plume-head, lithospheric stretching ... hypothesis.'

Both the Karoo (178 Ma) and the Siberian (248 Ma) events have also been attributed by White and McKenzie to major plume heads. Both these provinces are sufficiently old to have experienced some cooling of the affected lithosphere. However, the tomographic evidence shows that these regions are no different from other areas in the vicinity, which were not affected by the hypothetical plume head. So once again, for these two older events, Anderson *et al.* infer that there is no evidence to support the plume mechanism.

In their discussion, Anderson *et al.* state that tomographic results contradict the premises of all currently popular plume and flood basalt scenarios. Thus, we suggest, the fact that no plume signature can be found related to the plateau basalts to which White and McKenzie applied their model, renders their application to the several examples of plateau basalt development more than a little suspect.

4.4.3 Depth of plume and the melting problem

One can presumably judge the degree of support for a hypothesis by the number of authors of papers and other publications who use the concept to explain their particular geological problem. Using this criterion, support for the plume hypothesis, regarding the emplacement of LIPs, seems to be reasonably widespread.

Publications on the topic of plumes can usually be divided into two groups. The first of these are written by authors who accept that plumes exist and are capable of producing the required volume of melt to explain the particular eruptive body, or bodies, in which they are interested. The second, and much smaller group, are concerned with the mechanisms which can be invoked to explain the geological and geophysical evidence. With reference to this second group, there is little general agreement as regards the mechanism. Notwithstanding this apparently widespread support for the plume mechanism, some of those who advocate the use of various aspects of the emplacement mechanism are aware of its shortcomings.

There are three main hypotheses that have been proposed to explain the relationship between mantle plumes and flood basalts: namely the Campbell and Griffiths (1990) plume-head hypothesis, the White and McKenzie (1989) model, which combines the concept of lithospheric thinning and the emplacement of a plume, and the wet-plume hypothesis.

(a) Campbell and Griffith hypothesis

These authors consider that a large plume head originated at the core-mantle boundary, as indicated in Figure 4.8a, which is a photograph of a laboratory-induced plume. In Figure 4.8b, this plume head, which is assumed to have a diameter of 1000 km, rises to form, beneath the lithosphere, an oblate circular disk, with a diameter of 2000–2500 km. This leads to an uplift of the overlying lithosphere of 0.5–1.0 km, and the development of volcanic activity. Plume head melting occurs as the consequence of adiabatic decompression when the top of the plume reaches the top of the upper mantle – and continues to take place as the plume head continues to flatten. The final diameter of the plume head is considered to be approximately 2000 km (Figure 4.8c).

Melting, they contend, will start at the hot leading edge of the top of the plume, where the plume can melt to produce high-MgO magmas. As the plume head continues to rise and flatten, the cooler entrained mantle edge of the plume *may* start to melt if it rises to a sufficiently low pressure, i.e. shallow depths. However, it is difficult to quantify the amount of melt that may be generated by their model.

In their model they take the original plume head to have a diameter of 1000 km, so that it will have an initial volume of approximately 5×10^8 km^3 (Figure 4.8a). So it is interesting to compare the assumed initial volume of the plume head with the volume of melt that arrives at the surface in a continental or oceanic flood (or plateau) basalt. Continental flood basalts such as the Deccan Traps, may have a volume as small as 1.0×10^6 km^3, while a major oceanic flood basalt may attain a volume of about 60×10^6 km^3. Thus, the range of melt at the surface for a relatively small CFB may be as little as 0.2 per cent of the original volume of the hypothetical plume head. However, this volume of extruded material reaches 12 per cent for a large oceanic plateau basalt. Such a volume of melt is probably excessive, so that the application of this model to the Ontong–Java and other large oceanic eruptive bodies may require special pleading.

These authors also require their model to give rise to 'runaway extension', a necessary precursor to the splitting of continents which has been noted to coincide with, or closely follow, the emplacement of major continental flood basalts. They note that the uplift caused by the arrival of a plume can only generate small lateral extensile strains, and suggest the plume can enhance ambient tensile stresses in the lithosphere (induced by subduction) to give rise to continental splitting. As we shall see later in this chapter, this is not a viable mechanism.

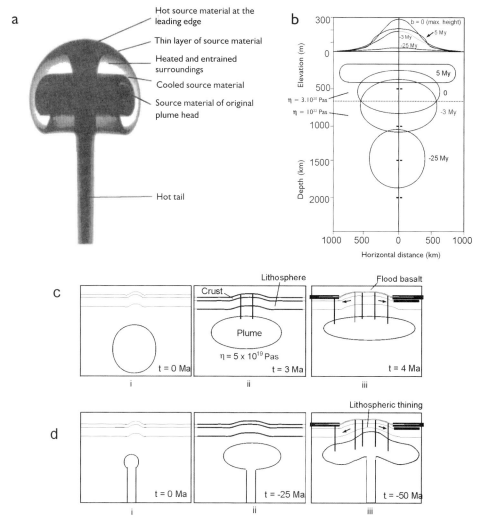

Figure 4.8 (a) Laboratory model of a thermal plume (after Griffiths and Campbell). (b) Diagrammatic summary of the predicted size and change of shape of a plume head, and the relationship of plume depth to surface uplift. (c) i, ii and iii indicating the progressive development of the plume as it rises (after Griffiths and Campbell). (d) i, ii and iii: stages in the White and McKenzie hypothesis for formation of flood basalts which was initiated by a plume head with an original diameter of 200 km.

(b) White–McKenzie plume/extension hypothesis

The starting plume assumed by White and McKenzie is much smaller than that taken by Campbell and Griffith and also it is assumed to have been initiated at a shallower depth (Figure 4.8d). In addition, these authors assume that the plume head has a diameter of 200 km, so that it has a volume of less than 1 per cent that of the Campbell and Griffiths model. Hence, it is inherently less able to generate large volumes of melt. Indeed, we have already noted that these authors are not greatly concerned with the depth at which the plume originated because, they argue, melting only takes place where the plume head reaches and abuts against the base of the lithosphere. To compensate for the small plume head they introduce the influence of lithospheric thinning, which we discuss in a later section.

(c) Super-plumes and wet plumes

An alternative approach is to invoke super-plumes or wet-plumes (Hawkesworth and Gallagher, 1992) which are initiated in a 'wet' transition zone between the upper and lower mantle. Providing the wet mantle material contains about 0.3–0.4 per cent water, the dimensions and rates of movement of mantle material could be increased to a degree that would provide the required volumes of melt at the base of the lithosphere. However, at this time, the assumed water content in the mantle is somewhat conjectural.

It has been suggested that chemical evidence can be adduced to support this hypothesis. For example, the chemistry of the Deccan Traps permits one to infer that melting took place in the mantle region immediately beneath the lithosphere. However the Paraña flood basalts are richer in silica and have a ratio of trace elements and isotopes which are different from those melt rocks which originated deep in the mantle. Indeed, the chemistry of the extrusive rocks of the Paraña area permits one to infer, in this instance, that melting took place in the relatively cold lithosphere, a unit which probably extended downward from the surface for well over 100 km. A significant amount of water in these rocks would certainly enhance the possibility of melting of the lithospheric rocks. However, as we shall see in Chapters 5 and 6, melting in the lithosphere can readily be generated by a reasonably well understood mechanism.

Debate of these different approaches has been vigorous. For example, Saunders et al. (1992) point out that the relatively short period between the initial contact, from below, to the generation of melt is likely to be less than 10 Ma. Despite the heat transfer that may take place between plume and continental lithosphere, they argue that large volumes of melt material are unlikely to be generated, and even that the melt that occurs may freeze in situ as heat is lost to the lithosphere.

Anderson et al. (1992) point out that Griffith and Campbell ignore aspects of plate tectonics such as rifting, the role of melting, the interaction of plumes with mantle and plates, the spacing of plumes, their initiation in a convecting, internally heated mantle and the expected geochemistry of source material in contact with the core.

Some other alternative hypotheses consider magmatism as a *passive* response to lithospheric spreading and rifting, coupled with the serendipitous unroofing of a plume. The reasons for the initial spreading and thinning in these latter hypotheses is usually left unstated, or are attributed to 'extension' in continents.

We have touched only lightly on the problem of the depth of origin of plumes. As can be inferred from the discussion so far, there is no evidence, nor is there likely to be in the near future, to indicate whether plumes are initiated at the core/mantle boundary or in the transition zone of the mantle, or even in the uppermost layers of the asthenosphere. Moreover, White and McKenzie (1989) argue that the depth of origin of the plume is of purely academic interest. These authors make the point that it is the production of melt material which is of paramount importance. As they note, the potential temperature of the plume is only 100–300°C higher than the surrounding area. Only in the LVZ are the pressure and temperature conditions such that the mantle is close to melting. As the increase in temperature caused by the plume is modest, the plume will only give rise to melting in a relatively narrow depth zone immediately beneath the LVZ. Consequently, they conclude that the depth of the stem of the plume is immaterial.

It is now generally accepted, White and McKenzie aver, that the melt material will have only limited lateral migration (< 30 km) in the plume head, so that the whole of the melt, except for a minute quantity retained within the plume, will be available to extrude at the surface.

Let us consider first the type of plume that is required to produce the modest quantity of melt-rock associated with hotspots. We take the area of the plume in Figure 4.4, at a temperature of 300°C above ambient as the area most likely to melt, and the radius of the hottest part of the plume in the stem to be 40 km; then, the area of this hottest part of the plume is about 5000 km^2.

White and McKenzie suggest that only about 30 per cent of the plume is likely to melt. So, if we further assume that the rate of upward migration of this inner core of the plume is 1–3 km Ma^{-1}, then it follows that the amount of melt generated, within the vicinity of the LVZ, would be about 1660–5000 km^3. Hence, such a sustained rate of upward movement of the core of the plume at (1–3 km Ma^{-1}) would be sufficient to supply the melt rock that could give rise to a typical, oceanic, volcanic ridge.

However, some LIPs, with a volume of more than 10^7 km^3 were emplaced in a period of about 1.0 Ma. To sustain such a rate of emission of magma, using the simple model given above, the central core of the type and dimension of the plume modelled in Figure 4.4 would need to have an upward velocity of several thousands of km per Ma. Alternatively, the cross-sectional area of the stem of the plume would need to be much greater than that indicated in Figure 4.4.

It will be clear from this very brief outline of current thinking that there is a general consensus that a plume model is best able to explain the emplacement of a large igneous province. However, it is equally clear that every model proposed has its supporters and its critics. Indeed, for the various parameters used and quantified in these models, it is only certain that plumes are only able to supply the quantity of molten material known to be associated with hotspots, ridges and chains of volcanic islands and seamounts.

White and McKenzie (1989) point out that 'when continents rift to form new ocean basins, the rifting is *sometimes* accompanied by massive igneous activity'. (The italics in this last sentence are ours, for we wish to point out that we are not dealing with a *universal rule*.) As we have noted above, these authors consider that the rising plume is aided by thinning of the lithosphere into and through which the melt from the plume rises to the surface.

In the light of these conclusions, it is interesting to quote White and McKenzie (1989, p. 7688). 'We note in passing that we can say little about how deep in the Earth the (plume) convection penetrates. However, the temperature in the upper mantle where melting occurs is insensitive to the depth at which the mantle instability is initiated. So it does not matter for our model.'

From this statement one can infer that, for these authors, whether or not a plume existed beneath each CFB is not critical. Rather, all that is required is some viable mechanism of lithospheric thinning which can cause extensive melting in the upper mantle.

4.4.4 Lithospheric thinning and the ß factor

McKenzie (1978) presented a model, the main elements of which are illustrated in Figure 4.9. This model is based upon the concept of simple *lithospheric attenuation* as represented in two dimensions, where the lithosphere is taken to include the crust and the upper mantle down to the 1333°C isotherm. (This figure is revised to a somewhat lower value in later publications.) McKenzie postulated a square section of lithosphere of length and depth *a* which is constrained to undergo *instantaneous* extension by a factor ß (Figure 4.9a), so that the length of section becomes aß, and the lithosphere experiences a concomitant thinning of a/ß. As a result of this stretching, the isotherms are packed more closely together and heat flow is increased (Figure 4.9b). Eventually, thermal equilibrium is re-established and the 1333°C isotherm returns to its original level.

It can be inferred (Figure 4.9c) that, because of the assumed crustal thinning, the mean density of the lithosphere and crust has increased and is greater than that in adjacent unstretc.hed areas. This results in subsidence of the attenuated region, which can then become a depressed trough, or zone, into which sediments may be deposited.

This model has been used to explain the development of grabens of the type shown in Figure 4.9d. The infill of the rift by clastic sediments is attributed to the early stretching phase. Once stretching is complete and the maximum development of the graben is attained, the thickness of the lithosphere slowly re-establishes itself as the 1330°C isotherm sinks. During this late stage of development, the on-lap sediments are deposited to provide what is termed a *steer-head*, or *Texas Long-horn* basin. This similarity in geometry of the graben to a steer-head is purely the result of vertical exaggeration. Without such exaggeration the true profile would be more akin to a cross-section of a *sting-ray!*

In studies of the North Sea the values of ß, inferred from refraction and thermal subsidence, are less than 1.5 (Barton and Wood, 1984). In the Great Basin, the value of ß approximates

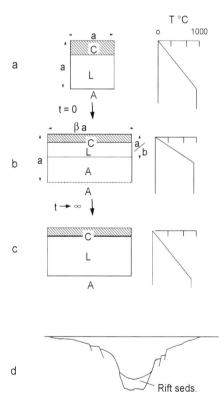

Figure 4.9 The ß strain model and thinning of continental lithosphere and subsequent migration isotherms (after McKenzie, 1978).

to 2.0, with small areas locally exceeding 3.0. However, much larger values of ß have been inferred from stretching of continental margins (White *et al.*, 1987).

This is an ingenious but also simplistic model, because it assumes, for mathematical convenience, that (1) extension is instantaneous and (2) that the lithosphere is homogeneous and isotropic, which (it is tacitly assumed) behaves mechanically as a ductile body. However, the McKenzie ductile model has been used to interpret grabens, which are inherently brittle structures, especially if they are required to develop at very high strain-rates.

4.4.5 Fracture-based evaluation of ß required to form a graben

Normal faults in the upper to middle levels of continental crust develop by brittle, or semi-brittle failure, and result from a lateral reduction in stress. Of course, the lateral extension model proposed by McKenzie will result in a reduction of stress, but, as we shall see, only quite small lateral extension is required to cause the initiation and development of normal faults with linear sections.

Consider the section of Figure 4.9d which is shown, simplified, in Figure 4.10. The reduction in lateral stress required to initiate faulting near the surface will be small, but will increase in magnitude at progressively deeper levels in the crust. The exact value of stress required will be controlled by the physical constants of the rock mass (which include the cohesive strength, the coefficient of friction and the fluid pressure within the rock mass (see Price and Cosgrove, 1990)). Let us assume that the rock mass is dry, for in such conditions the required reduction in magnitude of the horizontal stress is maximised and so will favour the McKenzie model. Let us further make the reasonable assumption that failure at depth takes place when the ratio of the maximum to least principal stresses is 4 : 1, and estimate the stresses required to cause failure at a depth of 20 km. At such a depth in the crust, the total vertical stress will be approximately 5.0 kb. Hence, to give rise to normal faulting (at a ratio of maximum to minimum principal stress of 4 : 1), the least horizontal principal stress must be 1.25 kb. As we have seen in Chapter 2, in conditions when the horizontal plane is unstrained, the initial ambient horizontal stress will be about 1/3 of the vertical stress (i.e. 1.7 kb). Hence, the lateral extension required is that which will reduce the lateral stress by approximately 0.45 kb. Let us now estimate the strain required to cause such a reduction in horizontal stress. If we take the average Young's modulus of strong rocks in the crust to be 10^5 bar, then the extensional strain (e) required to cause a lateral stress reduction of 450 bar is given by 450/100000 = 0.0045. That is, the line AA' (length 100 km) in Figure 4.10 need only be extended by 0.45 km to incur fault initiation (i.e. ß = 1.0045).

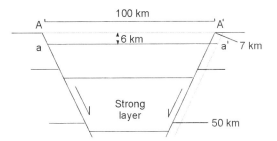

Figure 4.10 Section through an idealised, hypothetical graben, indicating the extension required to generate downthrow of 6 km in the graben.

We now further assume that two normal faults form facing each other (Figure 4.10), and at the surface the faults are 100 km apart and both dip inward at 65°. Hence, at a depth of 20 km the fault planes will be 77 km apart. To bring about a 6.0 km vertical movement on the normal fault, the lateral extension needs to be 7 km (ß = 1.07). Boundary faults to the graben, with this amount of downthrow, would cut and displace the strong layer of the continent, thereby giving rise to rifting which may potentially split a continent. Even if we have overestimated the value of Young's modulus by a factor of 2.0, this would only lead to ß = 1.14. Real grabens develop over long periods, so may incur a significant proportion of ductile strain. The simple ß model which has been applied to such 'natural' graben develop-ment, significantly overestimates the amount of stretching required to produce the structural effects caused by instantaneous brittle failure. To apply this concept to the instantaneous thinning of lithosphere by lateral stretching for ß= 20 to 50 is simply not valid.

4.4.6 Quantification of lithospheric thinning and enhanced melt production

The problem of lithospheric thinning and the quantity of melt that can be produced as a result of this thinning was addressed by McKenzie and Bickle (1988). Their argument, to which the reader is directed, is presented in a long and thorough paper which cannot be adequately summarised here. However, briefly, the authors point out that calculations of the volume and composition of melt induced by thinning of the lithosphere requires knowledge of the variations of melt fractions (X) with pressure P and temperature T. Such a study requires a detailed analysis of experimental data. (They note in passing that the simple model presented by McKenzie, 1978, which has been outlined in earlier paragraphs, omitted such analysis.) The approach used by these authors was empirical and quantitative; and consisted of three steps.

(1) The authors obtained analytical expressions for variations of the solidus T_s and the liquidus T_l temperatures. These expressions were based on experimental data derived from various sources.
(2) They then derived the melt fraction (X) as a function of P and T and
(3) established the melt composition as a function of X and P.

It was further assumed that the thickness of melt was generated by instantaneous, adiabatic decompression of asthenospheric mantle.

McKenzie and Bickle then expressed the relationship between melt thickness and temperature for different values of ß over an extended range of 2 to 50, as shown in Figure 4.11a. White (1992) later modified these findings by using what he considered to be the more correct value of 400 J (kg K^{-1}) rather than 250 J (kg K^{-1}) for the entropy change on melting. Accordingly, he considers that the relationship between melt thickness, temperature and ß is better represented by Figure 4.11b.

We note that two physical parameters enter into the calculations conducted by White and by McKenzie and Bickle; they are the (1) the thickness of the oceanic crust and (2) the entropy change on melting. The range and mean thickness of oceanic crust are well established and are cited in both papers by these authors. The difference between entropy values of 400 and 250 J (kg K^{-1}) is not trivial. It will be inferred from Figure 4.11a and Figure 4.11b that the White version requires significantly smaller values of ß = 20, rather than that of ß = 50, proposed by McKenzie and Bickle to generate oceanic crust. Thus, even if one accepts that the instantaneous stretching/thinning model gives rise to melting, the precise thickness of melt that will be produced at any given temperature for some specified value of ß is somewhat contentious. However, what is even more important is the fact that plate tectonic events occur slowly.

In the Great Basin, where ß is about 2, the stretching strain-rates are extremely small, so that even the development of ß = 2 has taken tens of millions of years. To attain ß = 50, the stretching would need to progress for 25 times longer. Clearly, plate tectonic controlled natural lithospheric spreading or thinning is very far from instantaneous. Hence, the thicknesses of melting for values of ß shown in Figure 4.11a and Figure 4.11b, would not be realised as the result of slow stretching and thinning.

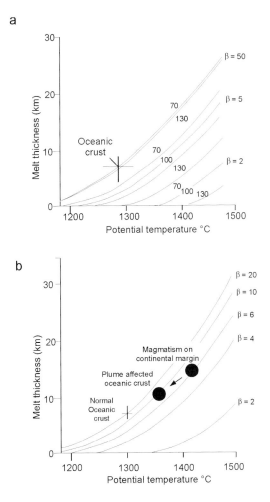

Figure 4.11 Indicating the degree of pressure-release melt, by instantaneous adiabatic decompression of asthenospheric mantle, for different thicknesses of lithospheric thinning, where the thinning is represented by ß. (a) is after McKenzie and Bickle (1988) and (b) is after White (1992). The difference in these diagrams derives from using the entropy change at melting of 250 J (kg K^{-1}) in (a) and 400 J (kg K^{-1}) in (b). Note the different values of ß required to 'produce' oceanic crust.

McKenzie and Bickle have stated that the ß values 'with a factor of 2' are typical of those found in intracontinental sedimentary basins which have not subsequently developed into ocean basins. A factor of ß = 5 or 6, they suggest, is roughly the point at which stretched continental crust breaks to fully igneous oceanic crust: and the factor of ß = 50 (or, possibly ß = 20, following White) is representative of the conditions that give rise to total continental rifting. If one translates a ß = 50, this means that it is necessary to envisage an initial 100 km square section of continental lithosphere to be instantaneously converted into a section 5000 km wide by 2 km deep. Such an instantaneous change in geometry of lithosphere by lateral stretching is physically impossible.

From our simple fracture-based model of graben development, we suggest that the value of ß is more likely to be 1.14 rather than a value of 2 cited by these authors. If we assume a square section of lithosphere of initial length 100 km, this means we would expect the development of a major graben with a depth of 6 km to develop if the lateral spreading is 7 km, or at most 14 km. If the value of ß were 2.0, this would mean that lateral extension has been 100 km. Thus, the lateral extension, which our brittle model requires, is about an order of magnitude smaller than that required for the McKenzie and Bickle or White model.

If, as we contend, the McKenzie ß model leads to about one order of magnitude exaggeration as regards the development of major graben basins in continental lithosphere, it is likely that a value of ß = 5 (i.e. 500 km of extension), which these authors required to cause the opening of oceans, is also heir to a comparable degree of exaggeration.

These large values of ß are important to the cited authors, for only with such magnitudes of (instantaneous) lateral stretching is it possible, by their mechanism, to generate the volume of magma necessary to give rise to LIPs with a volume of basalt in the range of 10^{6-7} km^3.

It is emphasised that the concepts proposed by these authors, namely that of instantaneous thinning of continental lithosphere cannot be explained by any conventional plate tectonics model. Consequently, they cannot justify the cited values of ß. These values have been specified because they are what are required to generate the observed, or estimated, volume of melt.It is apposite, therefore, to consider how continental extension and lithospheric thinning may develop.

4.4.7 Development of areas of extension in continental lithosphere

We have argued in Chapter 3 that any continental lithosphere which is driven by the gravity-glide mechanism of the oceanic lithosphere will be in compression in the direction of plate motion. The degree of compression in the continent will be greatest at the leading edge of the oceanic lithosphere, but will diminish in magnitude as the leading edge of the continent is approached (Figure 4.12a). If the continental margin is *passive*, the compressive stress is only that needed to balance the traction of the opposing down-going oceanic slab, which is probably less than 0.5 kb. However, if an Andean-type mountain chain has developed, as indicated in Figure 4.12b, then the compressive stresses are likely to attain values of 4–6 kb (4–6×10^8 Pa).

It is possible to draw such sections, which are parallel to the direction of absolute plate motion (and are normal, or at high angle to a spreading-ridge), through an adjacent continental mass in many, if not all, continental areas of the world. Hence, as we have argued earlier, in Chapter 3, since the maximum horizontal stress (which is usually aligned with the direction of plate motion) is almost always compressive, how may horizontal extensions occur in continental masses?

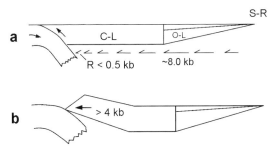

Figure 4.12 Compressive stress variations in oceanic and continental lithospheres in the direction of plate movement (a) for passive margins and (b) for continental lithosphere supporting an Andean-type chain.

One model that will give rise to such a biaxial horizontal stress field is indicated in Figure 4.13a, which indicates a re-entrant coastline of a mega-continent, into which one end of a spreading-ridge has migrated. As will be inferred from this diagram, adjacent areas in the mega-continent are acted upon by forces which result in a diminution of the horizontal stresses, where the radius of curvature of the re-entrant coastline is greatest. A real-life situation of such a scenario is shown in Figure 4.13b, where the S Indian Ridge abuts against the Austral-Antarctic landmasses prior to the break-up of Australia and Antarctica.

Such interaction between a spreading-ridge and a continent could give rise to the generation of new strike-slip faults, or grabens; or the rejuvenation of older, pre-existing fault systems. Let us first consider the development of strike-slip faults and regions of extension that can be associated with their development.

(a) Generation or rejuvenation of strike-slip faults

If the degree of extension is sufficient to make the horizontal stress in the direction of spreading smaller than the vertical stress, which results from gravitational loading, while the maximum horizontal stress, in the direction of absolute plate motion, is greater than the vertical stress, then the triaxial stress system that results is as indicated in Figure 4.14a. If the differential stress $(S_1 - S_3)$ is sufficiently large, then strike-slip failure can take place in the strong layer of the continental lithosphere (Figure 4.14b), which will usually give rise to the development of one dominant strike-slip fault (Figure 4.14c). These conditions of high differential stress in the strong layer, with the maximum horizontal stress greater than the vertical stress, are most likely to develop in the continental lithosphere nearest the interface between the continent and the 'pushing' oceanic lithosphere.

This latter figure indicates the orientation of the conjugate shears that could develop. Usually, only one fracture or set of parallel fractures develops. If the strong layer is relatively homogeneous and isotropic, the fracture or fracture-set will tend to make an angle of about 30° with the axis of maximum principal stress. However, continental lithospheres have a long history of deformation, so the strong layer may already contain vertical fractures. If the trend of such fractures falls in the range of 5°–55°, relative to the S_1-axis, the pre-existing fault or fractures may be rejuvenated in preference to the development of a new shear fracture.

Strike-slip faults in the strong layer may make themselves manifest at the surface in a number of ways. For example, a plate boundary fault, on the large scale, may be seen on the map as an approximately linear feature over a length of hundreds of kilometres. Strike-slip

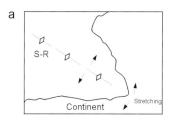

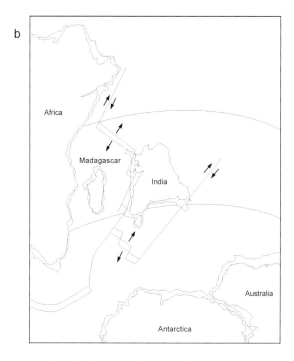

Figure 4.13 (a) Hypothetical model of a spreading-ridge that penetrates a re-entrant embayment in a continent. (b) Situation 75 Ma ago when Australia and Antarctica were still connected (after McKenzie and Sclater, 1971).

faults in four localities around the Pacific are shown in Figure 4.15a. In detail, however, the structure is seen to be more complex, with associated and adjacent anticlines and basins as well as secondary faults (Figure 4.15b and Figure 4.15c).

However, here we are primarily interested in the development of intra-continental structures, where strike-slip faults in the strong layer may give rise to *pull-apart basins*. If the strain-rates induced in the continental rocks are *extremely slow*, even the strong layer may begin to yield in a ductile manner, so that even the strong layer of the continental lithosphere gradually thins in a ductile manner, while the crust thins by the development of listric normal faults. Such a mechanism may well have played a role in the development of the Great Basin of the western USA.

Smaller basins may form in one of two other related ways.

The first of these is indicated in plan in Figure 4.16a, if two parallel, but off-set, strike-slip faults in the strong layer have the same movement sense. Should these faults also

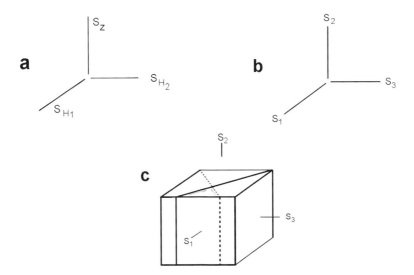

Figure 4.14 (a) Vertical and horizontal axes of principal stress. (b) Relative magnitudes of principal stress required to give rise to strike-slip faults. (c) Typical orientation of such faults which may develop in the stress field indicated in (b).

propagate upward until they reach, or are close to, the surface, a rectangular potential void space will tend to be generated that reaches down to the base of the strong layer. Such a 'deep hole' cannot, of course, be sustained. Pressure-release melting will tend to cause mantle material to flow upward, while, near the surface, vertical walls of the fault will fail, so that infilling of the void will occur from the upper crustal rocks. Nevertheless, a deep basin is likely to form. It is thought that the Dead Sea, in the Jordan Valley, and Lake Baykal, Central Siberia, may have been formed in this manner.

The second mode of basin development occurs when the cover rock above a single major strike-slip fault acts in a ductile manner. In relatively weak cover rock, normal faults can cause the development of elliptical *en echelon* basins (Figure 4.16b).

A major strike-slip fault may eventually cut through and split a continent. Indeed, the Red Sea, the Gulf of Akaba and the Jordon Valley probably show the transition from a relatively simple strike-slip fault to the early stages of development of a spreading-ridge.

(b) Generation of major grabens

It is a misconception that the existence of a major graben is evidence that a continental lithosphere is, or has been, subject to a regional 'tension'. Certainly, lateral *extension* can generate commensurate reduction in the least horizontal stress. However, it should be noted that such extension does not give rise to a general tension in the crust.

When the least horizontal stress is sufficiently reduced by extension, the maximum differential stress $(S_1 - S_3)$ will give rise to normal faults, which if they are 'opposed' can give rise to a *graben*. We have seen (Figure 3.21) that when a rock mass fails in shear it results in a decrease in the magnitude of the differential stress. Here we are concerned with normal faults where, at any depth, the magnitude of the maximum principal stress is determined by gravitational loading, so, at a specific depth, the vertical stress (S_1) will remain reasonably

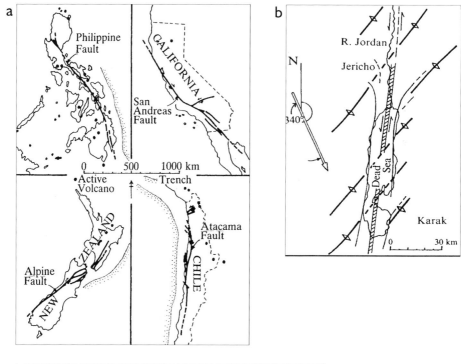

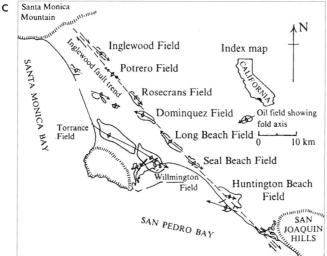

Figure 4.15 (a) Strike-slip faults from four locations around the Pacific rim (after Allen, 1969. © American Geophysical Union). (b) Details of types of superficial structures associated with deep strike-slip faults in the Jordan Valley (after Freund, 1965). (c) The San Andreas Fault (after Moody and Hill, 1956).

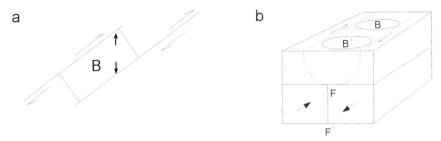

Figure 4.16 Plan-view, showing how deep basins may develop (a) between major strike-slip faults with the same trend and movement sense and (b) in relatively weak sediments above a major strike-slip fault in the basement.

constant for long periods. Once a normal fault, or a graben, has developed, the least principal horizontal stress is controlled by the criteria determining slip. Hence, when the normal fault or graben is established, the least horizontal stress near the fault plane is constrained and will remain as a compressive stress.

Away from the normal fault, or graben, extension will reduce the magnitude of the least horizontal stress. However, as soon as failure conditions are met, further normal faults will develop, so that over large areas continued extension will give rise to the development of more and more normal faults or grabens. Within a continent, very special conditions have to be met to generate internal extension. Even when these conditions are met, the least horizontal stress, at reasonable depths of a few kilometres, or more, will remain in compression.

A lateral reduction in horizontal stress can be induced, for example, where the divergent axes of greatest principal stress, generated by an arcuate spreading-ridge, will cause orthogonal extension and the possible development of a graben (Figure 4.17a).

Whether or not the extension occurs will depend upon the movements of adjacent plates. However, if this extension does give rise to failure of the continental lithosphere to form grabens, these structures will tend to die out inland from the coast. The reason for this can be inferred from Figure 4.17a, which shows that, 'inland', the continental stresses will decrease in magnitude. Near the coast, the compression will tend to displace a marker line A to A', as the result of elastic strain (the displacement is grossly exaggerated). The extension of arc A', relative to arc A, is given by *dl* of the hachured triangle. A similar, smaller triangle *dl* of arcs BB¢ shows that the extension is much smaller than for arc A. Also, as arc B is greater than arc A, the elastic strains in arc B will be much smaller than in arc A. Consequently, the lateral reduction of the horizontal stress normal to the S_1 direction will be commensurately smaller. Hence, even if a graben may form near the coast (Figure 4.17b), it is almost certain to die out inland. This mechanism could possibly explain the development of the graben system, shown on the Tectonic Map of Australia, that cuts the recent sediments in the Lake Ayer Basin.

In any event, this mechanism cannot be applied to explain the development of such features as the North Sea/Rhine/Bresse Graben in Europe, and the even more spectacular African Rift Valley. These major features, bounded as they are by inward dipping normal faults, present evidence that seems to demand that they be interpreted in terms of extension normal to the trend of the rift. But how can such, usually local, extension take place?

Normal faults are not the only features associated with major grabens. Such structures have outward sloping *shoulders* and often exhibit intrusive and eruptive activity (Figure

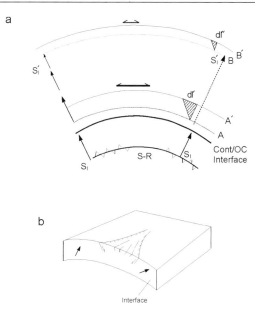

Figure 4.17 (a) Development of lateral extension in a continent driven by a curved spreading-ridge, which has also defined, in plan, the curvature of the continental lithosphere. (See text for details.) (b) Type of graben that can develop in such a continent as the result of such lateral extension.

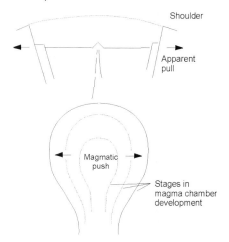

Figure 4.18 Diagrammatic representation of rifting at the surface as a consequence of emplacement of a magma chamber below.

4.18) (see Illies, 1977, 1981; Price and Cosgrove, 1990). From these features, one may infer that during its development the graben was underlain, in places, by magma chambers, which support the rock above and also provide the upward pressure necessary to generate the outward tilting shoulders of the graben. The magmatic pressure is equal to that generated by the gravitational loading of the superincumbent rocks in the graben. Magma is a liquid of relatively low viscosity, so that its pressure is hydrostatic and pushes normal to the graben at a pressure equal to the gravitational pressure induced by the rock cover beneath the graben and its shoulders.

The horizontal magmatic stress will be equal to that of the gravitational pressure, so will be equal to S_1 (i.e. S_2). Therefore, the magmatic pressure significantly exceeds the ambient horizontal stress S_3, and forces the wall-rock of the magma chamber to expand in the direction normal to the trend of the graben, or proto-graben, by pushing back the wall-rock of the magma chamber. It is this shouldering aside of the country rock at some moderate depth from the surface that causes the rocks above to stretch and develop into a graben. Therefore, to infer from this stretching in the superficial layers of the continental crust that the whole of the continental lithosphere has been subjected to stretching, may or may not, be correct. Indeed, where evidence of shoulders and igneous activity exists, such a conclusion is likely to be incorrect.

Rather, we contend that a major graben is more likely to be generated by one or more elongated magma chambers that cause superficial extension in the upper crust of the continent. How may linear or curved major magma chambers come into being, sometimes with a total extent of thousands of kilometres? This is a question that cannot readily be answered by currently held conventional concepts relating to plate tectonics. However, major grabens, especially those with shoulders at the rim of the rift valley, coupled with eruptive activity within the rift, should not be interpreted as evidence of continental tension. (The grabens that may form a 'triple point' at or near the crest of an upwelling hotspot dome do, of course, relate to tensional stresses, but these are not major features when compared with those like the African Rift system.)

4.5 Plumes and the splitting of continents

It has been noted for many years that the emplacement of continental flood basalts is closely associated in time with continental break-up. Consequently, there is general agreement that the two events are linked, but there is a lack of consensus as to which event came first in space and time. Was it the rifting of a super-continent? Or did the emplacement of the flood basalt give rise to the break-up of a super-continent?

It will have been seen that there are three main types of model. These are:

(1) A plume arrives at the asthenosphere/lithosphere boundary beneath an already existing zone of lithospheric thinning, which, until the arrival of the plume, had not reached the critical degree of extension necessary to cause continental rifting.
(2) A plume-head which is sufficiently energetic to give rise to significant thinning of the lithosphere arrives at the base of continental plate, prior to the eruption of the magma produced by the plume.
(3) The emplacement of the LIP is sufficient to induce continental break-up with the subsequent development of an ocean.

It was first suggested that a major plume alone could give rise to lithospheric rifting, leading to a passive margin and the development of an ocean. However, this concept did not gain general support because of the theoretical difficulty in providing the amount of melt necessary to form a CFB. To overcome this shortcoming, it was argued that the required amount of melt could be produced if the plume entered lithosphere which was already experiencing extension. The proposed erosion mechanism is merely a way of producing lithospheric thinning without relying on previous lithospheric extension. Let us consider the relative merits of these models.

4.5.1 Emplacement of a LIP at a pre-existing zone of stretching

(a) Thinning by chance

We are unaware of any *conventional mechanism* which can be put forward to explain how rifting can give rise to plume development beneath a rift area. One is, therefore, left with the factor of *chance*. This assumed relationship, that a major plume reaches the asthenosphere/lithosphere boundary beneath an existing zone of stretching (and therefore thinning of the lithosphere) leads to an obvious question: what is the probability of the stem of a plume rising beneath a rifting section of the lithosphere?

Let us assume that there exists a rift-graben without shoulders and devoid of eruptive activity in the rift zone (which may eventually develop into an ocean) with dimensions comparable with the East African Rift. We take this hypothetical feature to have a length of 5000 km and an average width of 100 km, so that the area of the feature is 500,000 km^2. The plume, we assume, can rise anywhere on the surface of the Earth, which has an area of approximately 500,000,000 km^2. Hence, the chances of the centre of the plume arriving beneath the rift area is 1 : 1000.

One can reasonably argue that more than one such rift of the specified size existed in the continents at any given period of geological time. However, if we assume the co-existence of 20 such major rifts, the chances are still about 1 : 50.

Of the major continental flood basalts that have developed in the last 250 Ma, i.e. (1) the Siberian Province, (2) the Karoo, (3) the Antarctic Province I, (4) the Antarctic Province II, (5) the Paraña-Etendeka Province, (6) the Deccan Trap, (7) the Ethiopian Flood basalts and (8) the Columbia River basalts, only the relatively small Columbia River Province and the massive Siberian Province are not related in space and time to the opening or evolution of an ocean or association with a major graben or former graben. If we add to this group the basalts of (9) the North Atlantic Province, which is held by White and McKenzie (1989) to be responsible for the opening of the N Atlantic, then out of the nine named sites, only two are not related in space and/or time to rifting and possible ocean development. The likelihood of a 1 : 50 chance event occurring seven times out of nine is very small.

Whatever the theoretical desirability of having extension-driven rifting forming before the emplacement of a major plume, common sense demands that, if rifting is indeed related to a plume, the plume arrived first.

(b) Thinning as a consequence of basal erosion of the lithosphere

It has been suggested that thinning can take place by erosion of the lower lithosphere by the arrival of an energetic hot plume. The mechanism is not well specified, or quantified, but erosion is thought to occur as the result of the emplacement of the mushroom-shaped plume head and its excess temperature relative to the lower lithosphere.

This type of behaviour, if true, should apply to all plumes. There are numbers of hotspots (presumably driven by plumes) which occur, with trails, in continental bodies (Figure 4.3). If all plumes can cause lithospheric thinning, why have the hotspots in Africa not given rise to splitting of that continent? It could be argued that this mechanism can only become important *in situations* where the continental lithosphere is very thin. In which case, see the argument presented in the previous section.

(c) Plume-induced rifting

It has been suggested by several authors that the potential of the lithosphere to slide down the dome would engender a significant compressive stress on the undeformed lithosphere surrounding the circular feature, which would tend to enable the graben to develop beyond the limits of the dome. For example, Price *et al.* (1988) and White and McKenzie (1989) argue that rifting will take place above a plume. The Price model (Figure 4.19a) is based upon the 'pie-crust' concept of a strong lithosphere which is capable of sliding on a LVZ. With this model, there will be a tendency for all elements to slide outwards along radii, but they will be restrained by the tensile strength of the lithosphere, the confining effect at the circumference of the uplift caused by the plume and the tensile strength of the lithosphere at the apex of the dome.

If one considers a segment of the pie-crust (Figure 4.19b), it will be clear that the mass of the lithosphere with a tendency to slide will be resisted at the apex of the pie-crust, along a vertical line, by a tensile force P (in combination with other resistive forces already mentioned). The tensile stress that is generated along the line is the magnitude of the tensile force, divided by the area of the vertical line (which of course is zero). Hence, however small the value of P may be, the tensile stress goes to infinity, so that fracturing of the apical region of the pie-crust must fail in extension.

Price *et al.* (1988) suggested that if the dome occurs in a lithosphere where the horizontal stresses at any specific depth were originally equal in all directions, the most likely failure pattern would give rise to a triple-point (Figure 4.20a). At the surface, and to shallow depths of a few kilometres (at most), the lithosphere may fracture along vertical tensile fractures, but below a few kilometres these fractures will give way to normal faults, so that the triple point is likely to develop into a junction of grabens. If this mechanism is significant, all topographical domes thought to be associated with a hypothetical plume would exhibit a triple junction of grabens.

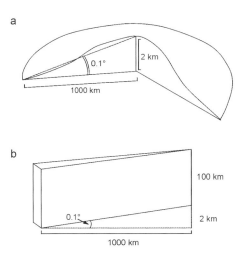

Figure 4.19 (a) Representation of lithosphere uplifted by a plume head which develops below the LVZ. Such a 'pie-crust model' has been used to explain the generation of triple points which can give rise to splitting of a continental body. (b) Shows a very thin sliver of the pie-crust which permits one to estimate the increment of horizontal stress that develops radially from the edge of the pie-crust as the result of gravity-glide on the LVZ.

A map of Big Island (Hawaii) has a roughly triangular plan (Figure 4.20b). This shape is consistent with a double triple-point fracture system centred on Mauna Loa (ML) and Mauna Kea (MK). One may infer from Figure 4.20b, that the length of the three arms of the triple point are about 100 km, or about 10 per cent of the radius of the plume. Hence, one may further infer that a plume is unlikely to generate a tensile fracture that can split the uplifted area into two or more major segments.

Let us quantify this effect. Here we will take a simple *flat-earth* model of the largest type of dome (Figure 4.19a) with a radius of 1000 km and a maximum relief of 2 km. It will be inferred from this diagram that the dip of the lithosphere is always small, so let us take the average dip indicated in Figure 4.19a of 0.1°. (We have already argued, in Chapter 3, that to substitute a straight-line conical section for a very low-angle, curved section will not introduce significant errors.)

Consider a thin, wedge-shaped sliver of lithosphere, completely in isolation, with a greatest width of 2 cm tapering to zero at the vertical line, marking the centre of the dome, and other dimensions as shown in Figure 4.19b. If we assume that the average density of this continental slice of lithosphere is 3.0 g cm^{-3}, it can readily be shown (see Chapter 3) that the maximum radial horizontal stress, which such a sliver will cause to develop in the adjacent horizontal lithosphere will be 300 bars. The wedge-shaped sliver will, of course, be surrounded by similar elements to which it is firmly attached. As well as viscous retardation at the base, tensional elastic strains, which result from the tendency of the lithosphere to slide down-hill and, therefore, to extend circumferentially, will place a further degree of restraint on free-gliding of such a hypothetical sliver. This would further reduce the magnitude of the increment of horizontal stress that may develop in the surrounding horizontal lithosphere. Hence, the increment of extra, average horizontal stress induced by the dome in the surrounding lithosphere is likely to be not more than about 100 bar.

The vertical stress gradient in the continental lithosphere will be complicated by the changes in density as one traverses downward from the surfaces. However, as already stated with an assumed average density of 3.0 g cm^{-3}, the average vertical stress gradient will be 300 bar km^{-1}, so that the vertical stress will increase from atmospheric pressure to 30 kb at

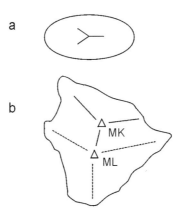

Figure 4.20 (a) Minor triple-point rifting possibly induced by a plume. (b) Map of Hawaii, showing its slightly irregular triangular outline, and the position of Mauna Loa (ML) and Mauna Kea (MK), with lines commensurate with a double triple-point junction, indicating that natural events are likely to be more complicated than our simple model shown in (a).

the base of a 100 km thick lithosphere. In the upper crustal areas, where the stronger rocks will be elastic up to the point of brittle failure, the horizontal stress in the flat-lying rocks around the dome will be about one third of the vertical stress (see Chapter 3).

At depths in the lithosphere where the rocks are ductile, the horizontal stress will be much higher, and even approach the value of the vertical stress. Hence, the average horizontal stress in 100 km thick, undisturbed, continental lithosphere is likely to be in excess of 5 kb, so that a negative increment of stress of -100 bar represents a stress change of less than 2 per cent, and so, in general, is unlikely to cause any significant effect. Only if the lithosphere were on the point of failure, prior to the emplacement of the plume and the resultant dome, would such a small increment of stress have a local influence, but even this would not extend much beyond the limits of the dome. Basal action of outward movement of the material of the rising plume has not been taken into consideration. However, McKenzie and Bickle (1988) considers that this uprising material does not migrate laterally for more than about 30 km, so presumably would have a small effect on a structure with an area of 3×10^6 km^2.

Without doubt, from time to time, vast volumes of molten basalt have reached the surface from depths in the mantle of several hundred kilometres. The mechanism(s) which may possibly drive molten basalt to the surface is, we contend, still speculative. For example, Anderson (1998) analyses the mantle problem and, specifically, the axioms underlying the various aspects of the plume paradigm. This paper is not easily summarised, so the interested reader is urged to study the original publication. He introduces his paper with a quotation. 'It is only after you have come to know the surface of things (i.e. the crust and strong upper lithosphere) that you can venture to seek what is underneath. The surface is already so vast and rich and various that it more than suffices to saturate the mind with information and meaning.' It also obfuscates those features and events which are more deeply buried.

Anderson points out that the core–mantle boundary is one of the most important internal surfaces. The changes in density and viscosity across the core–mantle boundary are larger than at the surface. Molten metal contacts crystalline oxide at this boundary and the dross of mantle and core processes collects there. Temperature variations are damped to near zero. However, because of pressure, buoyancy is hard to come by and heat is readily transmitted by conduction.

The plume hypothesis has influenced most fields of geochemistry, petrology and geo-dynamics. However, in order for narrow thermal plumes to slide almost 3000 km through a convecting internally heated mantle without deflection or cooling, they must be strong. In the laboratory, the plume material must be removed from the system, heated externally and then reinjected. Narrow plumes with low viscosity do not arise spontaneously in a heated fluid.

The plume hypothesis is axiomatic. Axioms, Anderson points out, are 'self-evident truths' so are rarely stated explicitly, however, when they are, they may not be at all self-evident. In Appendix 1 of his paper, Anderson briefly discusses the plume paradigm. 'A paradigm is the infrastructure of a culture of ideas and thought processes that are shared by practitioners in a given discipline. It includes assumptions, techniques, language and defence mechanisms. Every paradigm has paradoxes: as paradoxes multiply a paradigm can become vulnerable. It can only be overthrown from the outside and, ironically; only when the tenets become unfalsifiable. (Just-so stories are not falsifiable.)' Anderson concludes the paper with a series of Appendices, the first of which lists 20 key axioms of the plume paradigm. He states, 'Few readers will believe most of these axioms as stated, but unstated they underly much of the modern Earth science. The widespread belief in deep mantle plumes rests on the validity of these statements. Sometimes beliefs outlive the axiomatic framework they were based on. It is useful, therefore, to isolate the key assumptions and re-evaluate them as data and theory advances.'

There is yet another aspect relating to plate tectonics which is not taken into consideration by some authors dealing with plume theory. This has been succinctly expressed by Lithgow-Bertelloni and Richards (1998) who state: 'Unfortunately, we cannot reproduce the toroidal/ poloidal ratios observed from the Cenozoic, nor do our models explain apparently sudden plate motion changes that define stage boundaries. ... The most conspicuous failure is our inability to reproduce the westward jerk of the Pacific Plate ... implied by the Hawaiian-Emperor seamount chain. Our model permits an interesting test of the hypothesis that the collision of India with Asia may have caused the Hawaiian-Emperor bend. However, we find that this collision has no effect on the motion of the Pacific Plate, implying that important boundary conditions are missing from our models.'

These authors are stating that *conventional* methods of analysing plate motions are seriously flawed, in that there are known abrupt changes in track of islands and continents where the change appears to take place extremely rapidly (probably in a few tens of thousands of years). Moreover many of these events are associated with stage boundaries.

Whatever the pros and cons regarding the generation and emplacement of plumes, we can reasonably discount the possibility of such plumes arriving at or near the surface in such a short time-span required by the track data that can be inferred from programmes that permit the movements of islands and continents to be followed through time. We conclude, therefore, in the light of the evaluation presented by Anderson and the statement made by Lithgow-Bertelloni and Richards, that a mechanism exists (which has been ignored by the 'plumists') that explains the development of abrupt track changes and other events associated with stage boundaries.

4.6 Conclusions

The literature on the topics covered in this chapter is large, so that the treatment of the subject matter in this chapter is of necessity brief and incomplete. However, we suggest that the discussion has been sufficient to demonstrate that there is no general acceptance for any of the various mechanisms cited. There are obvious incorrect conclusions regarding the various models proposed for splitting of continents by domical uplifts or by tensile stresses. Moreover, none of the models set out address the problem presented by Lithgow-Bertelloni and Richards cited above, namely that the models used in analysing plate movements are currently imperfect, for they do not have a model which explains 'sudden plate motion changes that define stage boundaries'. (By 'sudden' we suggest that these authors mean that the period of change in plate motion is accomplished in a few tens of thousands of years.)

Even though the various concepts and studies discussed in this chapter have possibly advanced our understanding of some mantle phenomena, after decades of research no clear-cut and completely irrefutable mechanism has come to light that will satisfy the requirements set out by all authors cited above. Clearly, large volumes of basalt must reach the surface of the Earth. However, the uprising of hypothetical mantle plumes and the melting phase that takes place below the lithosphere requires several millions of years, followed by another million years or so in which basalt erupts on the surface of continents and the floor of oceans. None of the mechanisms cited above can explain the development of 'sudden plate motion changes that define stage boundaries'. Indeed, such sudden changes in direction and/or speed of plate motions simply do not come into the parameters considered by the various plume models.

Indeed, geoscientists have been ignoring a vital parameter. We have argued that the assumption of instantaneous lateral stretching is not valid. Nevertheless, we agree that the very large volumes of melt required to form LIPs can only be generated by dramatic thinning of the lithosphere. We are at odds with the authors cited earlier because they have only considered thinning of the lithosphere to be generated from the mantle upward. This has resulted in them requiring an impossible, instantaneous, stretching mechanism. However, in the following chapters, we shall demonstrate that the thinning of the lithosphere and the production of large volumes of melt can be generated by substantial thinning, or even complete removal of the lithosphere, from the surface of the crust downward, by the impact of large cometary bodies. In Chapter 5 we will introduce factual evidence regarding important experimental work which reproduced 'Moon craters on Earth', as well as presenting the pertinent theoretical concepts regarding 'impact tectonics'.

Chapter 5

Geometries and mechanisms of impact structures

5.1 Introduction

The Earth has a common feature with all the planets and satellites in the Solar System, in that it has endured extensive meteoritic and cometary bombardment. In the first half of the 20th century this 'Earth bombardment' was either given relatively little attention or relegated to the period prior to about 3800 Ma. However, in the second half of the 20th century, there has been a growing recognition of the importance of this process to more recent geological evolution. The circular features of the Moon have interested astronomers and philosophers for centuries. Smaller circular features on the Moon, it was initially thought, could be equated to volcanoes or calderas on Earth. The dark and apparently level areas of the Moon were likened to seas and oceans on Earth, and termed Maria. For a long period, the larger circular features were a complete mystery. However, it was eventually recognised that major circular features on the Moon could be attributed to impacts.

The surfaces of Mercury, Venus, Mars, the Moon and other satellites of the Solar System bear testimony to the importance of impacts in the developing of cratering. The Earth could not have escaped this process.

It will be seen in Figure 5.1 that in the first quarter of the 20th century the number of accredited *certain* impacts could be counted on the fingers of one hand. By 1950, the total number of certain impact features had doubled. However, in the period 1960–70 there was a dramatic increase in the number of acknowledged, certain, impact events. This rate of 'discovery' has been maintained, so that in the late 1990s, the number of certain impacts had increased to about 150. The trend-line in Figure 5.1 has been projected to the year 2000, where, the extrapolation indicates, the number of certain impacts is likely to number at least 180. This chapter is largely concerned with those characteristic features which enable impact structures to be recognised. Accordingly, we shall begin with a section entitled 'The Recognition of Impact Structures', which we start with a brief account of a few of the more recent events: for example, the Tanguska event, which occurred in late June in 1908 and some other recent 'burn-out' events in the lower levels of the atmosphere.

We then note the number of certain impact structures, and some which are at present a little less certain, and indicate their range in size. We refer to analyses which lead to a prognosis of the number of impact events over a range of crater diameters which can be inferred from Moon and Earth impacts. There appears to be a marked disparity between the estimated number of impact craters which are inferred from these analyses, as compared with the relatively small number of certain impacts which have been observed on Earth. To

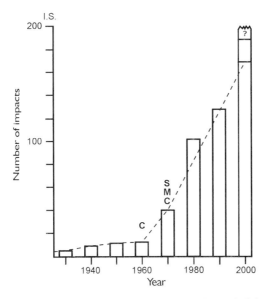

Figure 5.1 Number of *certain* impact features established up to the end of the 20th century.

indicate the probable magnitude of the bombardment on Earth, we shall cite, in the first section of this chapter, the analysis presented by Oberbeck *et al.* (1993), which enabled them to estimate the number of impacts greater than a specified size, for a given period. For example, they indicate that in the last 2000 Ma the Earth has been struck by 110 major impacts which gave rise to craters in excess of 100 km in diameter, while, in this same period, there have been over 1.4×10^6 impacts causing craters with diameters in excess of 1 km. This bombardment is not yet over. Indeed, there is no indication that, within the life expectancy of the Earth, the bombardment will ever come to an end. This disparity between the estimated and observed number of impacts on Earth is discussed briefly, but we shall also return to this topic, from time to time, in this and also later chapters of this book.

The inflection point in 1950–60 (Figure 5.1) marks the onset of the 'Cold War', a period in which explosive experiments, using nuclear devices and high tonnage TNT explosions, produced craters that revealed previously unknown, or unrecognised, features which developed in association with major impacts. These experiments, especially those conducted at Suffield, Alberta, Canada, produced data and geometries of craters that quickly revolutionised scientific appreciation of crater geometry. The multiplicity of structural elements that combined to form these experimental craters, and which presents a much clearer insight into the mechanism of crater development, are dealt with in the section entitled 'Man-made craters'. The geometry and type of structures produced in various explosive experiments, which are described at some length by Jones (1995), are presented in this second section.

The multiplicity of structural elements that combined to form these various types of craters, permitted a much clearer insight into the mechanisms of 'Impact tectonics', as presented by Melosh (1989), O'Keefe and Ahrens (1982) and others. These various aspects are discussed in the section which deals with 'Stresses and reaction of target rock as the result of impacts'.

5.2 The recognition of impact structures

Until recently, the importance of meteoritic impacts on the development of the Earth has received scant attention from most geologists. This is perhaps understandable, as geologists usually spend their professional lives trying to understand and explain geological phenomena wholly in terms of terrestrial processes. One of the reasons for neglecting the importance of meteoritic impacts is an uncritical application of the concept of *uniformitarianism*. We contend that *catastrophes* are an integral part of uniformitarianism.

It is fortunate for mankind that even quite a small impact event, which gave rise to a crater with a diameter of no more than about 1 km, has probably not occurred in continental areas during the period of recorded history. Statistically there should have been as many as four such events in this period. Possibly, of course, these modest-size meteorites fell into the oceans.

However, even this lack of activity may be more apparent than real. For example, Lewis (1996) and Vershuur (1996) cite historic examples of small impacts, while Tollmann and Tollmann (1993, 1994) present arguments for moderately large oceanic impacts that occurred around 7300 BC. Moreover, it is now being suggested that layers of burned vegetation, which had initially been attributed to grass fires, are now considered to be the result of more extensive and distant forest fires, which may have been initiated by impacts. Such events, which probably occurred in the period between 2200 and 1200 BC may have resulted in the destruction of Bronze Age communities in the Middle East, China and possibly the E Mediterranean area.

A potentially catastrophic event occurred in the morning of 30 June 1908 in the Tanguska region of N Siberia. Fortunately, this area was almost completely uninhabited, so there was no recorded loss of human life. However, two brothers, on a hunting expedition, experienced considerable shock when their tent was 'blown away'.

Scientific expeditions were not allowed into the area for 20 years, so the epicentre was not found until 1928. Despite repeated searches no evidence of cratering or of extra-terrestrial material has been found.

The cause of the 'wind' was an *air-blast* that flattened the trees from a central zone, throughout an area of 2000 km^2. However, the 'blow-down' pattern was not circular, but exhibited a 'butterfly shape'. From the bilateral symmetry of this butterfly pattern, Russian scientists set up an experiment using small, cylindrical, explosive charges, set at a small distance above a board which contained a 'forest' of small pegs which were initially set perpendicular to the peg-board. Each peg was attached to the baseboard by a thin ductile wire, which permitted the individual 'trees' to be blown down in a direction wholly determined by the explosive blast. They very quickly established that the line of symmetry of the butterfly pattern gave the track of the bolide and, by varying the angle of inclination of the cylindrical explosive and comparing the pattern of 'blow-down' in the models with the known pattern in the field, they were able to establish that the angle of track the bolide made with the horizontal was 10°. They also found that immediately beneath the charge, the model trees remained upright. This was exactly what had been seen in the field where, immediately beneath the focus of the blast, the trees in the Tanguska forest had been stripped of their branches but the trunks remained standing like 'telegraph poles'.

With these experimental results available, other scientists showed that the explosion was almost certainly caused by the vapourisation of a fragment of an icy comet which occurred at a height of 6–8 km above ground. The energy of the explosion has been estimated to be

1 megaton (1 Mt) of TNT (Toon *et al.*, 1997), over 60 times larger than the bomb that destroyed much of Hiroshima. The most probable cause of this event can be attributed to the impacting in the atmosphere of one unit of a comet stream, made up of fragments of a larger cometary body which orbit around the Sun. Between 40 and 50 such comet streams are known. However, the most probable 'culprit' is the ß-Taurus stream which makes an angle of 5° with the plane of Earth's orbit. The two orbits intersect in November and June, with the maximum likelihood of collision occurring on 30 June. Such collisions are thought to be probable every 100–300 years.

Had this event occurred above London, the devastated area would have covered almost the whole of Greater London within the M25 ring road. Indeed, if this air-burst had occurred above any of the cities on Earth, the devastation and loss of life would have focused the attention of mankind in general, and the scientific community in particular, on the study of what is arguably the most important of all the catastrophic dangers that face mankind.

Subconsciously, or even consciously, such relatively recent events are categorised as rare or remote occurrences and tend to be dismissed from our thoughts. This luxury of self-delusion, has now been swept away by *Earthwatch*. Toon *et al.* (1997) note that several objects, with energies comparable with the device which devastated Hiroshima, enter and are destroyed in the Earth's atmosphere every year. They further comment that 'the populace is not generally aware of these events'. However, they are routinely recorded by surveillance satellites operated by the U.S Department of Defense (Tagliaferri *et al.*, 1994). An incident occurred on 1 February 1994, when a bolide exploded over Micronesia, with an energy estimated as 20–100 kT, after penetrating the atmosphere to an altitude of 21 km (McCord *et al.*, 1995). As we shall see, impacts which produce a crater 1.0 km or greater in diameter, are only apparently rare events when measured in terms of the recorded history of mankind. They are far from rare events when measured by more usual geological time-scales.

The two most important, practical reasons for the lack of acceptance of the importance of major impact structures were, firstly, the lack of a sufficiently large 'world view', so that extensive circular features were not readily recognised. This difficulty has now, of course, disappeared with the arrival of satellite imagery. Secondly, relatively small features, which have now been shown to be of impact origin, were, in the main, for many years attributed to volcanism, or *crypto-volcanism*, where 'crypto' means secret, hidden or concealed (so that 'crypto-volcanism' really translates into 'unknown-mechanism').

Lists of impact events on Earth have been compiled by, among others, Grieve and Dence (1979) and Hodge, (1994). Hodge lists over 150 craters in 139 different sites, which he regarded as *certain*, up to the year 1992. He closely follows the compilation of Grieves and Dence, but adds eight new sites. However, he notes that there are many other structures around the world which have been proposed by various scientists to be probable impact structures; and some of these, with further study, are likely to be upgraded to *certain*. In order that such an upgrade from probable to certain can be achieved, evidence must be adduced to show that the impacted rocks of the structures have experienced shock-metamorphism, with the generation of exotic minerals and structures.

It is interesting to note that, currently, a crater may be classified as either *certain* or *dubious*. We suggest that this simplistic classification could now be expanded to (1) *certain*, (2) *probable*, (3) *possible* and (4) *unlikely or dubious*. There are certain features, some of which are illustrated in this chapter, which exhibit a variety of characteristics that render them as *almost certain* (i.e. *probable*). However, because such events have been only recently observed by satellite imagery in a remote part of the globe, or they are buried features which have been inferred from remote sensing and have not been surveyed on the ground or drilled to establish whether or not they contain diagnostic evidence of impact (which we discuss below), they would be termed *dubious*, a rating which is far too low.

The thoughts expressed in the previous paragraph are, of course, contentious. However, from what has been reported earlier, it is clearly not in question that the Earth has experienced bombardment by comets or meteorites. Nevertheless, when one compares the number and size of known impact craters on Earth with those of our satellite or of Venus it must be admitted that the Earth structures appear trivial, both as to number and size of crater. The reason for this, however, is obvious. The Earth is tectonically active with a surface that is rapidly eroded, or replaced. Other bodies in the Solar System record a longer history of bombardment.

5.2.1 Number of impacts on Earth and Moon

There is an increasing awareness among Earth scientists regarding impact events on Earth, as the result of the high-profile debate over the influence of one or more impacts which are thought to have given rise to 'extinctions' at or near the K/T boundary, 65 Ma ago. Moreover, as pointed out by Grieve and Dence (1979), there is an 'increasing maturity of our knowledge of the geological and geophysical effects of impacts'. Today, the Earth-science community more readily recognises the importance and significance of 'zone-fossils' such as shatter-cones, melt-rock and high-pressure shock-metamorphism.

Small circular, terrestrial impact features which are not buried, but have been eroded, have been correctly identified as impact structures. For example, Martin (1969) reported the circular *BP Structure* in a remote part of the eastern Libyan desert, and identified it as an impact structure, which is listed in Hodge (1994) as a certain impact, despite the low relief of the rim of the feature. At this time, i.e. three decades ago, most geologists would have referred to it as a crypto-volcanic feature. However, Martin (pers. com.) had the advantage of having previously visited Gosses Bluff, an impact feature in central Australia.

5.2.2 Certain craters

Hodge (1994) gives the location of specific 'certain' impact craters, their diameter (or other dimensions), together with maps, photographs and references, to which the interested reader is directed. He cites a total of 57 impacts with crater diameters of 10 or more kilometres. The number of events with a crater diameter of more than 10 km for a specific age-range are as given in Table 5.1.

Table 5.1

Age (Ma)	Number of events
0–100	22
100–200	10
200–300	8
300–400	6
400–500	2
500–1000	6
> 1000	3

These figures do not, of course, indicate that the rate of bombardment is necessarily increasing, but that sedimentation or erosion plays an important role in hiding or obliterating evidence of the older structures.

Eight of these features have diameters of 50 km or more. Three of these eight events, which occurred in the Phanerozoic, exceed 100 km in diameter, and one is 180–200 km in diameter. The structures listed by Hodge include Chicxulub, Yucatan, Mexico, with an estimated diameter of 180–200 km; the event to which some attribute the demise of most of the dinosaurs. Hence, one may infer that an impact that causes a crater with a diameter of over 100 km is not a trivial event.

The continental areas of the Earth, currently and for the last few hundred million years, at least, occupy a little less than 30 per cent of the Earth's total surface area. Statistically, if impacts were evenly distributed throughout its surface area, about 70 per cent of impact events would have occurred in the oceanic areas of the planet.

Even when an impact occurs in seas covering continental areas or shelves, considerable difficulties are encountered in discerning impact features, when the crater is obscured by later sediments. An example of such a buried impact structure (Figure 5.2a and b) in the Jurassic rocks of central Montana, USA has been described by Plawman and Hager, 1983). The sections in this figure respectively show the seismic reflectors and the interpretation of the structure.

It is known that, because of active plate motions, ocean floors rarely date back beyond 200 Ma. Consequently, all impact events that occurred in the oceanic areas prior to this time will be obliterated by subduction. Moreover, detection of submarine impact structures is even more difficult than finding corresponding continental features. As we shall see, such moderately large craters as may develop on the ocean floor are likely to be largely or wholly

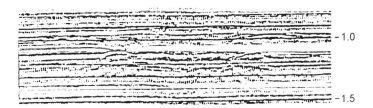

Figure 5.2a Seismic section through inferred buried impact structure.

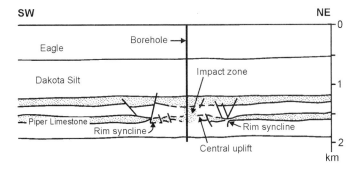

Figure 5.2b Interpretation of seismic section (after Plawman and Hager, 1983).

obliterated. Indeed, of the examples cited by Hodge, only one, the 45 km diameter Montagnais structure off the coast of Nova Scotia, is entirely submarine.

There are, however, other known features buried beneath the sea-floor of continental shelves. For example, Mjolnir, a 40 km diameter circular structure in the Cretaceous/Jurassic sediments beneath the Barents Sea, off the north shore of Norway, has been described by Gudlaugsson (1993). The structure is defined by a series of geophysical anomalies. Seismic reflection data permit one to infer the form of a complex impact structure, which exhibits gravity and magnetic signatures that are consistent with those associated with other, known, large impact structures.

Gudlaugsson considered possible alternative explanations for the structure, which included salt and shale diapirism or even an igneous eruption. However, he rules out these explanations on geological and geophysical grounds and infers that the structure, which developed about 140 Ma ago, was likely to have resulted from a meteoritic impacting body with a diameter of about 2 km. However, as pointed out by Grieve (1993), until the Mjolnir structure is drilled and it has been established that shock-metamorphic effects exist in the target rock, the Mjolnir will, strictly speaking, be classified at best as only a *probable* impact.

Two smaller hydrocarbon-bearing structures were reported in the previous year, namely the 16 km diameter, Ames structure in Kansas (Carpenter and Carlson, 1992) and the 12 km diameter Avak structure in Alaska (Kirschner *et al.*, 1992).

Off-shore submarine circular structures in the continental shelf are likely to continue to be discovered in the constant search for hydrocarbons. However, identification of large, deep ocean, circular structures presents a much more difficult problem.

Impact structures are only accepted as certain impact features when there is evidence of melt-rock in the form of tektites or small nodules, shatter-cones, deformation lamellae in minerals such as quartz and the existence of concentration of elements which are rare on Earth but more common in some meteorites. This demand for proof is, of course, understandable and a necessary guard against the zeal, and possible uncontrolled imagination, of some geologists. (However, to put this in context, one suspects that thirty years ago, pioneers such as Dietz, Shoemaker, Dence and Grieve would certainly be regarded by many of their 'peers' as being 'scientifically unreliable'.)

5.2.3 Quantification of large impacts on Earth

Even from the limited evidence presented above, one can conclude that the Earth has been, and continues to be, bombarded by meteorites and comets. However, the record of impacts on Earth, both as regards their magnitude and number (as represented by the list of impact events categorised above as being *certain*) appears trivial when compared with the evidence of cratering on the Moon. This has caused some Earth scientists to infer that the predominant bombardment of both Moon and Earth occurred early in their history (i.e. > 3800 Ma ago) and so they suggest that the bombardment is now over, or is of little geological importance. However, as may be inferred from Figure 5.3, which shows a non-dimensional curve for the total accumulation of impact craters on the Moon, such a conclusion is completely incorrect.

Clearly, from Figure 5.3, the greatest impact activity took place in the early Precambrian. But equally clearly, it can be seen that the rate of accumulation of craters over the past 3000 Ma approximates closely to a straight line, and that the line reaches zero at the present day. However, one must not unthinkingly conclude that this curve shows that current, impact activity is zero. The number of craters accumulated at any time (A_t) is the rate of impact (R)

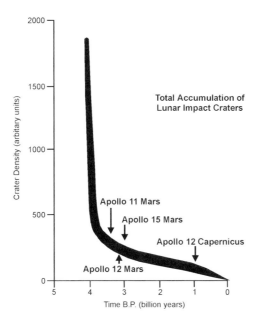

Figure 5.3 Envelope of (dimensionless) accumulation of impact craters on the lunar surface as a function of age. The approximate linear increase from the present back to 3.5 billion years ago implies a reasonably constant flux for that period.

multiplied by the duration of the chosen time limits (t). The present day would equate to $t = 0$. Hence, as is apparent from Figure 5.3, this equates to a zero rate of accumulation for the present day. However, because this figure indicates that the accumulation curve over the past 3000 Ma is approximately linear, it can be inferred that the average flux of impact over that period is approximately constant.

When data are scarce, most scientists turn to statistical methods of interpretation. Accordingly, let us turn to the question regarding the size and frequency of impacts in the past 2000 Ma. Also, let us now limit ourselves to 'certain' craters with a diameter in excess of 20 km, which were recognised and accepted as impact structures that formed in N American and E European cratons, as tabulated by Grieve and Dence (1979). From these data, they estimated the average rate of formation of terrestrial craters in the Phanerozoic; a period for which the dating was reasonably accurate.

Based on the work by these authors, Oberbeck *et al.* (1993) averaged their crater production rate for craters greater than 20 km and obtained a lower bound cratering rate of 2×10^{-15} km^2 yr^{-1}, a rate which, they asserted, is appropriate for at least the last 2000 Ma. These authors quote the conclusions of the Basaltic Volcanism Study Project (1981), namely that the number of craters produced with a diameter larger than d in kilometres was proportional to the -1.8 power of d. Moreover, they consider it reasonable to assume that this cratering flux has been approximately constant for the last 3000 Ma.

From the conservative estimates of those produced by Grieve and Dence and the functional relationship obtained by the BVS Project, Oberbeck *et al.* derived the general expression:

$$N(d) \ \text{km}^2.\text{a}^{-1} = kd \ \text{km}^{-1.8} \tag{5.1}$$

for production of the number (N) of craters per km² per year, larger than any observed diameter (d) from 2000 Ma to the present day (where k = 4.4 × 10⁻¹³).

We have noted that the estimates used to derive Equation 5.1 are based on a small database. In future decades, it is extremely likely that more impact features will be discovered in the areas considered by Grieve and Dence, hence the estimate of the 'crater production rate' will be upgraded. One can also infer that the estimated number of impacts predicted by Equation 5.1 is almost certainly conservative. The numbers derived from Equation 5.1 can be expressed graphically as the linear log-log relationship (Figure 5.4). Alternatively, examples of the number of cratering events of a specific size for the 2000 Ma period are listed in Table 5.2, which also shows the average time separation between such impacts, as inferred from this analysis.

In addition, Oberbeck *et al.* suggest that this equation is supported by astronomical observations of asteroids which currently intersect the Earth's orbit. Thus, from a statistical study of these Earth-crossers, Shoemaker (1983) estimated the probability of impact by Earth-crossing asteroids, which he converted to an 'equivalent cratering rate' on Earth. His estimated rates fall within the error bars of the rates determined by Grieves and Dence which, using the time-scale set up by Harland *et al.* (1982), only extend back to about 590 Ma (or, as is nowadays cited, 545 Ma).

If we assume that the impact flux has remained constant throughout the 2000 Ma period, it can be inferred from Table 5.2, that in the last 545 Ma, the period in which we are particularly interested, 1900 impact craters, with diameters equal to or in excess of 10 km, can be expected to have developed, with about 30 of these events having a diameter in excess of 100 km. There appears to be a marked disparity between observation and statistical evaluations!

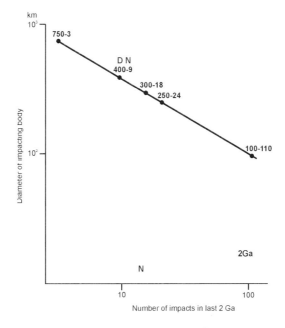

Figure 5.4 Log-log relationship between number and diameter of impact structures (based on relationship derived by Oberbeck *et al.*, 1993).

Table 5.2

Diameter (km)	Number of impacts per 2000 Ma	Average time interval between impacts of a given size
>750	3	1:670 Ma
>500	6	1:400 Ma
>100	110	1:18 Ma
>20	2003	1:1 Ma
>10	6974	1:287,000a
>5	24,283	1: 82,000a
>1	1,430,000	1:1400a

If we consider the impact flux in the inner Solar System to be constant, then over a period of 500 Ma, the number of impact events with diameters equal to or greater than 10 km will number 1500. In the last 300–500 Ma, the number of impacts on Venus has been in excess of 900. The thickness, density and temperature of the Venusian atmosphere would have eroded and reduced the speed and dimension of an incoming body more rapidly than a comparable body passing through the Earth's atmosphere. Hence, a proportion of the incoming bodies, especially those composed mainly of ice, would probably be destroyed before they hit the Venusian surface. The age of the Venusian surface may be as little as 300 Ma, so that the degree of correlation between the Venusian impact density and that inferred by Oberbeck *et al.* for Earth, is quite remarkable.

For those readers who are unfamiliar with the variety of diagnostic structural features that are associated with craters, we suggest that the evidence for a greater degree of cratering on Earth than is indicated by Figure 5.1 is best appreciated by the study of the craters and related features brought about by man-made explosions.

5.3 Man-made craters

During the last few centuries, the military learned the 'art' of creating small craters with, initially, relatively small quantities of gunpowder. As with many other aspects of technology, mankind's understanding was greatly enhanced by war, or the threat of war. The effects of cratering became more generally appreciated during World War I, but it was not until World War II, in 1940, that British scientists conducted a series of tests to determine empirical relationships that would permit estimates to be made of the size and shape of craters caused by high-explosive (HE) charges. These tests, carried out in soil and representative rock types encountered in England, permitted relationships to be established between the diameter of the crater and the energy of the charge. The range of tests carried out later by the U.S. Army Corps of Engineers in the 1950s on soils and lightly cemented soils was impressive. They too established relationships between the various parameters.

At the end of and immediately following World War II, the advent of atomic and hydrogen bombs enabled the relationship between the diameter of the crater and the size of charge (i.e. the energy of the explosion) to be extended through several orders of magnitude. Quite a number of experiments were conducted in the USA, Canada, Australia and islands in the Pacific (and, presumably, also in the former USSR and elsewhere). This work which was, of course, largely directed at purely military objectives, has now been mainly superceded by computer modelling. The geometry of the crater produced by these explosive experiments depended upon the position at which the explosion was initiated relative to the surface.

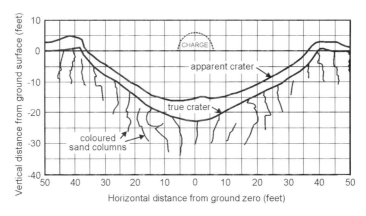

Figure 5.5a Section through crater which developed in weak sediments following a 20 ton TNT explosion. Such craters are very close to circular in plan.

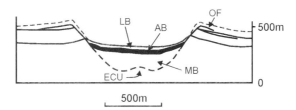

Figure 5.5b Diagrammatic section through Meteor Crater which is very similar to that of Figure 5.5a (after Shoemaker).

Figure 5.5c Oblique view of Meteor Crater reveals that the rim of the structure shows 'straight-line' sections which are probably determined by the orientation of pre-existing joint or other forms of fractures.

However, for tests where explosions of relatively small to moderately large TNT tonnage took place at the surface, the craters which developed were always of a simple basin-like geometry. The section through a crater which developed as the result of a 20 ton TNT explosion is shown in Figure 5.5a and a natural, but very much larger structure, the Barringer or Meteor Crater, Arizona, is shown in Figure 5.5b and in vertical view in Figure 5.5c.

These structures, although termed *simple* craters, show a degree of evolution. It was soon realised that, initially, excavation formed a short-lived, *transient crater* (Figure 5.6a) which then rapidly evolved into a simple crater by slumping of the transient crater walls (Figure 5.6b). It will be noted that, as a consequence of this slumping, the simple crater is wider and shallower than the transient crater. An important feature of the transient crater is that, in section, it approximates to the nose of a parabola with the ratio of the depth to diameter of the transient crater approximating closely to 1 : 3.

Grieve and Pilkington (1996) show, in an elegant diagram (Figure 5.7), how, by adjusting the scale of individual craters, the sections of simple craters can be superimposed on an idealised section, so that the degree of erosion can be inferred.

In small, 20 ton TNT tests, the surface around the crater is thinly covered with earth and debris derived from the crater. The nature of this surface cover did not become immediately clear from such small energy explosive experiments. Indeed, it was not until much larger explosions of about 100 tons of TNT, set off in the test-sites of the USA and Canada, that it was recognised that the near-surface rocks were 'blown upward and outward to form a recumbent flap' (Figure 5.8) which can also be seen in Figures 5.5b and 5.7. Indeed, it was Shoemaker's experience with the results of moderately large tonnage explosions that enabled him to demonstrate that the Barringer Crater, Arizona, was a *certain* impact crater. As a result of this identification, the structure was renamed Meteor Crater.

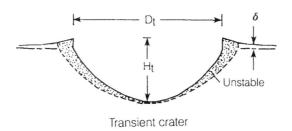

Transient crater

Figure 5.6a Section through a *transient crater*, which rapidly evolves by slumping to (see Figure 5.6b).

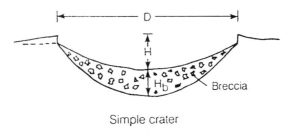

Simple crater

Figure 5.6b A *simple crater* (after Melosh, 1989).

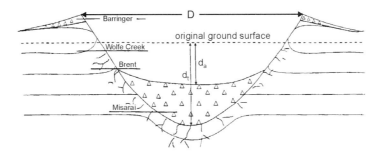

Figure 5.7 Superposition of simple craters, by adjusting the scales of a number of natural impact events (after Grieve and Pilkington, 1996).

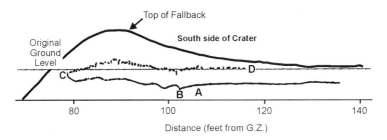

Figure 5.8 Flat-lying recumbent fold outside the crater formed by a 100 ton TNT explosion (after G.H.S. Jones).

It required a further series of major explosive experiments to provide the information that permitted one to infer the evolution from the simple crater to the more complex features, that are now recognised as exhibiting the characteristics of natural major impact craters. These experiments were conducted at the Defence Research Establishment, Suffield, Alberta, Canada in the period from about 1960 to 1970. Little of the experimental work was reported except in internal reports. A brief outline of some of the important results was presented by Jones (1977, 1978). However, in his retirement, he was persuaded by the present author to prepare an extremely valuable document (entitled *The Suffield Craters as Analogues of Impact Structures*; Jones, 1995) which gives an overview of the range of experiments conducted in the decade 1960–70. It was only when charges in the range of 100–500 tons of TNT were exploded on poorly indurated sediments that the more complex crater geometries developed which, hitherto, had only been observed on the Moon; but whose mechanical significance had not been understood.

5.3.1 Geometry of real and 'model' craters

The morphology of craters has been studied in natural examples on the Earth, the Moon, other planets and satellites, and also by means of theoretical and physical models. The theoretical models and many of the physical models (especially those using small energy sources) have used a target material which is usually homogeneous and isotropic.

The results of such calculations or experiments are usually in the form of a basin crater (Figure 5.5a). Most natural craters on Earth with a diameter of less than about 3.5 km (e.g.

Meteor Crater, Arizona, and Brent Crater, Canada; Figures 5.5 and 5.7) exhibit this geometry. Such craters, whether they be natural, experimental or theoretical, represent the geometry of the initial phase of excavation and development of large real craters. Hence, understanding this initial phase is of critical importance in interpreting the mechanisms involved in cratering, which will be considered later in this chapter.

To understand how natural, circular, impact features can be detected, even after deep erosion, and how they are related to folds and faults which are also generated during impact, let us consider the morphology that develops as the result of explosions of known energy, and the structural development at deeper levels within and outside the crater, which were revealed by detailed survey coupled with quite deep trench excavations across the experimental area.

We will be mainly concerned with the model TNT explosive experiments conducted at Suffield, Alberta, which gave rise to complex craters and various forms of structures which were produced by 100–500 ton TNT explosions. These craters and associated structures produced in explosive experiments constituted a breakthrough, that was recognised by Dence, Dietz, Grieve, Shoemaker, Roddy and others as exhibiting important diagnostic features which could be seen in many tens of natural impact structures on Earth, as well as on the Moon.

Cratering experiments, which were being conducted in the USA as well as in Canada, were designed to establish the probable behaviour of ground movement when subjected to megaton nuclear explosions. Some of the tests in the USA were in excess of 500 tons of TNT (or TNT equivalent). In December 1997, the US government released information regarding their various tests using nuclear weapons. One such explosive test, in Nevada, with an energy of 104 kilotons TNT equivalent (approximately seven times more powerful that the Hiroshima bomb) gave rise to a 'simple' crater with a diameter of 1280 ft (390 m) and a depth of 320 ft (98 m).

Those tests in the USA that had the explosive charge on or close to the surface of the test site, produced only relatively simple basin craters. In Canada, explosions of 500 tons of TNT produced completely different results.

King-Hubbert (1937) published an extremely important paper on the theory of scale models, as applied to the study of geological structures. He showed that if one wishes to imitate large geological structures by using scale models, then the model material must be proportionately weaker than the rock in which the full-scale structure has, or will, develop. Thus, it can readily be inferred from this *scale-model theory*, that the overall size and form of the crater depends critically upon the magnitude of the explosion and the strength of the target rock.

The tests carried out in the USA between 1950 and 1960, in which the explosive charge was configured to rest on the surface, were conducted mainly in desert areas, where the surface rocks were generally indurated sediments and/or volcanics which were dry, and in some instances approached or even exceeded the strength of the rocks in which the full-scale structures would develop, as the result of a multi-megaton explosion. Also, in these desert test areas, the water-table sometimes tended to be at a depth comparable to the diameter of the crater. Hence, as the majority of the US tests were scale-models probably in the range of 1 : 1 to 1 : 10, it is not surprising that the craters that developed in these tests were simple basins.

There is no implied criticism of the tests conducted in the desert areas in the USA for the sites were, of necessity, mainly determined by social factors. Indeed, small-scale experiments were conducted by Davis (1967), of the US Army Engineers, to establish the effects of a near-surface water-table on the dimensions of the crater which developed. However, in these experiments, the charges were limited to 250 lb wt, so that, again, only simple craters were formed.

The explosive tests conducted at Suffield, however, were carried out on the site of a dried-out postglacial lake, where the sediments were poorly, or non-indurated, and the water-table was at no great depth beneath the surface. The strength of such sediments is of the order of one hundredth to one thousandth that of well-indurated sediments or metamorphic or igneous rock. The resulting structures that developed in the Suffield explosive tests deviated sharply from what, at that time, was regarded as the normal pattern. The larger of the Suffield craters did not conform to the normal, simple basin, but instead they appeared remarkably similar to structures which were then only known on the lunar surface and a few terrestrial structures which were, at that time, attributed to 'crypto-volcanic' agencies.

It is, of course, difficult to infer precisely the scaling factor of these Suffield explosive experiments. Tests in the 20–100 ton TNT range probably gave good modelling of megaton, or multi-megaton, nuclear explosions. However, the 500 ton tests, which produced a crater-rim diameter of about 87 m, probably represented real large-scale impact structures with a diameter of 20–100 km, or possibly even larger. Thus, the nature of the test site, which comprised a sequence of flat-lying, lacustrine sediments, with the water-table at a depth of about 8–9 m, was sufficiently uncompacted and weak that, at the scale of the detonation energy, the modelling was more accurate for large-scale natural impacts than for the results one could expect from megaton nuclear tests conducted on strong rock.

Fortunately, members of the scientific community beyond the military, were alerted to the production of 'lunar-type' craters at Suffield, so that this work had an important influence and enabled the pioneers in this field to identify a number of natural structures which, at that time, had not been firmly identified as impact structures. Indeed, these Suffield craters were closely studied by the astronauts destined to land on the Moon.

The first 500 ton TNT experiment was code-named Snowball. In this experiment, the TNT blocks were constructed to form a hemispherical configuration, so that Half-Snowball would have been a more accurate description. The diameter of the hemisphere of explosives was 34 ft (10.46 m).

The Snowball explosion (see Frontispiece) gave rise to a crater with a diameter of 87 m. The overall view of the water-filled crater (Figure 5.9) is, of course, instructive, but tends to diminish the magnitude of the structure. For this reason we have included Figure 5.10, which shows a 6-foot-tall man (Gareth Jones) on the internal slope of the crater, from which one can infer its depth and also the arcuate form of the crater rim and the main depression. This photograph was taken soon after the explosion, well before the upward migrating water buried most of the crater area.

Over a period of several days, water poured into the crater from a pronounced *central, vented uplift*, which, eventually, just cleared the surface of the lake which developed in the crater, as can be seen in Figure 5.9.

It was immediately realised that the extra energy of the 500 ton TNT explosion had given rise to a dramatic and unforeseen change in the pattern of crater-floor movement. This central uplift morphology had been seen on the Moon. Also, it was soon realised that the central uplift morphology was matched by several of the impact structures in the Canadian Shield, hitherto considered by some to be *crypto-explosive* features. This was the first Suffield cratering experiment that enabled moderate size, natural impact craters to be recognised as such, and greatly stimulated the pioneers in this field of study.

Figure 5.9 Panoramic view of Snowball crater showing central uplift in a moat of groundwater. Diameter of crater is 87 m (after G.H.S. Jones).

Figure 5.10 Internal shot of portion of crater taken soon after the explosion when ground water was beginning to seep into the base of the structure (after G.H.S. Jones).

The reader will perhaps be surprised by the degree of 'contrast' in the reproduction shown in Figure 5.9 and in other photographs shown later in this chapter. This can be attributed in part to the fact that the original photographs were taken 50 or so years ago, when the standard 'black and white' film emulsion was orthochrome, which inherently tended to produce high-contrast images. However, there was a second and more important cause for the high-contrast photographs. This was the intensity and extent of the fire-ball set off by the 500 ton TNT explosions. For a distance from ground zero, a zone of black 'scorched-earth' extended out beyond the crater rim for a distance of 3–4 times the diameter of the crater (i.e. about 250–320 m). The temperature of this 'fire-ball' was not only responsible for the scorching, the pressure wave and the high temperatures generated by the 500 TNT explosions were sufficient to fuse grains of quartz and produce small hollow spheres which Jones termed 'silica pop-corn'.

In his list of *certain* impact structures, Hodge (1994) cites several natural examples of such craters which exhibit the central uplift, e.g. Gow Lake, Saskatchewan (4 km diameter), Haughton, Northwest Territories (20.5 km diameter) and Gosses Bluff, Australia (22 km diameter). It is interesting to note that the Gow Lake structure is just outside the diameter limit for simple basins, suggested earlier, of about 3.5 km. Such a limit is better considered as marking a transition zone between simple and more complex craters. This follows from looking at the Suffield cratering experiments with hindsight.

The first recognised central 'bump' was noticed after a 100 ton explosion. However, it was not until the Snowball experiment took place that it was realised that the central 'bump' in the floor of the 100 ton crater was an embryonic central uplift. It is interesting to note that Meteor Crater (Figure 5.5b) also exhibits such an embryonic central uplift.

Military dissatisfaction with the observed structural form exhibited by the Snowball crater had most illuminating consequences; it resulted in the configuration of the charges used to produce subsequent craters to be changed from the hemisphere to that of a sphere which was tangential to the ground (see Frontispiece). The detonation with this *'Prairie Flat'* configuration produced the remarkable ringed structure shown in Figure 5.11. Later detonations, such as Dial Pack, extended and confirmed the pattern of ringed structures.

In the second of these 500 ton explosive experiments, the blocks of explosive were arranged to form a sphere which was in contact with the surface at ground zero. The lower half of the sphere was supported by slabs of polystyrene foam. The sphere of explosives had a diameter of 24 ft (10.46 m). The explosive gave rise to a crater with a diameter of 64 m and a disposition of structures within the crater which was completely different from that of Snowball. As will be seen in Figure 5.11, the crater floor contained multiple, circular uplifts, with associated 'volcanic cones'.

A comparable 50 km diameter circular feature is shown in Figure 5.12, in the Sahara Desert area of Mauritania. This ringed structure has been photographed from a space capsule and also from a Space Shuttle, but we are not aware that the feature has, so far, been studied on the ground. Consequently, it has not yet been classified as a certain impact structure.

The reasons for the change from central uplift to ringed uplift, which resulted from the new configuration, is not completely understood (Melosh, 1989). However, it can be inferred that, because of the spherical charge, Prairie Flat primarily resulted in an air-blast. This developed a smaller diameter crater than did the same tonnage of TNT of hemispherical form for Snowball. As we shall see in later chapters, Snowball can possibly be likened to an impacting comet.

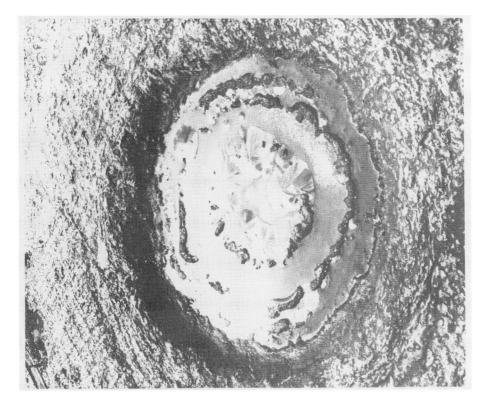

Figure 5.11 Oblique aerial view of crater with concentric ringed uplifts which developed after the 500 ton TNT Prairie Flat explosion. Crater diameter is 65 m (after G.H.S. Jones).

Figure 5.12 A view from space of concentric rings in a 50 km diameter impact structure in the Sahara of Mauritania, caused by a 'direct' impact or 'low altitude' explosion (c.f. Prairie Flat 'bomb' (Frontispiece) and aerial view (figure 5.11)). The picture was taken by a member of NASA prior to 1979.

While there is much that is common to the Snowball and Prairie Flat craters, the details vary enormously as regards the location of the secondary pseudo-volcanism. For example, in Snowball, most of the observed pseudo-volcanic cones lay beyond the crater rim, along the circumferential and radial fissures. In Prairie Flat, there was an intense concentration of such cones on what appear to have been fractures associated with an inner circumferential structure on the inner floor of the crater.

A detail of the turn-back of strata, which occurs beneath the rim of the Prairie Flat Crater is shown in Figure 5.13. This structure leads to an inverted limb of the flap outside the crater, as indicated in the section of the 100 ton TNT experiment shown in Figure 5.8.

Such inversion of strata is now well documented for natural craters, e.g. Barringer/Meteor (Shoemaker, 1960), Flynn Creek (Roddy, 1968) and Vredefort (Dietz, 1963). It required only a few years, before this type of feature had become sufficiently well-documented to become a diagnostic feature of moderate to large impacts in sedimentary and even basement rocks.

Very small-scale experiments, not necessarily involving explosives, have also been conducted in laboratories. These usually involved firing a small metal sphere at a weaker target area. However, even a lead pellet fired from a powerful air-pistol, when the target is weak clay, will provide comparable results, as indicated in Figure 5.14. In such experiments, the target material often takes the form of a 'frozen' incomplete recumbent 'flap'. Such flaps in model work represent 'arrested' development because (a) the energy of the impact is not sufficient to spread the flap further and (b) the model is of such a scale that the flap material is sufficiently strong to support itself without collapsing to form a completely recumbent feature. Flaps with comparable geometry may develop in basement rock covered by sediments, as the result of a major impact. In such instances, the flap is rooted in the basement and spreads to cover the sediments around the crater, which, of course, supports the weight of the flap.

Dietz (1963) realised that such model profiles (Figure 5.14) were reproduced on a far grander scale in the Vredefort structure in S Africa. Moreover, he recognised that minor conical shear features, now termed shatter-cones (which we shall discuss later in this chapter) were almost certainly a manifestation of an impact event. This enlightened insight was altogether too much for the vast majority of geologists. Indeed, it is only in the last decade of this 20th century that the majority of S African and other geologists have been persuaded by his thesis.

5.3.2 Structures external to the crater rim

A characteristic of these Suffield tests was the time and effort expended in determining the deformation and displacement of the target material. To enable the ground motion in and around the crater to be established, boreholes were drilled, prior to the explosion, and infilled with sand. In early tests, the study of the deformation was restricted to the area outside the crater. In the major 500 ton experiments these sand columns (now containing marked cans, filled with sand and set at specific depths in the sand columns) were emplaced all the way to the bomb and later, following the explosion, their movements were extensively surveyed.

(a) Nose of recumbent fold

An important feature that soon became evident from the TNT explosion craters was the nose of a *flat-lying recumbent fold*, which lay under the *crater rim* (Figure 5.13). The existence of such a flat-lying fold had been detected in earlier explosive tests, in which 100 tons of

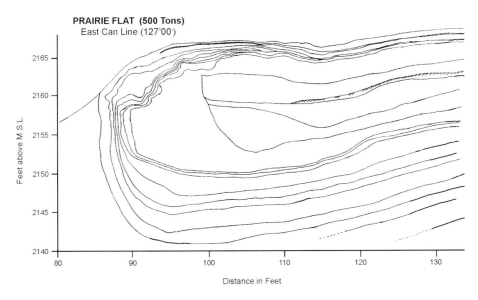

PRAIRIE FLAT (500 Tons)
East Can Line (127°00')

Figure 5.13 Detail of the nose of the recumbent fold formed by the Prairie Flat 500 ton TNT explosion (after G.H.S. Jones).

Figure 5.14 Section through a small-scale model in which the overturned flap is frozen. A section through the Vredefort impact structure exhibits such a flap.

TNT had been used. These recumbent folds are *not* the result of fall back of ejecta, but result from a coherent overturning of the flap, in which even incoherent sand strata retained complete positional integrity, even though a bed with an original thickness of about 20 cm was thinned to less than 0.5 cm (Jones, 1978).

(b) Peripheral graben

Beyond the nose of the flat-lying recumbent fold, the next major features to be encountered beyond the rim of the crater formed by the 500 ton explosion, are concentric, vertical or steeply inclined fractures (Figure 5.15a). These fissures and fractures were not everywhere immediately visible, for they were sometimes covered by ejecta and sometimes obliterated by sand washed up from depth. These fractures were sometimes open at the surface (Figure 5.15b). At depth, a section through these fractures showed that they formed a sand-filled dyke (Figure 5.16a) which sometimes gave rise to sand-volcanoes at the surface (Figure 5.16b). These fractures formed a pair of circumferential, high-angle normal faults, which define a *graben*.

Figure 5.15 (a) Trace of one of the concentric fractures at the surface with out-flow of mud. (b) These fractures are sometimes quite wide at the surface. Gareth Jones is shown in such a fracture for scale (after G.H.S. Jones).

Figure 5.16 (a) Section of concentric, vertical, or steeply dipping fractures, sometimes covered by mud flats or ejecta. (b) Sedimentary volcanoes also form above these concentric fractures (after G.H.S. Jones).

(c) Peripheral folds

Concentric with the peripheral graben were series of folds of small wavelength and amplitude. These folds were best seen in the tarmac paths (Figure 5.17) that were constructed radially to ground zero to facilitate the surveying and building of the 'bomb'. The folds indicate that, at the time of their development, the maximum principal stress acted radially, while the concentric fractures permit one to infer that when the graben formed, it was the least principal stress that acted radially.

(d) Radiating vertical fractures

Less obvious major structures were the *vertical radial fractures*, which were somewhat irregular in their surface manifestation (Figure 5.18a). These, too, were utilised for the outflow of water, mud and sand from below, which sometimes formed sedimentary 'volcanoes'. One may infer that when this latter group of fractures developed, the greatest principle compressive stress (S_1) acted radially from ground zero, while, normal to the extension fractures, the least principal stress (S_3) was tensile. These extension fractures were not detectable in the vicinity of the rim or in the crater area, because of the fallen ejecta. Indeed, they were only clearly visible when they disrupted the surface, usually accompanied by the outflow of silt-laden waters.

Because the true extent of individual radial fractures was often masked, it is difficult to give a precise indication of their average length. However, from the mapped surface features which developed around the 500 ton Snowball explosion, it was clear that the extent was at least one crater diameter beyond the rim. Also, one may infer from the quantity of sand and silt brought to the surface, that in some areas, at least, the fractures reached down to and penetrated below the water-table, which was at some 28 feet (8.5 m) below the surface.

Major extension, i.e. tensile, fractures are normally limited to the uppermost regions of the crust. At depths of a few kilometres, the gravitational loading causes the rocks to change their mode of failure to normal faulting. However, such changes occur when the rock-mass is deformed at normal geological strain-rates. Here we are concerned with high-magnitude

Figure 5.17 Sections of small-wave length folds that developed in tarmac paths, outside the crater rim (after G.H.S. Jones).

pulse stresses, the duration of which is exceedingly brief. Such stresses can rip rock apart in milliseconds, before normal shearing can develop.

One may compare the relative dimensions of the radiating tension fractures, in Figure 5.18a, with those apparent around a 200 km diameter circular feature, in Nevada, USA (Figure 5.18b). Sufficient evidence has been adduced by Warne (see Willar, 1991) to demonstrate that the circular feature is probably the result of an impact some 375 Ma ago. (We shall give evidence which supports the existence of this event in the following chapter.) An earlier study of the lineaments in and around this circular feature by Rowan and Wetlaufer (1978), showed that these fractures may well have influenced mineralisation. We note at least three fractures A, B and C, in Figure 5.18b, which appear to be radial tensile fractures. The length of the main radial fractures is as much as 1.5 diameters of the circular feature. That is, the length of the main fracture is in reasonable accord with that obtained from the 500 ton explosive experiments.

Features on the Moon lend further support to the considerable extent of such vertical fractures. Working from the basic map of the Copernicus crater and surrounding area, prepared by Shoemaker, Shotts (1968) inferred from the linear alignment of groups of three or more small craters, that vertical fractures radiated from the Copernicus crater and extended for several crater radii beyond the lip of that crater. Shotts suggested that the alignment of these minor craters was the result of 'exhudation of volatiles'. Jones (1995) inferred that radiating dykes, which connected and fed the small craters, represented an immediately post-impact event.

a

b
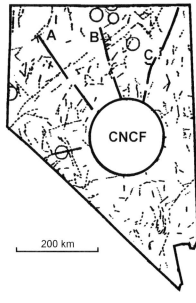

Figure 5.18 (a) Vertical radial cracks (revealed by series of stakes) traversing flooded area beyond Snowball crater (after G.H.S. Jones). (b) Map of radiating dykes (A, B and C) around Central Nevada Circular Feature (after Wetlaufer).

(d) Thrusts

The final important features that developed below, and well beyond, the crater rim were low angle shear fractures. The importance and extent of these features were only discovered as the result of the post-explosion excavation and survey that followed the major tests.

In the literature relating to terrestrial craters, there has been scant mention of the existence of *thrust faulting* associated directly with the creation of the crater, though such observations do exist. For example, the existence of a thrust which is visible on the inside of the Meteor Crater wall, on the 'Astronauts Path', is pointed out by Shoemaker (1960). At Suffield, even on one small (5 ton) crater, a thrust fault was detected that breached the surface at a distance of a little over one crater radius from the rim. In the larger trials, including both the Snowball and Prairie Flat 500 ton explosions, excavation beneath the rims revealed *anomalies* which could be explained as thrust planes, but the visual evidence was controversial.

It was not until the present author postulated the existence of such thrusts (for reasons that will become apparent in Chapter 7) that he examined the original data for the Snowball experiment (Jones, 1978). From analysis of the field data supplied in one section, it was clear that sets of adjacent cans in a single hole exhibited a similar displacement vector, but that above and/or below such sets of 'can-vectors', there were other sets which exhibited markedly different displacement vectors (Figure 5.19). From this difference, it can be inferred that there were one or more discontinuities in any single borehole, between one set of can-vectors and the adjacent set. These discontinuities can be traced laterally to adjacent sand-columns, thereby permitting an inclined shear-plane to be defined. It was found that the lines of discontinuities clearly represented planes of differential movement which dipped inward (towards a vertical line, projected below ground zero) at an angle of dip of approximately 25°. The movement sense on some of these low-angle fracture planes is that associated with normal faulting. However, as we shall see, explosions, whether they be from

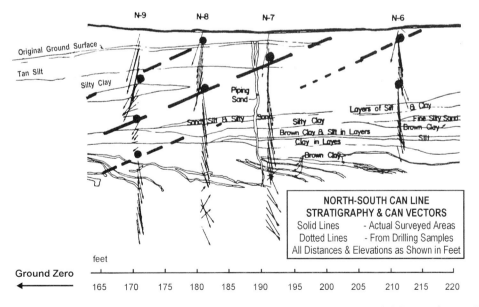

Figure 5.19 Shear plane inferred from displacement of can-vectors excavated following Snowball explosion, dipping at about 25° toward a vertical line at ground zero (after G.H.S. Jones).

TNT or the result of meteoritic impact, first give rise to a compressive phase which is then followed by elastic rebound. It was, therefore, inferred from these observations that the low-angle fractures formed as thrusts, which then experienced re-shear in the normal mode during an elastic rebound phase.

Analysis of data from the north side of the Snowball crater shows that this slumping had not occurred, and the residual movement along a similar gently dipping, inwardly inclined, shear plane was consistently *outward* in the thrust mode. This was also the situation along the excavated radii of the Prairie Flat Crater (Figure 5.20), where the thrust faulting is clearly demonstrated. From the measurements, it can be inferred that the thrusts crop out at about twice the radius of the crater.

5.3.3 Correlation between experimental and real impact structures

The essential relationships between the structures which develop in central, or ringed, uplift craters which formed at Suffield are indicated in Figure 5.21. It was almost immediately recognised that these explosive structures presented important diagnostic features which would permit real impact features to be identified.

The importance of these Canadian experiments can readily be inferred. For example, Jones (1965) showed a direct correlation by scaling up the features obtained in the Snowball experiments and superimposing this pattern on an area in W Africa. It can be seen (Figure 5.22), that the drainage pattern around Lake Bosumtwi was almost certainly controlled by the inferred rim and peripheral graben. Thus, Jones demonstrated the probability that Lake Bosumtwi marked the site of a reasonably recent impact event, prior to its authentication as a certain impact structure. (The almost straight line feature AA′, that extends to the right of the structure, represents the best exposed vertical tension fracture that formed in the Snowball experiment. It will be noted that this fracture extends for a distance at least equal to the diameter of the crater rim beyond the crater.)

It was soon realised that the peripheral graben which developed around the Snowball crater was a diagnostic feature of a major impact. However, the outer fault plane may be

East Section

Distance in feet from Ground Zero

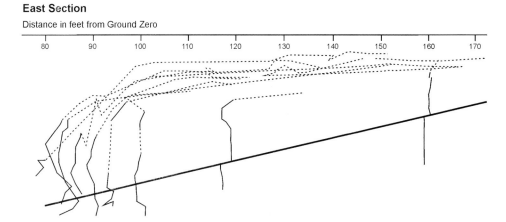

| 80 | 90 | 100 | 110 | 120 | 130 | 140 | 150 | 160 | 170 |

Figure 5.20 Thrust displacement along an inward-dipping shear plane after the Prairie Flat explosion (after G.H.S. Jones).

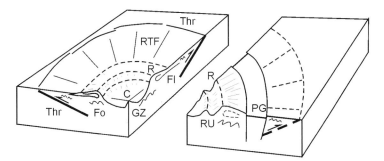

Figure 5.21 Schematic representation of structures developed by impacts. A simple crater is shown to the left. The peripheral graben is omitted, for it may not be an obvious feature of a small impact, but would certainly be expected to occur around a crater more than about 5 km in diameter. The block-diagram to the right shows the ringed uplift of the type generated by the Prairie Flat 500 ton explosion.

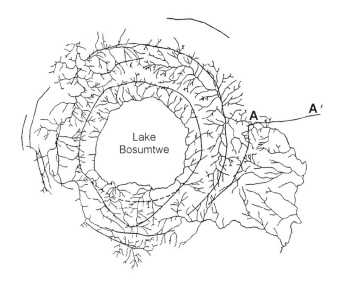

Figure 5.22 Snowball fracture pattern superimposed on Bosumtwi drainage system. The line AA marks the position of the best exposed vertical tension fracture, which, it will be seen, extends at least 2.0 crater diameters from ground zero (after G.H.S. Jones).

dominant as in the particularly well-exposed, 100 km diameter Manicouagan structure, in Quebec, Canada (Figure 5.23a).

We have shown that natural impact features can be either central uplift, or ringed-uplift features. It is interesting to note, therefore, that the twin, Clear Water Lakes in N Canada show signs of a hybrid impact form. The larger of the lakes exhibits a concentric ridge around a small island that almost certainly represents a central uplift feature. We have already noted that simple craters do not usually exceed a diameter of 3.5 km. Hence, it is unlikely that the smaller ClearWater Lake, with a diameter of about 10 miles (16 km) is a simple crater. We suggest that, if soundings of this lake were available, they would show the structure to be a central uplift crater, but where the central uplift has not reached the surface level of the lake.

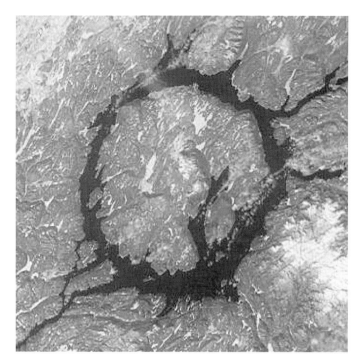

Figure 5.23a Manicouagon impact structure with a diameter of 100 km, showing well-developed peripheral graben as defined by two lakes (courtesy of NASA).

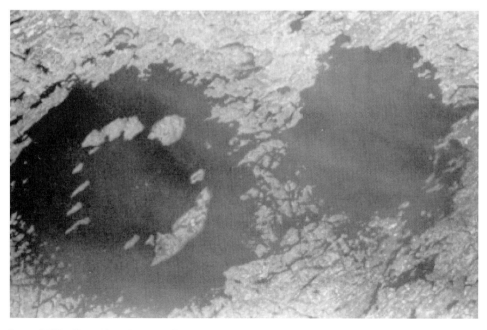

Figure 5.23b Clear Water Lakes N Canada (courtesy of Canadian Geological Survey).

Thus, the Suffield experiments, because they were scale-models, provided a unique wealth of data on a variety of structures generated by man-made explosions. Without this information, our knowledge of structures that result from natural impacts would be significantly diminished.

5.3.4 Small-scale and microscopic structures

As a result of the energy of the explosive or impact event, some melting or even vaporisation, of the rock mass can take place. In some nuclear experiments with energy equivalents in excess of 1000 of tons TNT, *tektites* (solid glass globules) were ejected, and modest quantities of glass were found in the crater rocks. In moderate-sized, natural, terrestrial craters, such as the Ries Structure of S Germany, *suevite* (an impact breccia with a shock-produced glass matrix) is a common feature. The identification of high-pressure minerals (e.g. polymorphs such as *coesite* and *stishovite*, high pressure forms of silica, as well as *diamond*) provides important and supportive evidence that the structure under investigation was formed by impact. To these can be added the identification of shock-induced features within individual mineral grains, particularly quartz and feldspars.

Remnants of meteorites are only rarely found associated with impact features. Those which are most easily recognised are the iron meteoritic fragments, such as are found in and around the Meteor Crater, Arizona. However, it has recently become possible to relate the geochemical signature of the impacting body by identifying the siderophile elements, such as iridium, osmium and platinum, within melt-rocks and breccias associated with the impact structure.

Intermediate-scale structures in the rock, as opposed to the mineralogy, which are diagnostic of impact events, are *shatter cones* (Figure 5.24a) with characteristic converging striations. These cones are features which result from shear failure, as indicated in Figure 5.24b. Such structures were formed in high-energy experiments conducted in the USA It was inferred that the differential stress necessary to produce such shatter cones is equal to or in excess of 20 kb.

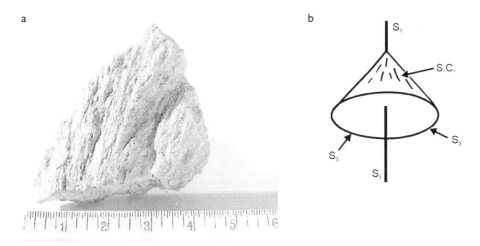

Figure 5.24a Shatter cones. (b) Stress configuration required to produce shatter cones, provided the differential stress exceeds 20 kb (2 GPa).

The presence or absence of these various small to microscopic-scale features is the diagnostic evidence that proves (if they are present) that the structure under investigation is the result of an impact. However, it must be emphasised that the finding of such evidence in or adjacent to terrestrial craters requires careful study. When the crater is on the ocean floor, the task becomes many orders of magnitude more difficult. Hence, if such high-stress minerals are not found, this does not prove that the structure under investigation is a non-impact feature.

These by-products of a moderately large impact event are usually much easier to detect, or infer, than it is to establish the site or even the existence of the impact scar. Based on a study by Alvarez *et al.* (1980), specimens taken from close to the Cretaceous/Tertiary (K/T) boundary were obtained from various wide-spread sites. These small samples contained quantities of tektites, quartz with deformation lamellae and high concentrations of iridium. The authors claimed that such a concentration of items near the K/T boundary was evidence that the boundary itself was the direct or indirect result of a major meteoritic impact or impacts that must have had an influence that was world-wide. Indeed, evidence for one or more events ranges from the Southern States of the USA to Italy far to the east, and the Pacific Ocean far to the west. However, Robin *et al.* (1993) suggest that the evidence they found in the Pacific points to separate events and that several events are necessary to explain the global distribution of spinel-bearing nodules. The postulated event(s) soon attracted attention, for it was also suggested that it contributed to the extinction of the dinosaurs. The hunt for the site of the event which caused such world-wide havoc was immediately set in motion. Currently, the favoured site is that of the Chicxulub crater in N. Yucatan, Mexico, which has a diameter of about 180–200 km. (Hildebrand, 1991). However, as will be seen in subsequent chapters, we concur with Robin *et al.* and suggest that other events contributed to, or even exceeded, the influence of the Mexican event at the time of the K/T transition.

5.3.5 Geophysical techniques

We have already noted that geophysical techniques have been used to establish the existence of buried impact structures. There are, of course, a number of techniques which are available for such studies. However, we feel that a discussion of such techniques is outside the scope of this book, which has largely been based on experimental and geological field studies of impact events.

Australia has, for some years, been the happy hunting ground for those studying craters. The Australian Geological Survey Organisaton (AGSO) published an important journal (*Themic Issue: Australian Impact Structures*, Vol. 16, 4, 1996) which was fittingly dedicated to the honour of R. Dietz (1914–95) and Eugene and Carolyn Shoemaker for their major contribution to the study of impact structures.

Naturally, several of the papers are dedicated to the study of Australia's most famous impact structure, Gosses Bluff (Figure 5.25a). These articles deal not only with the traditional geological mapping of the structure (Milton *et al.*, 1995a), but also with seismic, magnetic and gravity studies (Milton *et al.*, 1995b). A section of Gosses Bluff based on seismic data is diagrammatically represented in Figure 5.25b. It can be inferred that the circular wall formed by the bluff may be an artifact caused by differential erosion, though, of course, fundamentally controlled by the mechanics of impact.

Those readers who are interested in the application of geophysical techniques to the study of impact structures are directed to the papers cited above and that of Grieve and Pilkington (1996).

Figure 5.25a Oblique view of Gosses Bluff.

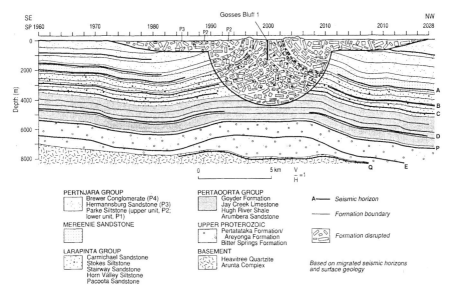

Figure 5.25b Section through Gosses Bluff impact structure. The ringed-uplift feature is probably the result of differential erosion (after Milton *et al.*).

5.4 Moon v. Earth

Despite the number of probable and certain impacts we have mentioned, and the abrupt and important changes in atmospheric and marine environments which are registered in the geological record, probably the most important single element which has helped geologists to discriminate against the importance of impact tectonics on Earth is the apparent lack of circular or inter-linked features on Earth, which are such an obvious feature of the Moon. Even when maps and aerial or satellite images of Earth are studied, such obvious and abundant evidence of circular features is not readily apparent.

As we have already noted, there are several causes for such an apparent difference in the degree of development of such impact features on the Earth relative to those of its satellite. However, these can sound like excuses. So let us see what evidence exists of Moon-like crater patterns on Earth.

5.4.1 Moon-like circular assemblages on Earth

For several decades of the 20th century, the main source of visual data by remote sensing came from aerial photographic images. As many such photographs were desperately lacking in contrast, it took a trained and practised eye to see and interpret the subtle details, faintly in evidence, in such images. Because of the subtlety of the data, a person not trained in such interpretation does not have to be a natural sceptic to doubt the validity of the interpretation. However, when presented with an image-intensified version of the same view, or enhanced satellite images, the practised interpreter is usually vindicated, though the intensified image may show details that even the practised observer did not detect.

Evidence which indicates a concentration of impact activity on Earth which could be considered significant, seems to have evaded discovery. But such evidence that crater density almost comparable with that on the Moon, does exist and was 'brought to light' (literally) in a study by Saul (1978). His investigation showed that there are numbers of moderate-sized circular features in Arizona, USA (Figure 5.26).

These features were not inferred from field observations or photogeological techniques, but are the result of studying the geomorphology of this portion of Arizona, as represented by an accurate three-dimensional model of that state. This model was viewed with illuminating lighting which was restricted to a horizontal beam. By changing the direction of illumination, high spots on the topographic model could be readily identified and plotted on a map. It was noted that metalliferous mines were conspicuously related to these circular features, especially when the circular traces intersected.

These are simple empirical observations which, one would suppose, require at least an attempt at an explanation. However, Saul, who presented this survey, offered no explanation of these features. We consider it reasonable to suggest, from the size and frequency of these circular features, that they relate to impact phenomena in the Precambrian basement and that traces of these basement structures have been inherited by the cover rocks. (We suggest that an alternative explanation, that these circular features represent 'nests' of caldera, lacks plausibility – and requires as much, or even more, verification than the impact hypothesis.)

Norman (1980), who carried out a study of Saudi Arabia using aerial photographs and satellite images, was less inhibited. He too found evidence for an array of quite large arcuate

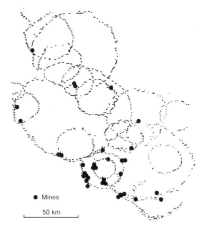

Figure 5.26 Showing mines in relation to ancient circular features in Arizona, USA (after Saul).

and inferred circular features, most of which, he suggested, were traces of ancient impacts which had probably been 'inherited' by the flat-lying sediments from the underlying archean rock mass (Figure 5.27).

It will be noted that, in both these examples, we are not observing direct evidence of cratering, but the influence of structures in basement rocks of circular or arcuate fractures, which make themselves evident in much younger cover rocks. Hence, the traces in the cover rock can be related to impact events, even after the crater itself has been deeply eroded.

It has been shown that the major meteoritic impacts occurred on the moon prior to 3800 Ma. By far the preponderance of basement rocks on Earth yield isotopic dates less than 4000 Ma, therefore circular features shown in Figure 5.26 and Figure 5.27, are likely to have occurred when the lower or residual rate of impact existed. Nevertheless, these figures show that the Earth's basement rocks are still capable of exhibiting a quite respectable record which can be interpreted as being the result of bombardment.

The work of Russian workers (Zhukov, Murav'yev and Popsuy-Shapko) which is presented by Zaychenko et al. (1982) is shown in Figure 5.28. The original papers are not quoted, but we can reasonably assume that the figures in this paper showing circular features are accurate copies of the originals.

Zaychenko et al. suggest that the evidence which militates against the circular features being impact features (*astroblemes*) include the following:

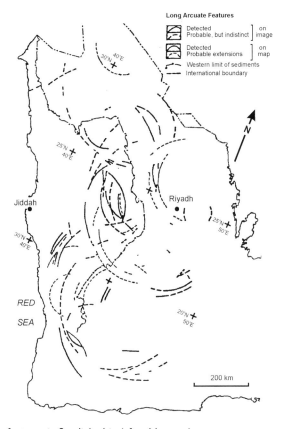

Figure 5.27 Circular features in Saudi Arabia (after Norman).

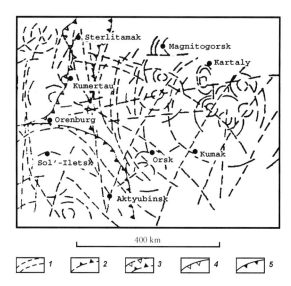

Figure 5.28 Circular features in the Orsk area of the former USSR.

(a) their spatial location at an intersection of two deep linear faults;
(b) their occurrence at location of dome uplifts;
(c) the occurrence of annular and *en echelon* normal faults with normal bedding of cover rock;
(d) several formation stages of volcanic rock; and
(e) the presence of allied tuff-containing explosion pipes outside the structure.

We do not wish to go into detail regarding these points, but would refer to the results of the structures which developed in the Suffield experiments, and suggest that many of these features given in (a)–(e) could reasonably be expected in a large, old impact structure. On the basis of the evidence of circular features, presented above, we suggest that there is evidence which supports the conclusion that impacts have occurred on Earth, in sufficient numbers to resemble impact patterns on the Moon.

The question is not *whether* the Earth has been bombarded, but *to what extent* could we expect circular impact scars to be a feature of the surface morphology of the Earth. The pattern of arcuate and circular features, shown in Figures 5.26 to 5.28, we suggest, are so similar to those found on the moon that there is an *a priori* reason to accept that the arcs and circles on Earth, shown in the various cited figures in this chapter were mainly caused by the same mechanism that caused the features on the Moon.

5.5 Stresses and reaction of target rock as the result of impacts

Impacts give rise to a variety of structures which extend from small-scale pitting of the surface to the development of craters several hundreds of kilometres in diameter. In this section, we shall deal only with the simpler elements of deformation mechanisms involved in the generation of impact craters. The reader interested in a more detailed evaluation of this topic is referred to the excellent text by Melosh (1989).

5.5.1 Compressive and tensile stress-waves

Let us consider an impacting body, be it an asteroid experiencing a degree of break-up, or a comet undergoing even greater disruption, which strikes the ocean or the continental crust. A relatively small proportion of the energy at impact will be converted into blast-waves, tsunami or seismic waves. However, most of the impacting energy is converted into thermal energy, which heats the bolide and target material, or into kinetic energy of the ejecta from a crater, which forms as the result of the impact.

The simplest situation to be considered, namely that of a homogeneous, isotropic spherical meteorite, making contact along a vertical trajectory, at a velocity V with the surface of a flat portion of a homogeneous isotropic target area on Earth, is represented in Figure 5.29a. At the point of impact (i.e. ground zero, or GZ), two stress waves are generated. One moves upward into the impacting sphere at a velocity V_a. The second wave radiates downward and outward from GZ at a velocity V_b. The velocities V_a and V_b are related to the density and elastic constants of the meteorite and the target rock in the vicinity of GZ. If we assume that both the meteorite and the target rocks have the same composition (e.g. basic rock), then $V_a = V_b$, which we take to be about 8 kms^{-1}. If the impact velocity is high, a compressive wave of considerable magnitude is generated that travels away from the impact point. In the target rock, the stress intensity decreases with distance from GZ until, at a distance many times greater than the diameter of the meteorite, the stress values in the target rock become of negligible importance. Meanwhile, a comparable wave generated in the meteorite progresses upward and outwards from GZ. However, whereas the target rock represents a large area (as viewed in section), that of the meteorite is relatively modest. Thus, if we assume that the diameter of the meteorite is 8 km, the compressive stress will reach the rear surface in 1 second, during which time the stress-wave will experience only modest attenuation.

At the rear free-surface of the meteorite, the stress-wave is reflected back into the sphere but with a change of sign, i.e. as a tensile rather than a compressive stress-wave. In compression, rocks are strong (especially in an ambient confining stress field as in the target rock); in tension, however, the same rock will be relatively weak. Consequently, as the tensile stress-wave rebounds into the impacting sphere, it exceeds the tensile strength of the rock

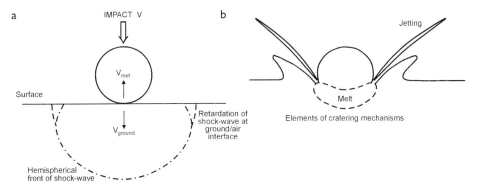

Figure 5.29 (a) Early phase of stress development as a spherical meteorite, with a trajectory which is perpendicular to the Earth's surface, strikes a flat area of target rock, indicating the initiation of cratering. (b) Assuming the impact has sufficient energy, the generation of melting of the target rock and part of the impacting body takes place, thereby giving rise to jetting of the molten material.

in the meteorite and causes continual spalling of the rear surface of the meteorite, until the tensile wave reaches the frontal limits of the sphere. At this stage, the whole of the meteorite is 'consumed' and ceases to exist.

As we have noted, the frontal part of the meteorite is being flattened, melted and possibly ejected, so that the return journey of the tensile pulse is shorter than that of the original dimensions of the meteorite. Hence, during the (say) 1.5 seconds that it takes the stress-wave to complete the outward and return journey through the meteorite, this body will have buried itself within the target rock. If the velocity at impact was high (say 40 km s^{-1}), the frontal part of the meteorite may have buried itself several tens of kilometres into the target rock.

In this 1.5 second period, the target rock has been subjected to a propagating compressive stress-wave. However, once the reflected tensile stress-wave in the meteorite hits the front of that body, the generation of the compressive pulse diminishes or even ceases.

The tensile wave in the meteorite will have done work in breaking up the impacting body, so that the magnitude of the tensile wave will be attenuated. It then propagates into the target rock as a tensile stress-wave, though it will certainly have a lower initial potential than that of the original compressive wave.

The decrease in the magnitude of the compressive stress-wave and the effects it has on the target rocks are illustrated in Figure 5.30. The compressive stress pulse will travel rapidly through the target rock along all radii and, provided the initial impact stress is sufficiently high, will result in the target rock becoming gaseous, molten, plastic or comminuted by brittle fracturing at progressively greater distances from GZ. The weakened target-rock will be moved by the compressive stress pulse along an outward and upward path. The ejection of this material erodes the limits of the crater, so that this, in part, contributes to the widening of the crater.

Near the surface, rarefaction reduces the maximum intensity of the pulse stress in a zone of interference, in which those compressive stress-waves that hit the rock/air interface are reflected as tensile stresses. This rarefaction causes spalling of the surface to take place. This enhances the ejection of target rock around the periphery of the crater, thereby contributing to the development of the parabolic section of the crater, together with the concomitant development of an overturned flap, possibly coupled with jetting. This mechanism continues until the compressive stress pulse decays to a small magnitude.

At this point, the movement pattern of the target rock changes. Prior to impact, the target rock was in a state of horizontal compression, which has been induced by gravitational loading and any tectonic influences which may exist. During the period in which the compressive pulse acts, the *in situ* compressive stresses are swamped by the compressive pulse stress which builds up large, compressive, elastic strains in the zone around the crater. However, as soon as the transient compressive pulse comes to an end, the combined *in situ* stresses and the release of that proportion of the pulse stress that engendered the elastic strains in the rock around the crater, possibly coupled with the influence of the newly arrived tensile pulse, causes a *rebound*, which is particularly effective in the upper regions of the crater wall.

This rebound will be extremely rapid, so that the inward-moving rock mass behind the free-surface of the crater will engender tensile stresses that will give rise to spalling and faulting of the 'aureole' of the transient crater. The falling away of the crater walls, as the result of the spalling, results in a flow of blocks and fragments of the country-rock downward and inward towards the centre of the crater. Thus, the transient crater rapidly widens and becomes shallower, to form a *simple crater* (Figure 5.6). The transition from the transient

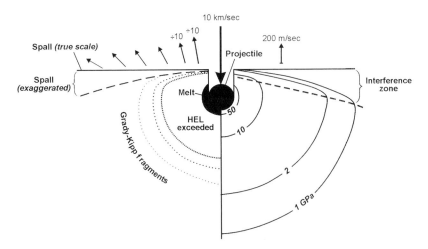

Figure 5.30 Schematic representation (in the left half of the diagram) of the decrease in maximum magnitude of the pressure induced by an impact. The 2 GPa (20 kb) contour has been interpolated to indicate the stress level, below which shatter-cones do not develop. The right half of the diagram represents the physical state of the target rock (after Melosh).

crater to that of a simple crater results in an increase in the crater diameter of about 19 per cent, with a commensurately shallower depth.

The form of deformation of the target rock depends upon the mass and velocity of the impacting body. Melosh presents a diagrammatic representation of the stresses and the types of deformation they engender if an impacting body (size and composition not specified) strikes the target rock at 10 km s^{-1} (Figure 5.30). In the left side of this figure, it will be seen that the magnitude of the pulse stress adjacent to the impacting body is over 50 GPa (500 kb), which reduces to 1 GPa (10 kb) approximately 4 diameters of the impacting body away from the point of impact (GZ). The minimum differential stress required to enable shatter cones to form is not less than 2 GPa (20 kb). This limiting stress has been included in this figure.

It will be seen in the right side of the diagram that, adjacent to the impacting body, the target rock becomes molten and can readily be ejected as a *jet* (Figure 5.29a). The zone beyond the melt exceeds the HEL (Hugoniot elastic limit), i.e. it behaves as a plastic material. Rapid deformation of such a plastic material results in the generation of heat. The outer zone marks the development of the Grady-Kipp fragments. This is a zone in which deformation is dominated by fracturing (the majority of which will be shear-fractures). Rapid shear movement on such fractures may cause melting, if the rock is dry (e.g. granite) and may cause significant heating in other situations. Hence, from the contact between the impacting body and the target rock, out as far as the 1 GPa stress contour, the target rock will increase its temperature very quickly. Throughout much of the volume of the pulse stress shown in Figure 5.30, this could give rise to a significant increase in average temperature.

Figure 5.30 relates only to one impact velocity; namely 10 km s^{-1}. A more general treatment has been given by Ahrens and O'Keefe (1977), who show (Figure 5.31) the decline in peak shock pressure (compressive stress pulse) for three specified impact velocities of 45, 15 and 5 km s^{-1}. The scale of stress magnitude is shown on the left, with the type of deformation corresponding to a given stress-range on the right. The basal margin gives the ratio of the depth below the original surface relative to the projectile radius.

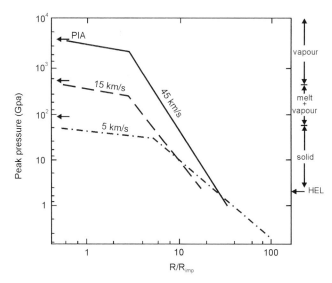

Figure 5.31 Attenuation curves for an impacting body striking at a velocity of 45, 15 and 5 km s⁻¹. The horizontal scale represents the depth below the surface of the pulse stress magnitude at distances from ground zero relative to the radius of the impacting body. The left-hand scale gives the stress intensity and the right shows the mode of deformation of the target rock (after Ahrens and O'Keefe).

It will be seen that there is a near-field of low attenuation rate which extends a little way beyond the limits of the impacting body. The far-field attenuation is significantly faster. The relationships shown in this figure are important and we shall return to them in the following chapter. However, it should be noted that they are based on the assumption that the target rocks not only lack relief but, even for events which give rise to large craters, the Earth was 'flat' (this is the planar impact assumption, or PIA). Also it assumes that the projectile is spherical.

5.5.2 Computer models

Even if the impacting body is initially completely spherical, it does not remain so, following impact. Ahrens and O'Keefe (1975) indicate four stages in the deformation of the impacting body and the target rock (Figure 5.32a). (It should be noted that the scale in the four 'snapshots' becomes progressively larger.) We have noted that the target rock adjacent to the impacting body may become molten. Melting may be even more important in the impacting body, for the meteorite has passed through the Earth's atmosphere, so that the frontal area may be close to, or actually, melting before it makes contact with the target rock. On impact, it is therefore to be expected that the meteorite will experience progressively more significant melting. Much of the melt formed, both in the meteorite and in the target rock, will meld, and is likely to be ejected in *jets* at the junction between the two bodies.

A more recent computer simulation by Boslough *et al.* (1995) (Figure 5.32b) shows that after a few seconds following impact, the shape of the impactor has little effect upon the magnitude and profile of the stress-wave front. These same authors (Figure 5.32c) indicate how seismic waves generated by the impact may be focused. They further estimate that such an impact would give rise to peak displacements at the antipode of the impact. They find that the seismic energy is most strongly focused in the upper mantle, within the asthenosphere, at a depth of about 200 km, where it is most readily converted to heat, thereby causing a degree of melting. Seyfert and Sirkin (1979), earlier discussed this antipodal affect in qualitative terms, while Shaw (1994) has also suggested that a major impact could give rise to an antipodal hotspot.

It should be noted that to this point, we have tacitly been concerned only with a meteorite that has a trajectory which is perpendicular to the 'flat' impact rock. This must be a very rare happening. In general, a meteorite or comet will strike the Earth's surface obliquely. Such oblique impacts are likely to generate a preponderance of the melt 'jetting' down-range (i.e. in the direction of travel of the meteorite), with perhaps a smaller quantity of melt being directed up-range (Figure 5.33).

It should also be noted that cratering is determined by the shock-wave generated in the target rock, so that the circular plan of the crater is usually little influenced by the track of the meteorite. Only if the track of a meteorite is at a very low angle to the surface of the target rock does the plan of the crater become markedly elliptical and, at very low angles of impact, give rise to the elongated 'trenches' of Rio Cuarto Craters in Argentina (Hodge, 1994).

5.5.3 Fractures and folds external to the crater

Let us now consider events of such energy that they can give rise to a central uplift crater with its attendant external fractures and folds. On the time-scale of crater development, the evolution of a simple crater into one which is more complex requires a somewhat longer period of development.

Jones (1995) reports on Snowball that 'the crater was approached within less than 5 minutes of the detonation'. A slow trickle of water started flowing into the crater within a few minutes of the explosion, but the surface of the crater remained dry and firm enough for surveyors to traverse the crater with ease. However, after some quarter of an hour there was an almost explosive venting close to GZ. A vent somewhat more than a foot in diameter was opened, and from that vent gushed sand-laden water which rose to a height of about 3–4 feet. The flow slowly abated and finally stopped after about a week.

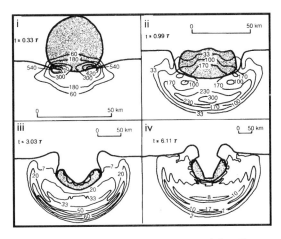

Figure 5.32a Four stages in an early computer simulated impact (after Ahrens and O'Keefe, 1975).

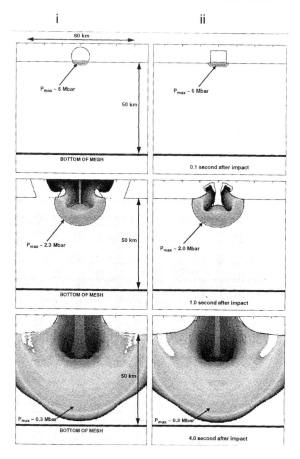

Figure 5.32b More recent (1995) computer simulation showing a series of pulse sections generated by a 10 km diameter (i) spherical and (ii) near-cubic asteroid with an impact velocity of 20 km s^{-1}. These sections show that after a few seconds the shape of the impacting body has little effect upon the magnitude and shape of the stress pulse (after Boslough *et al.*). Note the well-defined axial jet that develops early in the life of the spherical impact and perhaps a couple of seconds later in the rectangular body. This computer-simulated 'jet' was later seen to be an important feature of the larger impacts of the Shoemaker-Levy 9 event on Jupiter.

However, the fractures and folds beyond the crater rim were probably fully developed within seconds, or at least minutes, of the explosion, though the development of the sand volcanoes and other outflows of muds and sands may have gone on for many tens of minutes.

We shall first deal with the mode of development of the concentric and radiating fractures and the minor concentric folds. As noted earlier in this chapter, one is able to indicate the orientation of the least and greatest principal stresses which gave rise to the different structures that developed beyond the crater rim.

From the peripheral folds that can be seen in the tarmac path (Figure 5.17), it can be inferred that the maximum principal stress (S_1) acted normal to the fold axes. Fold development is controlled by the anisotropy of the layer, so that either S_2 or S_3 may be parallel to the axis of a fold. However, from the radiating, vertical fractures, it can be inferred that the circumferential stress near the surface is the least principal stress (S_3). This is substan-

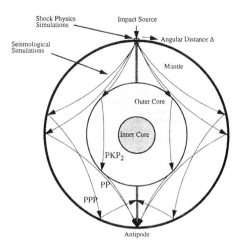

Figure 5.32c A section through the Earth shows how the shock simulation is limited to the region immediately around the impact point, but that the seismic waves are generated throughout the entire interior. Various families of compressional seismic rays are drawn to show how they focus at the surface or, in particular on the impact antipode (after Boslough *et al.*).

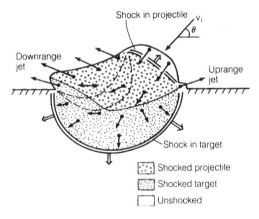

Figure 5.33 Asymmetrical generation of jetting as the result of an oblique impact (after Melosh).

tiated by vertical, strike-slip fractures which developed in a tarmac path which was deformed in a 20 ton explosion (Jones, 1995). Moreover, from the sands, clays and water that were brought to the surface along the radiating vertical fractures, it can be inferred that these extended well below the level of the original water-table. It is also reasonable to infer that the explosion caused permanent compressive volumetric strains in the sediments. These little-indurated, post-glacial sediments would have a relatively high porosity with low cohesion between grains. Hence the pore-spaces were easily deformed, which automatically resulted in the development of high fluid pressures in the rock-mass. Consequently, we suggest, the vertical, radial fractures resulted not only from the hoop-stresses, but were also aided in their development by hydraulic-fracturing, for which the fluid pressure (p) must be greater than the least principal stress (S_3). This argument can also be applied to the zone in which the thrusts occur.

It can be inferred, therefore, that both the radiating vertical fractures and the concentric thrusts formed during the compressive phase. However, the reasons for the change-over of the horizontal circumferential stress from S_2 to S_3 are not clear, and may reflect changes in the magnitude of the compressive pulse.

The thrusts were only revealed when, subsequently, deep cuts were made through the crater and outlying area. These structures developed when the maximum principal stress (S_1) acted radially from GZ, and the intermediate principal stress (S_2) was horizontal and concentric about GZ (Figure 5.34a). One can therefore conclude, from the orientation of these structures and the inferred orientation of the axes of principal stress, that these thrusts developed during the *compressive stress-pulse phase*. From the sections presented by Jones (1995), it is clear that these thrusts extended to a depth of at least 40 feet (12.3 m) below the surface.

When the peripheral graben developed, however, the least principal stress (S_3) acted radially from GZ, so one can infer that the bounding fractures which define the graben formed during the *tensile stress-pulse*, or possibly as a purely *rebound affect* (Figure 5.34b).

5.5.4 Gradation from simple to central, or ringed, uplift craters

It will be recalled that the diameter of the Snowball Crater was approximately 87 m. Consequently, from the relationship indicated in Figure 5.6 (Melosh, 1989, p. 129), the transient crater has a diameter some 19 per cent less than the simple crater. Hence, if we take the limit of the Snowball Crater to be 1.19 times larger than that of the transient crater, from which it developed, we can infer that the transient crater had a diameter of $87 \times 0.81 = 70$ m. Melosh also argues that the ratio of the depth of the transient crater relative to its diameter 1 : 2.7, so the depth of the Snowball transient crater was likely to be about 26 m. As the average depth of the Snowball crater floor is an estimated 4.25 m, the maximum depth of infilling of the central part of the Snowball transient crater is over 21 m, and this does not take into account the height of the central uplift. Jones (1995) does not actually quote the depth of the crater, because he realised that the Snowball explosion had given rise to what he described as a 'depressed rim, central uplift crater'. This description

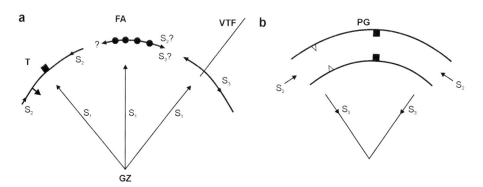

Figure 5.34 (a) Radiating compressive pulse giving rise to concentric folds and deep thrusts. Hoop-stresses coupled with high fluid pressure that results in the development of vertical radiating fractures. (b) Rebound phase, possibly coupled with a tensile pulse that results in the development of the concentric graben.

derived from the fact that, although initially flat, subsequent to the explosion, the whole of the surveyed site dipped gently inwards towards the outer limit of the crater rim. From the data presented in an E–W section (Jones, 1995), it follows that there was a gradually increasing dip from a point at the limit of the surveyed line of 2168 feet at 225 feet from GZ to 2160 feet, near the limit of the crater at 105 feet from GZ. The average slope was therefore 3.8°. Prairie Flat and Dial Pack (both 500 ton TNT experiments) exhibit a comparable inward dip of what had been, prior to the explosion, flat ground.

The change in dip of the original surface can be attributed to one of two mechanisms, or a combination of both, possibly acting in sequence. The first of these would be attributed to the intensity of the explosion, which would have caused excavation of the transient crater and would have generated high fluid pressures in the sediments surrounding the crater, by deforming the pore-spaces.

The second effect comes into play as soon as the pressure pulse has passed. The high pore-fluid pressure would reduce any cohesion with each other, which the grains may have originally possessed, to zero, thereby giving rise to an upheaval of the floor at the centre of the crater. Slumping of the walls of the crater would have occurred simultaneously, as well as a slower migration of the weaker elements of the sediments more distant from the crater walls (Figure 5.35a). As the walls slumped into the crater, they would tend to impede the upward progress of the diapir rising from the deepest part of the crater.

As regards Snowball, it is apparent that inward sliding of the wall rock completely covered the rising diapir, for a while. However, because of its liquidity, the central diapir continued to rise and soon became a prominent feature of a flooded crater (Figure 5.9). Elsewhere around the crater, the liquefied sediments from below the original water-table came to the surface to form sand and silt volcanoes and mud-flats.

Let us now consider this as a model for a quite major natural impact feature (Figure 5.35b). In this diagram, the metres have been changed to kilometres, the bars to kilobars and probable temperatures have been added. The liquefication mechanism caused by the Snowball explosion is substituted here by the flow of hot, relatively weak rock aided by the upward adjustment of the asthenosphere, which will assist in the production of the diapiric action. Concomitant slumping from the walls of the crater took place, giving rise to gravity-glide and tumbling of boulders and rubble towards the central part of the transient crater. Natural erosion and sedimentation in the crater will eventually modify and flatten the crater floor. From this section, because the mantle becomes involved with the diapiric intrusion, one may reasonably expect that such major natural impacts may well generate gravitational anomalies.

At least some multi-ringed craters are thought to be the result of the development of concentric normal faults that form within the crater, and dip inwardly towards the impact point. In other craters, however, the rings are considered to be the result of buckling of the inward-flowing debris, to cause variations in surface elevation of the crater floor. The interested reader is directed to the appropriate chapter in Melosh (1989).

5.5.5 Size of craters and the energy of the impact event

Energy is the primary measure of an impact's magnitude. In early literature, the unit of energy used is the *erg*. More recently energy is cited in *Joules* or in Megatons TNT equivalent. The relationship between the size of a specific crater and that of the impacting body that gave rise to the structure is imponderable. A crater's diameter is directly related to the

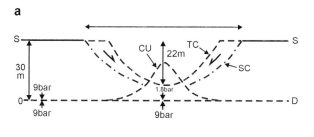

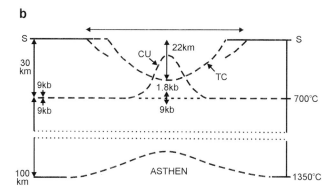

Figure 5.35 (a) Crater, diameter 87 m, with central uplift developed by 500 ton TNT explosion, indicating rebound and slumping features. (b) Comparable natural crater 87 km diameter with rebound and slumping features and upwelling of the asthenosphere. (The magnitude of stresses indicated in these diagrams are only approximate as are the temperatures indicated in (b).) CU = central uplift; TC = transient crater; SC = slumped crater; S = surface.

energy of the impact event. However, it is also influenced by the angle at which the impacting body hits the Earth's surface.

The energy (E) generated during the period of the disintegration of the meteorite or comet, which gives rise to the crater of specific dimensions, is given by

$$E = M.V^2/2 \tag{5.2}$$

where M is the mass of the impacting body and V the impact velocity.

The mass of the body relates to its volume, density and structure. The composition of bodies which enter the Earth's atmosphere range from iron, through chondrites to stony meteorites and comets.

Energy of impact, crater diameter and volume of melt

The energy of impact that gives rise to a natural impact crater of a given size can be inferred from the known energies of TNT or nuclear explosions that result in craters of specific diameters. Such a relationship between energy of explosion or impact, for two specific impact velocities, is given in Figure 5.36. Dence *et al.* (1977) showed that the relationship between the impact energy E_n and the diameter D of the crater is given by

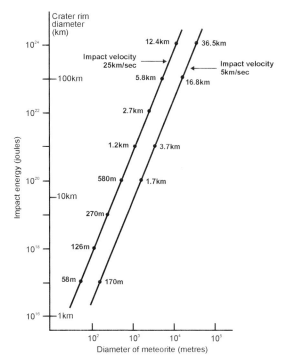

Figure 5.36 Scaling relationships between impact energy, and diameter of spherical meteorite for two specific impact velocities (after Dence *et al.*).

$$D = E_n^{1/3.4} \tag{5.3}$$

It is clear from Equations 5.2 and 5.3 that one cannot arrive at a unique solution regarding the mass and the velocity of the impacting body for any given crater.

Also, the explosive experiments are only comparable with vertical impacts upon a 'flat Earth'. As it is known that the angle of impact gives rise to variations in the energy required to form a crater of a specific dimension, the relationship between the diameter of the crater D and the energy E_n, as given by Equation 5.3 and Figure 5.36, should be considered as approximations.

Theoretical analyses show that, when melt conditions are attained by an impact event, the ratio between the mass of melt or vapour to the impacting body is proportional to the square of the velocity of the impact. However, Melosh (1989) notes that the effects of oblique impacts are relatively poorly understood, but the *jetting* caused by such an oblique impact may actually enhance the mass of melt or vapour produced.

The theoretical equations used to establish these relationships have not yet been verified experimentally, because current apparatus is not yet able to accelerate the impacting body to a velocity which is sufficiently high to cause significant melting. The predicted theoretical relationships between the ratio of the mass melted to the mass displaced from the crater are given by the diameter of transient craters formed (on the Moon and on Earth) for the impact velocities of 80, 20 and 15 km s⁻¹, (as shown in Figure 5.37). The points in this figure relate to four terrestrial craters, for which the minimum volumes of melt have been established by Grieve *et al.* (1977) and the Popigay crater in the former USSR (Ivanov,

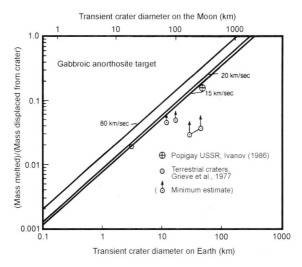

Figure 5.37 Relationship between melt and the speed of impact (after Melosh).

1986). It will be seen that one of 'Grieve's impact craters' (with a transient crater diameter of approximately 3 km) falls on the 15 km s^{-1} line. One can infer that the impact velocity for the Popigay crater and two of the 'Grieve craters', with transient crater diameters in the 10–20 km range, indicate impact velocities in the 5–10 km s^{-1} range, and the remaining data, for transient crater diameters in the 40–50 km range, exhibit a velocity well below 5 km s^{-1}.

As pointed out by Melosh, these data, although very limited, appear to be consistent with the predicted theoretical curves which have been plotted in Figure 5.37. The vertical arrows attached to several of these points indicate that the plotted circles are based on the minimum estimate of the ratio of melt to mass displaced from the transient crater. Hence, the tip of the arrows indicate that the probable velocities are even closer to the predicted relationships. Even with this caveat in mind, it is interesting to note that two points would indicate impact velocities of no more than 15 km s^{-1}, two of the points indicate an impact velocity somewhat smaller than 10 km s^{-1} and the remaining data still indicate impact velocities of less than 5 km s^{-1}.

5.5.6 Asteroids or comets?

Modern astronomical networks can detect kilometre-sized bodies in Earth-crossing orbit. Shoemaker made early studies of the relatively small asteroids which originally formed part of the Asteroid belt. A relatively small number of asteroids in this belt have, over time, been dislodged from the Asteroid belt, to become 'Earth-crossers'. Shoemaker spent years studying these bodies and initially estimated that there were perhaps 1200 such Earth-crossers. More recently, the number has been upgraded to about 2000. As of August 1994, 296 were known to exist. However, of these, only 172 have a diameter in excess of 1 km (Shoemaker *et al.*, 1990; Rabinowitz *et al.*, 1994). The impact energy of a 1 km diameter stony meteorite is likely to be about 10^{4-5} Mt. The vast majority of such bodies are less than 1 km in diameter. Consequently, should these bodies eventually impact on Earth they are not likely to give rise to craters greater than about 15 km in diameter.

We have seen that the majority of theoretical analyses relating to the effects of bodies impacting on Earth are based on the assumptions that they are rigid spheres (though shape is now known to be relatively unimportant), and that they follow a vertical trajectory and strike the target rock at a flat horizontal surface. In reality, none of these assumptions is likely to be met, although some asteroids may approach the 'ideal' as regards rigidity and shape. However, because of the inherent difficulties of mathematical modelling of even the most idealised of impact events, most sympathetic readers are ready to accept the inherent errors of such analyses. This acceptance of the inherent possible errors in modelling was understandable up to the early 1990s for, prior to this date, it was generally held that most impact events on Earth could be attributed to asteroids.

Although comets, which are spectacular visitors from the outer regions of the Solar System have traditionally been considered to be harbingers of death and distruction, their importance relative to meteorites has, until recently, been underplayed. However, we should note that the total mass of cometary bodies is certainly in excess of 50 times the mass of the Earth. Also there are comets and there are super-comets. The latter may have a diameter in excess of 1000 km (Taylor, 1992), but as far as is known, these fortunately do not intercept the Earth's orbit.

Until recently, the importance of comets as impacting bodies on Earth haves been little emphasised. However, as Shoemaker (1997) pointed out, the available evidence suggests that, in relatively recent geological time, comets have produced more than half the impact craters on Earth greater than 20 km in diameter, and probably all craters greater than 100 km in diameter. Also, Bailey and Emil'yanenko (1997) state that the predicted number of long-period comets from the inner Oort cloud is of the order of 4000. This is at least 100 times more than the observed number of de-iced Halley-type comets, so there should be many undiscovered inert Halley-type 'asteroids' or compact streams of cometary debris, still to be found in the inner Solar System. However, as noted above, since about 1995, it has been realised that the number of 'returning' comets, which make multiple passes around the Sun, may be in excess of 4000. Such comets may have diameters measuring up to 40 km. Accordingly, it is now held that all craters of 100 km or more in diameter are almost certainly the result of impacts by comets, rather than asteroids.

A modified and extended version of a diagram given in Toon *et al.* (1997) and shown in Figure 5.38a, represents the relationship between energy and crater diameter, based on an assumed impact speed of 15 km s^{-1} for a stony meteorite and a typical long-period comet (LPC) with impact speed of 50 km s^{-1}. The horizontal lines indicate the probable upper and lower limit of the Chicxulub event. Toon *et al.* present an equation on which this figure is based. It includes a ratio of the nominal to actual impact to the power of 0.087, so that an error of as much as 50 per cent is possible. It also assumes that the probable impact angle is 45°, measured from the vertical. Any given impact may easily differ by as much as 25°. (One may therefore infer that the estimated crater dimensions shown in Figure 5.38a are best regarded as order of magnitude calculations.)

A slightly modified version of a diagram by Toon *et al.*, which illustrates the cumulative frequency of impact, is shown in Figure 5.38b. This figure is based on the work by Shoemaker (1983) and Shoemaker *et al.* (1990) and data reported by Spacewatch. These curves indicate that at energy levels greater than about 10^6 Mt, the influence of comets is greater than that of asteroids, while above 10^8 Mt, comet impacts completely dominate the scene. As we have noted, Bailey and Emil'yanenko (1997) support this conclusion.

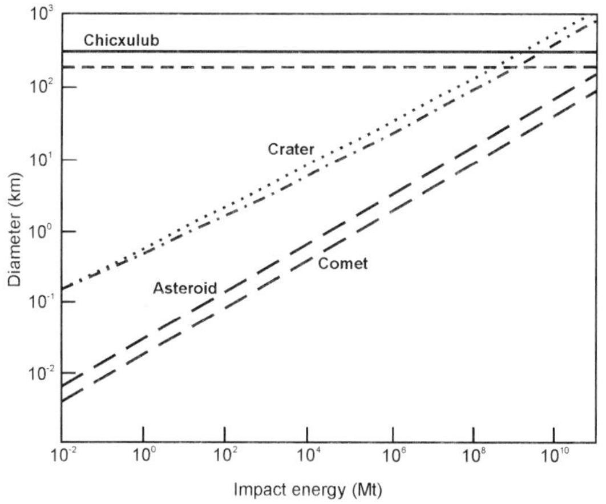

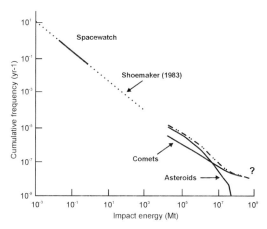

Figure 5.38a Estimate of the relationship between impact energy and crater diameter for asteroids and comets which are assumed to have an impact velocity of 15 km s^{-1} and density of 2.5 g cm^{-3} and 50 km s^{-1} and 1 g cm^{-3} respectively (after Toon *et al.*).

Figure 5.38b Estimate of the frequency with which objects of a given energy impact on Earth. The comet data for energy values above about 10^8 Mt has been extended as a dashed line. This would result in a concentration of large impacts at a cumulative frequency around about 10^{-8} yr^{-1} (after Toon *et al.*).

It is, therefore, necessary to attempt an assessment of the mechanisms associated with impacts of comet bodies. First and foremost we need to know the general structure and composition of comets.

We immediately encounter a major problem in that so little is known regarding the structure of these bodies!

The composition and structure of comets is very largely based on inference of their origin (mainly in the Oort cloud) and the behaviour of the comet's tail as it processes around the Sun.

It is thought that, initially, comets are composed mainly of ice, which perhaps acts as a weak cement for a loose agglomerate of stony blocks, boulders and smaller particles of unknown size. Not only may the stone aggregates be bound together by interstitial ice, but they may also be covered with an ice carapace. Such ice and stone aggregates may be up to 40 km in diameter. With time, such comets which orbit the Sun may lose their surface ice, leaving only a dark carapace of a relatively loose agglomerate of stony material held together, possibly, by a small matrix of ice. Such an ice-stripped agglomerate (Figure 5.39a) may have a 'porosity' of as much as 30–40 per cent. With such a high porosity and the inner agglomerate held together by ice, the body may have little or no structural rigidity (i.e. blocks will readily move relative to one another), causing it to deform as it enters the Earth's atmosphere. Alternatively, the inner core of the larger asteroids may consist of rock ranging in size from large blocks to small particles bound into a relatively strong body by ice, with perhaps a weak carapace of blackened dust and rock fragments.

A large comet with such a composition, which has spent almost all its life at close to zero temperature, will be relatively strong. However, heating by the Sun will weaken the outer zones of the body and create a significant thickness of weakened carapace. However, how deeply this weakened material extends into a large (40 km diameter) comet is conjectural. Indeed, much of ideas regarding the structure of comets is based on conjecture. However, it seems probable that large comets may have a reasonably strong core and a thick carapace of weakly connected or unconnected blocks and rock dust held in position by a weak gravitational field.

In smaller comets the whole core beneath the carapace may consist of blocks of stone, permafrost and ice which are perhaps 250 m in diameter. Such blocks beneath the carapace (Figure 5.39a) may have a porosity of 30–40 per cent. Also, the body may have little or no structural rigidity (i.e. blocks will move relative to one another), enabling the comet to deform or disintegrate as it enters the Earth's atmosphere.

When a comet or asteroid enters the Earth's atmosphere, aerodynamic forces cause it to slow down. However, a significant amount of retardation occurs only if it encounters a mass of air that is comparable to its own mass. The mass of a column of the Earth's atmosphere is about 1 kg cm^{-2}. From these statements, it follows that an infinitely rigid, spherical object with a radius greater than about 100 m will not be stopped by the atmosphere. However, few bodies, especially comets, are sufficiently strong to withstand the stresses induced by atmosphere; for, as a body travels through the atmosphere, it generates a frontal shock wave which, if the body's velocity is about 20 km s^{-1}, will generate a differential pressure of several kilobars across the bolide. This would be sufficient to disrupt the component elements of the core of a comet. An asteroid will be inherently more rigid than a comet-core. However, practically all known asteroids show evidence of pock-marks caused by smaller impacting bodies. As we have noted earlier, impacts generate tensile and shear fractures in the target. Such fractures will represent planes of weakness, into which the highly compressed atmosphere will rapidly invade and give rise to rapid break-up of the near-surface areas and progressively deeper zones of the bolide. As the body experiences partial or total break-up, it spreads laterally, thereby increasing the effective cross-section relative to that of its original dimensions.

As regards comets (especially those which have been stripped of their ice carapace through successive passages around the Sun), the resistance of the atmosphere will cause the smaller size particles in the agglomerate to slow down more rapidly than the larger blocks. They will be winnowed out of the main cluster and many well be consumed as 'shooting stars'.

Air pressure will tend to separate the larger blocks laterally from each other, so that the loose spherical agglomeration shown in Figure 5.39a will probably deploy rapidly into a form similar to that indicated in Figure 5.39b. This body could be modelled as an oblate spheroid.

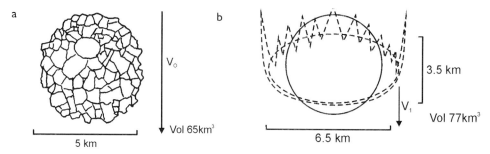

Figure 5.39 (a) Schematic representation of a 5 km diameter ice-free comet travelling at velocity V_0 which may be of the order of 30 km s^{-1}. The body is likely to have an initial porosity of about 30 per cent and a volume of about 65 km^3. (b) The possible change of sectional form as the comet enters the Earth's atmosphere and changes from a sphere to a flattened front with trailing elements containing winnowed and incandescent smaller particles. The body approximates to an oblate spheroid, perhaps 6.5 km in diameter and 3.5 km front to back, with a volume of about 77 km^3. The increase in volume can be attributed to spreading of the major rock units because of atmospheric resistance.   The 'continental' impact velocity (V_1) may be significantly less than 10 km s^{-1}.

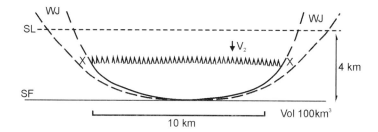

Figure 5.39c If the comet impacts in the ocean, the increase in resistance of the ocean is likely to cause a further significant flattening and broadening of the impacting ice-free comet, causing the body to have a width of perhaps 10 km and a front-to-back thickness of only 2 km, so that the volume of the oblate spheroid increases to about 100 km^3. (SL = sea level, SF = sea floor, WJ = water (or steam) jet.)

We noted earlier that we considered that Operation Snowball, with a hemispherical configuration, would result in a shock-wave comparable with that of an impacting comet. Our reasons for this statement are as follows. The generation of the explosive within the hemisphere would, within microseconds, radiate outwards to consume the TNT in a uniformly progressive manner. Hence, the explosion at the periphery of the circular base would occur slightly later than at GZ. Consider the section shown in Figure 5.39c. If we take this to be a reasonable envelope for a meteor about to strike a 'flat' oceanic floor, then it is obvious that the central portion of the section, GZ, will strike a little before the impacting of the outer portions of the meteor. If we take the horizontal zig-zag line to represent the effective limit of maximum impact energy, then the ends of this line are approximately 2 km above the first impact point. If we assume the impact velocity to be 5 km s^{-1}, the impact at X and X will occur at about 0.4 seconds after the initial contact at GZ. This, we suggest, makes Snowball a reasonable representation of a cometary impact.

The rate of slowing of a body in the Earth's atmosphere will mainly be determined by the mass and velocity of the body and its cross-sectional area. For a 5 km diameter stony meteorite travelling at an initial velocity of 15 km s^{-1}, the ratio of the momentum to air-resistance will be relatively large. However, if a block 100 m in diameter, which forms part of a de-iced comet, enters the atmosphere, the ratio of momentum to resistance is very much smaller, so that this and comparable components of the meteor will tend to reduce their velocity quite appreciably.

Indeed, one can specify minimum dimensions of ice, stony and iron bodies that are able to penetrate the Earth's atmosphere and reach the surface. Thus, an ice body must have an initial diameter in excess of about 150 m, while stony and iron meteorites require diameters respectively in excess of 60 m and 20 m.

A *solid*, stony meteorite, with a diameter of about 5 km, travelling at a velocity of about 15 km s^{-1}, will generally survive this passage through the Earth's atmosphere and experience relatively little diminution in its dimensions, so is likely to give rise to a crater with a diameter of about 40 km.

In contrast, the core of a de-iced comet with an initial diameter of 5 km and entering the Earth's atmosphere at 50 km s^{-1}, because it is comprised of relatively small elements, would be significantly retarded in the atmosphere prior to making impact on a continental area. If it were to impact into the ocean, it would briefly meet with an even higher degree of deceleration, so that it would make contact with the ocean floor at a relatively low velocity. Also, as indicated in Figure 5.39b, the oblate spheroid axes would increase and decrease respectively. Thus, for a continental impact, the maximum frontal dimension will be significantly greater than its initial dimension in space.

Were the impact to occur in the ocean, the oblate spheroid may further extend its length of major axis, possibly by 100 per cent, and decrease its minor axis by about 60 per cent (Figure 5.39c). It will be noted that the evolution from a sphere to the oblate spheroids of the given dimensions gives rise to a marked increase in volume. This merely means that the packing of the original, agglomeratic sphere becomes looser, as the body is deformed by the atmosphere and the ocean.

We have noted that impacting bodies are likely to follow a track which is oblique to the Earth's surface. An incoming ice-free comet with an angle of incidence with the Earth's surface as shown in Figure 5.40, would be 'refracted' as it hits the oceans's surface at an angle, so that within a fraction of a second the final impact effect could approximate to a vertical impact.

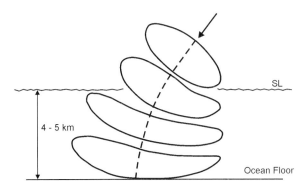

Figure 5.40 Indicating stages in the refraction of an equivalent oblate spheroid of an ice-free comet making oblique contact with Earth.

Shotgun model

It is interesting to compare the performance of a shotgun with that of a comet-body. Firstly, the muzzle velocity of a shotgun (full choke) is commonly about 1050 f s^{-1} (320 m s^{-1}). If the comet enters the Earth's atmosphere at a modest 15 km s^{-1}, the ratio of the initial velocities of comet/shotgun-pellet is therefore about 50 : 1.

The pattern of shot at full choke is such that, at a distance of 10 yards (9 m), 100 per cent of the pellets fall within a circle of 30 inch (76 cm) diameter, so that the angle of spread is 4.75°. This pattern of shot is established in about 0.1 second. If we take the bore of the shotgun to be 1.5 cm, the ratio of the target/bore diameter is 50 : 1. These data relate to discharge of gun at or near sea-level. At high altitude, because the atmospheric resistance would be smaller, the circle of spread would be correspondingly reduced.

As regards the passage through the atmosphere of a comet-body, its path length may be as much as 1000 km and the average velocity of transit may be reduced to 12.5 km s^{-1}, so that the transit time through the atmosphere may be 80 seconds. The density of the higher levels of the Earth's atmosphere is very low. Even so, the heating-up, during the passage of small particles through the outer layers of the atmosphere, is sufficient to vaporise these small bodies by the time they descend to within about 50 km of the Earth's surface. The larger component blocks that we infer to constitute a comet-body, which may exceed several hundred metres across, will be warmed and probably detached from one another by the 'hot' shock-wave in the atmosphere engendered by the passage of the component blocks of the comet body. This will give rise to the parting of the blocks one from another. This separating of the larger blocks will become progressively more marked as they enter the more dense, lower levels of the atmosphere.

The mechanisms involved in the scatter of shotgun pellets will involve the explosion that ejects the particles from the cartridge, the interaction of pellets with the walls of the shotgun barrel and with each other and, when they leave the barrel, the interaction with each other and especially with the atmosphere. The retarding and deflecting effect of the atmosphere will be related to the speed and the cross-sectional area of the pellet.

In Figure 5.39 we assume that a spherical comet body, measuring 5 km in diameter, is composed of a large number of 200–300 m component elements, to be loosely disaggregated at a height of 50 km above the Earth's surface, and that as it travels through the lower levels of the Earth's atmosphere, it changes its shape into an oblate spheroid with a horizontal diameter of 6.5 km. This is equivalent to an average angle spreading of 1.7°, which is approximately one third that of a full choke shotgun.

The shotgun scatter develops in uniform atmospheric pressure. A comet-body experiences passage through an ever-increasing density of atmosphere. Hence, the envelope of scatter will not be conical, but will have an increasingly larger angle as the Earth's surface is approached. However, for our purposes, the errors involved in assuming a constant angle of conical scatter are not considered. One may use the discharged shotgun cartridge as a very approximate model of a comet-body, but the modelling must be done with care. (The dimensions of the hypothetical comet represented in Figure 5.39 are based on the details of the 'shotgun model' given above.)

5.6 Summary, conclusions and inference

Bombardment of the Earth by large meteorites is not a thing of the distant past, to be relegated to the early Archean. There is ample evidence, to be seen on the Moon, that our satellite has an approximately linear accumulation curve of craters over at least the past 2000 Ma.

The Earth's atmosphere has guarded our planet against the host of minor events which are so evident on the Moon. The frictional resistance generated by the atmosphere, as small bodies enter at high speeds, consumes relatively small in-coming bodies so that they become *shooting stars*. The limiting dimensions which these incoming bodies must initially possess in order to survive and strike the Earth's surface depends upon their composition. For bodies which are composed of (a) ice, (b) stone and (c) iron, the diameters must exceed 150, 80 and 30 m respectively. If the Earth had no atmosphere, such small meteorites would have generated a host of craters with a diameter of perhaps 1–3 km.

It was only in the first half of the 20th century that a very small number of craters on Earth were considered to have resulted from meteoritic impacts. Even as late as 1960, less than ten such structures were recognised. However, between 1960 and the end of the 20th century, the number of features which have been designated as *certain* impact structures increased to more than 150. Many of these take the form of simple craters, which may exhibit some degree of erosion or have become partially, or even totally, infilled by sediments. Other recognised impact structures exhibit a more complex structural form, which include a crater with a central, or ringed uplift, together with a number of features external to the crater rim which include radiating fractures, concentric grabens and inward-dipping, concentric thrusts that formed below the crater rim and cropped out at the surface at about twice the radius of the crater.

Such complex 'impact structures' were seen in the explosive experiments conducted at Suffield, Alberta, Canada, between 1960 and 1970. These experimentally developed features enabled pioneers in the field to 'diagnose' the probable existence of ancient impacts, when erosion had uncovered the deeper structures of such an event.

The explosive experiments, coupled with the recognition that the stresses generated by a natural impact was so intense that melting and generation of high-pressure polymorphs and high-stress deformation features developed in some minerals, together with the development of features which became known as *shatter cones*, gave rise to a sharp increase in the number of impact structures which became designated as *certain*.

It is known that very ancient lineaments, which can be traced for thousands of kilometres, have remained periodically active throughout very long periods, influencing the emplacement of eruptive rocks and even sedimentary patterns. Therefore, there is no reason to suppose that major curved fractures would not also exhibit 'longevity'. We cited three studies where such circular or arcuate features exhibit the appearance of clusters of craters seen on the Moon. Moreover, studies of the number of *certain* impact structures in the shield areas of N America and Europe have resulted in the prognosis that in the last 2000 Ma at least 7000 impact events have occurred on Earth, causing the development of craters which exceed 10 km in diameter; and over 100 of these exhibit diameters greater than 100 km.

We have outlined some of the pertinent aspects of the mechanics of cratering and seen that the mode of development of major impact features is reasonably well understood, when considered from an experimental and theoretical viewpoint. However, we noted that there are intrinsic difficulties in applying these theoretical and experimental findings to specific natural events. Thus, the size of a crater is related to the energy of the impact. The magnitude of the energy that gives rise to a specific impact structure can be estimated from explosive models and the use of scaling laws. However, the energy of a natural impact is determined by the mass and the velocity of impact of the meteoritic body. A natural crater of a given size may result from a relatively small, high-velocity bolide, or a much larger, but relatively slow-moving impacting body. Also, in general, theoretical analyses simulate vertical impacts, while, in natural events the track of the impacting body will be oblique, but at an unknown angle to the Earth's surface. In addition, the impacting body is generally assumed to be spherical. Again, in natural impact events the shape of the meteorite, or asteroid, may differ quite markedly from a sphere (see Chapter 1).

Fortunately, Boslough *et al.* (1995) have shown that the initial shape of the impacting body is not particularly important as regards the stress wave, 2–3 seconds after impact. Nevertheless, theoretical evaluation of real impact events can only be regarded as approximations. Despite these caveats, if the impacting body is a single, solid and coherent unit, the inherent approximations may be acceptable. However, quite recently, it has been inferred that a large proportion of the larger impact structures on Earth have resulted from the impact of comets or from the core of comets which have become de-iced by multiple passages close to the Sun. The type of cratering to which such impacting bodies will give rise may well be different from that associated with a single-body impact. But our understanding of the variables which are introduced by impacts from a loose agglomeration of individual bodies of different size and shape leaves much to be desired.

Finally, as we shall see in Chapter 8, it has recently become evident that the flux of impacts in the latter half of the Phanerozoic (i.e. in the last 250–300 Ma) is probably at least double that which existed in the early Phanerozoic and late Precambrian. Hence, one can infer that the estimated number of impacts in the last 250 Ma, as inferred from analyses by Grieve (1981) and Oberbeck *et al.* (1993) is likely to be quite conservative.

In the next chapter, we shall begin to apply some of the concepts we have outlined in this chapter to known major continental impacts and the eight continental flood basalts as well as an ocean plateau basalt and show how impact events can be recognised even when the impact site is completely covered by basalts.

Chapter 6

Impacts and plate motion

The development of flood and plateau basalts, and continental rifting

6.1 Introduction

In this chapter, we look at a number of effects and structures which are not readily explained in terms of orthodox arguments or mechanisms; namely, abrupt changes in direction of motion of plates, the emplacement of continental flood basalts (CFBs), oceanic plateau basalts (OPBs) and the splitting of continents.

We have noted that Lithgow-Bertelloni and Richards (1998) state, 'Unfortunately, we cannot reproduce the toroidal/poloidal partitioning ratios observed from the Cenozoic, *nor do our models explain apparently sudden plate motion changes that define stage boundaries*. The most conspicuous failure is our inability to reproduce the westward jerk of the Pacific plate … implied by the Hawaiian-Emperor seamount chain. Our model permits an interesting test of the hypothesis that the collision of India with Asia may have caused the Hawaiian-Emperor bend. However, we find that this collision has no effect on the motion of the Pacific plate, implying that important boundary effects are missing in our models.'

The importance of this statement cannot be ignored. These authors are stating that the conventional methods of analysing plate motions leave considerable gaps in our understanding.

To study the motion of plates, through time, it is necessary to use a suitable computer program. There are a number of such programs available which permit the position of continental bodies to be plotted back through the Phanerozoic and even earlier periods. We used the Atlas system (Version 3.3), described as 'a mapping and global reconstruction system for the personal computer', which has been developed by Alan Smith and co-workers at Cambridge University, England, over the last two decades, or so. We shall present maps and tracks of the movement of continental bodies or individual islands during various periods in the Phanerozoic in this and in the remaining chapters of the book. We also comment on the strengths and some of the idiosyncrasies of this program, so that the reader will be able to assess the limitations of this and similar computer programs.

In the light of the experimental, explosive studies, reported by Jones, cited in the previous chapter, I had long realised that some of the *sudden* changes in plate motion could be the result of major impacts. However, Seyfert and Sirkin (1979) were probably the first to publish the important concept that abrupt changes in track of continents could be attributed to major impact events. Also, Negi *et al.* (1993) claim that the massive eruptive body which gave rise to the Deccan Traps was the result of a major impact in the vicinity of Bombay.

There are only a few major 'certain' Phanerozoic impacts, with crater diameters of 100 km or more. Are these large craters associated with track changes?. We use the Atlas system to show that each of these known impacts is indeed associated with sudden changes in direction and speed of movement. Moreover, in each case, the date of change of track coincides closely with the date of the associated basaltic eruption as inferred from radiometric dating of these events. In addition we show that two 'near-certain large impact events' also show a sudden marked change in track.

We then follow the suggestion by Seyfert and Sirkin (1979) and also by Negi *et al.* (1993) and show that all eight of the continental flood basalts exhibit changes in direction and/or speed of motion at times which are consistent with the dates of emplacement of the various flood basalts.

Of the 13 events studied, all are shown to be associated with a significant change in speed and/or change in track which are closely linked in time to the radiometric dating of the various features. Thus, we present *prima-facie evidence* that, like the five major impacts cited, all major continental flood basalts are the result of major impacts.

Using the impact mechanisms that give rise to major transient craters, discussed in Chapter 5, we show how such a major impact on a plate so changes the stress system acting on the plate that changes in direction and/or the rate of movement are to be expected.

We then argue that the amount of molten basaltic material that can be produced at the surface depends not only upon the size and energy of the impact but also upon the thickness of the lithosphere. Oceanic impacts result in a different mode of development of emplacement of basaltic material relative to continental impacts.

We show how a large volume of melt may be generated, which is comparable with the estimated volume of the huge Ontong-Java igneous body. In comparison, the volume of erupted rock of the Deccan Traps is only about 2 per cent of that of the Ontong-Java structure. It is shown that the disparity in volume of erupted rock, for a comparable size of impact event, can be related to its impact environment, and argue that the rate of emplacement and extrusion of basic material from the asthenosphere below a continental impact is much slower than in some oceanic areas.

We note in passing that the Ontong-Java plateau basalts are also associated with track changes which are closely linked to the dates of the erupted lavas, so that the correlation of impacts to events is 14 out of 14!

It was mentioned in Chapter 4 that several major continental flood basalts (CFBs) are associated with the splitting of continents. The emplacement of the massive Paraña igneous province occurred a little after a spectacular change in track and increase in speed of S America, which occurred about 135 Ma ago. We indicate how the major impact structure which caused the extensive Paraña flood basalts could have given rise to the opening of the S Atlantic. However, because we are going to present a series of 'tracks', it is appropriate that we first describe the use of the Atlas system.

6.2 The Atlas system

In the user's guide, the *Atlas* system is described as a comprehensive map-making package comprising two FORTRAN 77 programs that permit the plotting of coastline and bathymetric data. It contains Euler rotations and a variety of utilities for creating paleographic reconstructions.

The main aim of this program is to represent the relative positions of the continents and major islands during the last 600 Ma. The coastlines of continents and islands are ephemeral features, and earlier coastal limits can only rarely be inferred from the stratigraphical record.

However, as Smith *et al.* (1981) point out, most of us depend upon recognising the coastlines drawn on maps. Consequently, in this program, the current coastlines are displayed for convenience of recognition and are assumed to remain unchanged back through the ages. Hence, the Atlas program draws maps which are the estimates of the past positions of the present-day shaped coastline.

This requirement of 'recognition' leads to some interesting features. For example, Hawaii is an important current feature of the Pacific map. However, Hawaii only came into existence relatively recently. Yet, if one traces the evolution of the Pacific back to 600 Ma, the island is still shown on the map. The position of 'Hawaii' 600 Ma ago merely marks the position which the current Hawaii would have occupied, had it existed 600 Ma ago. Hence, one can use the position of Hawaii (and 'Hawaii') back through time to shown how the plates migrated, which makes the 'immortal Hawaii' a positive advantage.

There are other examples in which continental landmasses, which are currently quite separated, are shown, say 135 Ma ago, as being superimposed upon one another. This, of course, is an impossible situation which requires resolution. One such example is the superposition of the Antarctic Peninsula upon Patagonia. As we shall see, this particular problem can be resolved by 'straightening' the Patagonian Orocline prior to 65 Ma ago.

As regards the reliability of the maps, there are several sources of error. These include errors which are related to the uncertainties in the history of ocean-floor spreading. The problem regarding the position of the various continental and island elements becomes progressively more important as one regresses in time. In the first 100 Ma, the palaeomagnetic striping in the ocean floor provides reasonably accurate results. Further back in time, the ocean-floor spreading anomalies are less well developed or less well defined. As regards the continental fragments, when they collide the boundaries are changed by tectonic processes, thereby introducing further uncertainties.

Another source of uncertainty lies in the estimate of the mean paleomagnetic poles when 're-assembling' the continents. The accuracy of the reassembly depends upon the number and accuracy of the estimates of individual magnetic poles. Other sources of errors relate to the interrelationship between the fossil, magnetic and isotopic time-scales. For example, Atlas 3.3 is tied to the time-scale set out by Harland *et al.* (1982), who put the base of the Cambrian at 590 Ma, while more recent time-scales, e.g. by Gradstein and Ogg (1996) put the base of the Cambrian at 545 Ma (see Table 6.1).

It will be seen that, to the base of the Paleocene, the disparity of ages in the two time-scales are negligible. For the dates of the base of the Cretaceous to the base of the Triassic, both sets of dates show appropriate error bars, which overlap. From the Permian to the base of the Cambrian, Gradstein and Ogg do not cite error-bars. However, it will be seen that the positive error bar for the Harland estimates covers the cited Gradstein–Ogg figure of 290 Ma. As regards the Harland 360 Ma date, the negative error bar extends to within 1 Ma of the Gradstein–Ogg date. For the 408 date, the positive error bar falls short of the G. and O. date by 3 Ma, while for the Harland 438 Ma date, the negative error bar reaches within 2 Ma of the Gradstein–Ogg date. The negative error bar for the Harland 505 Ma date comfortably covers the Gradstein–Ogg date, so that it is only for the basal Cambrian age of 590 Ma (for which there are no Gradstein–Ogg error bars) that Harland *et al.* are seriously in error as regards an age, which is now generally accepted as 545 Ma. We have compiled a Compatible age column in which, except for the Cambrian date, the ages cited are compatible with both the Harland *et al.* and the Gradstein–Ogg ages. Accordingly, because the earlier versions of Atlas were based on the Harland time-scale and the 'Compatible' ages, except

Table 6.1

Time-scales by: Present day	Harland et al. (Ma)	Gradstein and Ogg (Ma)	Compatible (Ma)
Base of:-			
Pliocene	5.1	5.3	5.2
Miocene	24.6	23.8	24.2
Paleocene	65	65 +/- 0.1	65.0
Cretaceous	144 +/- 3	142 +/- 2.6	143.0
Jurassic	213 +/- 6	205.7 +/- 4.0	209.0
Triassic	248 +/- 10	248.2 +/- 4.8	248.0
Permian	286 +/- 7	290	288.0
Carboniferous	360 +/- 5	354	355.0
Devonian	408 +/- 6	417	412.0
Silurian	438 +/- 3	443	440.0
Ordivician	505 +/- 15	495	500.0
Cambrian	590	545	545.0

for the base of the Cambrian, are all within or close to 2.0 per cent difference, we shall continue to cite ages as presented in the earlier time-scale version.

Smith *et al.* (1981) comment that the total effect of the sources of error in the Atlas program is not known and is difficult to estimate. However, as will be seen later in this chapter, where we discuss the position of Grahamland and Patagonia 135 Ma ago, once the timing of the generation of the Patagonian Orocline is taken into account, the positional errors of the two landmasses are probably less than 100 km. At earlier times, of course, the errors probably become progressively larger.

In this and the following chapters, we use the Atlas System to obtain the tracks of easily recognised inlets, or capes, in the coastline of islands or continents, back in time from the present-day for perhaps 250–300 Ma, and in some instances, as far back as the base of the Cambrian (i.e. 545 Ma ago). Also, to obtain the maximum accuracy, we plot series of detailed tracks which may represent a duration of only a few Ma. The Atlas system has the facility of plotting these tracks directly 'by instruction'. However, a track, over a long period, may combine translation with rotation of the landmass. Consequently, in order that we can differentiate between the two modes of movement, we have usually printed the position of the landmass by a series of individual print-outs of specific ranges of age. The details of successive prints enable one to see the proportion of the movement that can be attributed to translation and rotation respectively. The presentation of the data has been simplified by presenting the final version of the tracks on different scales and different spread of time between points. For complex movement patterns, it became necessary to place points nearer together, in time and distance.

There is one further aspect of the Atlas system that requires comment. This relates to the fact that the tracks, from time to time, exhibit an 'instantaneous' change in direction of movement and/or change in speed of plates. These instantaneous changes are the result of extrapolation, or interpolation of data. As we have noted in Chapter 1, instantaneous changes in speed and/or changes in track cannot be induced by any mechanism associated with *normal* plate movements. However, as was pointed out by Seyfert and Serkin (1979), extremely rapid changes (judged by normal plate tectonic standards) can be induced by major impacts.

6.3 Tracks of certain impacts, continental flood basalts and an oceanic plateau basalt

Throughout the remainder of this book the reader will encounter a number of maps which show tracks that may extend as far back as 600 Ma. As already noted, tracks refer to an easily identified reference point on a map (usually a cape or inlet on the coast of a continent, or an island) which can readily be compiled by using the Atlas program. If, for example, we trace the migration of an island back from the present-day to 250 Ma ago, we shall find that the track will be, dominantly, a straight line or a gentle curve. However, from time to time, the track will exhibit, on the time-scale we use, an apparently instantaneous change in direction of motion and/or a change in the rate of plate movement.

In Section 6.3.1 we present the evidence of track changes associated with three known 'certain', one 'highly probable' and one inferred impact events and show that each event is associated with a sharp change in track at a date which is very close to the age of the impact structure, as determined from radiometric dating or paleontological evidence.

It has been suggested, by Negi et al., (1993) that the Deccan Traps are the result of a major impact in the vicinity of Bombay. As we shall see in Section 6.3.2, a case can be made to support this assertion. It is reasonable, therefore, to ascertain whether the eight known major continental flood basalts exhibit sharp changes in track. In Section 6.3.3, the track of the largest oceanic plateau basalt is briefly discussed, and it is shown that this major oceanic event is also associated with sharp changes in direction of motion and changes in rates of motion, at times which are closely related to the two inferred ages of emplacement of this oceanic plateau basalt.

6.3.1 Tracks associated with known or inferred impact events

Of the 150 certain impact structures listed by Hodge (1994), only three have a diameter of 100 km or over. These three craters are:

(1) Popigai, in Russia, N70°,111°E, diameter 100 km and age 35 Ma;
(2) Chicxulub, Yucatan, N21°,89°W, diameter 180–200 km and age 65 Ma;
(3) Manicouagan, Quebec, N51°,69°W, diameter 100 km and age 212 Ma.

However, there is a fourth example which we have considered, for decades, to be a probable impact structure; and which now, as the result of the investigations by members of the US Geological Survey and the Colorado School of Mines, is likely soon to be recognised as a fourth 'certain' large impact. This is known as the Central Nevada Circular Feature (CNCF) with a diameter of about 220 km. and estimated age of about 370–380 Ma.

The tracks of these four events are shown in Figure 6.1. Each event shows an abrupt change in direction. Only Popigai shows no appreciable change in rate of movement, but its change in direction is dramatic. All other tracks exhibit an increase in speed from about 5 per cent (for Chicxulub) to a threefold increase for Manicouagan. The dates of the impact event as determined by the tracks, compared with the dating of the event by more conventional methods, are in very good accord and well within the probable range, inherent in current dating methods.

Evidence, which is becoming more readily and generally accepted, that an impact event has occurred, is the sudden obliteration of certain marine or terrestrial life forms. For example,

Wang *et al.* (1994) adduce evidence from S China of a late-Devonian (Femmenian, 360–367 Ma) impact event. The location of the impact site is not known with certainty. A paleo-reconstruction by McKerrow *et al.* (1990) has shown that South China was very close to, and facing, W Australia in the late-Devonian. Microtektites are found in South China with siderophile anomalies (including Ir) spread over South China and Western Australia. Wang *et al.* also refer to a thin boundary clay which marks a drastic faunal change. 'A high-diversity benthic community below the boundary clay is abruptly replaced by a layer of sediments devoid of biological content ("a Dr. Strangelove Ocean") which was only gradually repopulated by brachiopods 2 m above the impact-event horizon'. From the palaeontological evidence, the authors put the date of impact at 365 Ma.

Accordingly, we constructed the track of an inlet on the Western Australia coast for the period 350–385 Ma, to see if this method would produce corroboration of the dating of the event. As will be seen from Figure 6.2, the impact date inferred from the track is only 0.2 Ma different from that established by Wang *et al.* Moreover, it will be seen that there is a change in track, at this date, of approximately 30° and the plate speed increased by 60 per cent. The date of the change in track is 364.8 Ma, exactly the same as that of the Central Nevada Circular Feature (Figure 6.1d).

It cannot, of course, be guaranteed that a change in track of an identifiable continental feature must be associated with an adjacent impact feature. There may be another, possibly larger and as yet unidentified, coeval, impact feature that has determined or modified the track associated with the four examples shown in Figure 6.1.

As we have noted earlier, the Shoemaker-Levy 9 event gave rise to multiple impact events on Jupiter, within a period of a few days. Such events, that may be almost simultaneous,

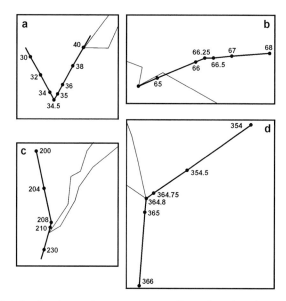

Figure 6.1 Tracks related to four known impact structures. Examples a, b and c are of *certain* impacts. (a) is that for the Popigai crater, diameter 100 km and age 34.6 Ma. (b) is that of Chicxulub, diameter about 200 km and age 66.25 Ma. (c) is that of Manicouagan, diameter 100 km and age 208 Ma. (d) is for a 'near certain' CNCF impact, diameter about 220 km and date 364.8 Ma.

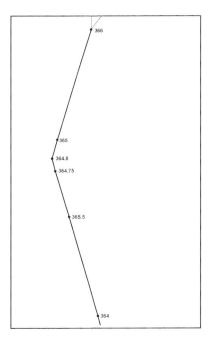

Figure 6.2 Track in W Australia inferred to be related to an impact in the S China Sea. Location of impact unspecified, but accurately dated from sedimentological evidence at 364.8 Ma.

can also be inferred on Earth. (See for example the twin Clear Water Lakes, Canada.) The S China Sea event and that which gave rise to the CNCF are possibly the result of impacts which took place within hours (rather than seconds) of each other.

In the light of the arguments presented above, it is suggested that the evidence presented so far clearly permits one to infer that abrupt changes in track and/or rate of movement are indeed associated with major meteoritic impacts.

However, this leaves one with an important problem. The four tracks shown in Figure 6.1 can be related to specific and certain, or near certain, craters of known diameter. The least 'dramatic' track is that for the Chicxulub Structure (Figure 6.1b). The diameter of this feature is probably in the range 180–200 km. The increase in plate speed after impact is 5–6 per cent and the change in track direction is 17°.

The CNCF structure (Figure 6.1d) has an estimated diameter of about 220 km. The increase in speed after impact is about 11 per cent with a change in the angle of track of 50°.

The track of Manicouagan (Figure 6.1c), with a diameter of 100 km, shows a four-fold increase in rate of movement after impact and a change of track of 30°.

Finally, the Popigai (Popigay) track (Figure 6.1a) exhibits a reduction in speed of 4–5 per cent after impact, but exhibits a change in the direction of track of 120°.

Clearly, the existence of a sharp change in rate of movement and/or direction of track can reasonably be associated with a major impact. However, one can also infer that the post-impact track itself is influenced, or even determined, by the pre-existing boundary conditions of the plate in which the impact occurs, and the degree to which they are changed by the impact. These parameters we do not always know and, possibly, may never know with any precision.

6.3.2 Continental flood basalts

As we have seen in Chapter 4, the location, extent and volumes of erupted rocks and the age and duration of emplacement have been widely studied (Coffin and Eldholm, 1992, and the many authors cited by these collators). The major continental flood basalts are known to have a volume which may range from about 0.2×10^6 km^3 to well in excess of 10^6 km^3.

Although eruptive activity, at any one site, may extend over a period of up to 10 Ma, it is generally held that the majority of these various flood basalts were emplaced in a period of 1.0 Ma, or even less (Gallett et al., 1989; Baksi, 1994; Courtillot, 1990). Thus, the rate at which melt is brought to the surface in CFBs and OFBs is as much as 3 to 4 orders of magnitude faster than the rate of supply of melt at a hotspot/plume.

The ready acceptance of the Chicxulub structure as being the result of an impact was predicated on the finding of small particles of iridium and stress-metamorphosed minerals in the sediments, of near K/T age, around the world. Such material is prime evidence pointing to the existence of meteoritic impact. However, the distribution of such material near the K/T boundary is so extensive that, it has been argued, there is almost certainly more than one source of this exotic material.

A *prima facie* case can be put that the Deccan Traps (and possibly certain other features that we shall discuss in the next chapter) which developed about 65–67 Ma ago, are the result of impacts; and one can reasonably conclude that rapid changes in speed and/or direction of movement of plates can be attributed to major impacts.

The major continental flood basalts that have developed in the last 250 Ma are listed in Table 6.1. In addition to the name, the location and approximate age of the main continental flood basalts (CFB) which have developed within the past 250 Ma, are also listed in Table 6.2.

As pointed out in Chapter 4, if a continent passes over a major hotspot it will be heated to an abnormally high temperature in the lower regions of the continental lithosphere. Consequently, it is reasonable to suppose that a large, and relatively recent, thermal event must have left a readily detectable *heat-signature*. Anderson et al. (1992) state that no such signature exists for six of the eight listed CFBs. (These authors did not include the Antarctic features in their study.) Their conclusion is based on their study of high-resolution, upper-mantle tomographic models in terms of plate tectonics, hotspot and plume theories. They state that there is no evidence for plume head or lithospheric thinning under any of the major continental flood basalts listed in Table 6.2.

Anderson et al. further state that tomographic results contradict the premises of all currently popular plume and flood basalt scenarios. Moreover, the fact that no plume signature

Table 6.2 Location and age of continental flood basalts.

Name	Location	Age (Ma)
Columbia River	USA	17
Aden and Ethiopian	Africa–Arabia	39
Deccan Traps	India	65
Paraña	S America	135
Antarctic	I	170
Antarctic	II	180
Karoo	S Africa	193–178
Siberian	N Russia	248

can be found related to the plateau basalts to which White and McKenzie (1989) applied their model, Anderson *et al.* suggest, renders their application to the several examples of plateau basalt development more than a little suspect.

Let us, therefore, turn to the possibility that the events listed in Table 6.2 are the result of impacts. We comment briefly on the various continental flood basalts and illustrate the tracks of these bodies, to demonstrate that they all may be attributed to impact events. These we shall present in the order shown in Table 6.2.

(a) Columbia River flood basalts

The Columbia River Basalt Group (CRBG), was originally estimated to have an area of about 200,000 km^2 and a volume of about 350,000 km^3 (+/– 20–30,000 km^3). A more recent estimate, together with aspects presented in Tolan *et al.* (1989) puts the area at 163,700 +/– 5000 km^2 and a volume of 174,700+/–31,000 km^3. The maximum volume of an individual lava flow exceeds 2000 km^3, and some flows are known to have advanced more than 750 km, which makes them the largest known terrestrial lava flows.

Because of their plate tectonic setting, the structures of the Central Columbia Plateau and its surrounding area are undoubtedly difficult to interpret (Reidel *et al.*, 1989). The Yakima folds developed under N–S compression that has persisted for the last 50 Ma (Zoback and Zoback, 1980). However, a generalised section of the structures through the Rattlesnake and Horse Heaven Hills in the central Columbia Plateau shows a series of spectacular thrusts (Figure 6.3), where it will be seen that the peaks and valleys define a series of thrust blocks. Reidel *et al.* further state that the axial trend of these particular thrusts produce 'a crude fanning' of anticlinal ridges across the fold belt, that changes from N50°E in the west to N50°W in the east. These curved thrust structures are unique in the area and the authors, understandably, did not present any comment on how such a set of radiating thrust structures may have developed. For example, it is possible to envisage a small cylindrical plume being placed beneath the central point of the arc of folds. As we have seen in Chapters 3 and 4, an updoming caused by a small plume would not be sufficient to generate lateral stresses of sufficient magnitude to induce such arcuate thrusts. Alternatively, one could suppose that a narrow pipe of igneous melt was introduced at the centre point of the arc which then expanded in diameter, thereby causing the surrounding country rock to be pushed back. Such a mechanism could conceivably produce arcuate thrusts. However, it is equally probable that such an increase in radius of the pipe would generate radiating tensile fractures which became dykes. In either event, one would expect there to be evidence of such an igneous pipe at the surface.

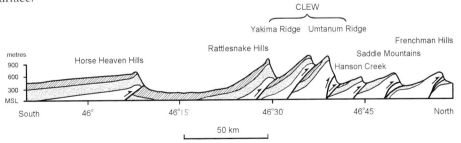

Figure 6.3 Generalised geological cross-section through Rattlesnake Hills and Horse Heaven Hills in the central Columbia Plateau (after Reidel *et al.*, 1989).

I am confident that the reader, on referring to the thrust structures that occurred after the 500 ton Snowball explosive experiment (Chapter 5), will be able to find a more plausible explanation for these Columbia River thrust blocks.

These thrusts are thought to have been initiated prior to the emplacement of the Columbia River FB. However, they continued to develop as the lavas were erupted, from about 16–17 Ma and more recently.

The track for the period between 25–15 Ma, in which it is inferred that the impact event in central Washington State, USA, took place, is shown in Figure 6.4. The changes in the track are better shown in Figure 6.5, where it will be seen that the early part of the track was constant in direction and speed from 25 to 19.5 Ma. From 19.5 (the inferred impact date) to 15 Ma, the track changed direction by almost 10° and increased its rate of movement by 4.4 per cent.

The volume of basalt erupted as the result of this event is much smaller than the Deccan Traps. Indeed, the Columbia is the smallest of the major FBs. It can also be inferred that the track change of this event predates the range of dates in which general eruptive activity took place.

(b) Afar plateau basalt (or Arabia-Ethiopia-Africa plateau basalt)

This plateau basalt is, perhaps, the only one listed in Table 6.2 which could be explained wholly, or at least mainly, in terms of conventional plate tectonic mechanisms. Thus, it can be argued that, when the Indian Continent made contact with the area which was to become the Tibetan Plateau, this northward push was sufficient to open the Gulf of Aden and then cause the generation of the Red Sea. Just how the Red Sea was initiated (using conventional plate tectonic mechanisms) may require some special pleading, but we are quite prepared to accept that the collision of India with the southern limit of Asia did contribute to the development of the Gulf of Aden and the opening of the Red Sea.

However, we would point out that hotspots occur at either side of the Straight of Bab el Mandeb, at the southern limit of the Red Sea and also at opposite sides of the Red Sea further north (Figure 6.5). We have already suggested that hotspots may be induced by a meteor, or a comet, or to an antipodal event: so an impact could be involved in the development of the Afar PB.

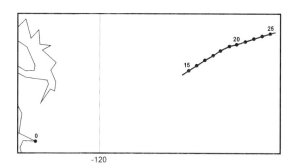

Figure 6.4 Track from 25–15 Ma, showing position relative to the coast and present-day reference point 0.

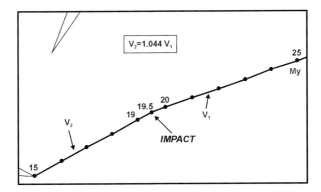

Figure 6.5 Detail of track from 25 to 15 Ma, with sudden change at 19.5 Ma.

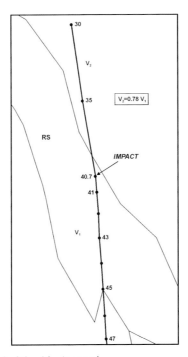

Figure 6.6 Track for the period of the Afar 'impact'.

The track of the point to the east of Beilul is shown in Figure 6.5(??6.6??). It will be seen that from 47 to 40.7 Ma, the northward track is constant in direction and speed. At 40.7 Ma the track direction changed through an angle of 6° and the post-impact motion (V_2) slowed abruptly by 22 per cent relative to V_1. Such behaviour is consistent with that of an impact. Hence, we conclude that although conventional plate tectonics may have played a role, the opening of the Red Sea can best be attributed to an impact 40.7 Ma ago. Again, the impact event is in good agreement with the radiometric dating of the onset of eruptive activity, with impact pre-dating the eruptions. (Also, as we shall see below, the Paraña event left no evidence of track deviation in Africa, so it is perhaps not surprising that the Afar event also gave rise to relatively little deviation of track and rate of movement.)

(c) Deccan Traps

It has been known for several years that India suddenly doubled its rate of northward movement some time in the period 72–60 Ma. However, as far as we are aware, except for the general hypothesis presented by Seyfert and Sirkin (1979), this change of pace has not been explained.

The track of India over the period 72–60 Ma, which represents the plate motion during that period, is shown in Figure 6.7a. This track is derived by identifying a specific geographical feature on the coast of India, near present-day Bombay. As we have noted, the Atlas system assumes that the present-day geographical shapes of islands and continents are maintained back in time, so that the position of 'Bombay' can be identified for specified ages. In Figure 6.7a, the positions of 'Bombay' from 72 to 60 Ma are given in steps of 1 Ma. Here we are particularly concerned with the sudden change in speed of India, so in Figure 6.7b we show the change of position of 'Bombay' in the period from 67.4 to 67.15 Ma. We plotted the positions of Bombay for the period 67.15 to 67.4 Ma, with the points between 67.30 and 67.20 Ma (i.e. a period of 100,000 years, in steps of 10,000 a, Figure 6.7b).

Let us *assume* that the change in the rate of motion could be accomplished in 5000 a. In the period between 67.235 and 67.250 Ma, the slowest possible rise-time is represented in Figure 6.7b by the straight line. However, because of the mass of the Indian lithosphere, the inertia of the system will demand that there is a period of increasing acceleration, starting in 67.235 Ma, followed by a steady-state speed, which then gives way to a deceleration before the plate settles down to a near-constant plate speed following 67.230 Ma. Thus, the rise-time path is likely to follow the S-shaped dashed line shown by the insert in Figure 6.7b. It may be inferred, therefore, that the maximum acceleration may have had a duration of only a few hundred years. The total rise-time of 5000 a is, as we pointed out, an assumption (the rise-time could, perhaps be as long as 50,000 a). However, whichever figure is assumed, such acceleration is far too fast to be explained in terms of mechanisms normally associated

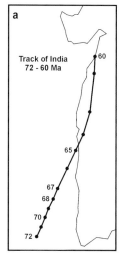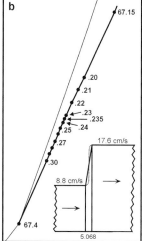

Figure 6.7 (a) Track of 'Bombay' over the period 72–60 Ma. It can be inferred from the distances between the points representing specific times that there was an abrupt change in velocity of the plate at about 67 Ma. (b) A detail of the track shown in (a) reveals that the velocity of plate motion doubled in a period which is assumed to be about 5000 years.

with plate movements. For example, we have noted that the emplacement of plumes and the attendant hotspot may take at least a million years. Hence, a postulated plume cannot explain the acceleration-time that is to be inferred from Figure 6.7b. We merely note here that change of speed can reasonably be attributed to a major impact. We also note that this track is unusual in that it does not show any obvious deviation following impact. The inferred impact date is 67.4 Ma, while the dates for the erupted Deccan Traps are usually about 65 Ma.

(d) Paraña

The track (Figure 6.8a) associated with this continental plateau basalt is the most dramatic we have so far seen; and arguably was the event that initiated the opening of the S Atlantic. Between the years 138–135.1 Ma, it will be seen that S America moved generally northward at a rate of 6–8 km Ma^{-1}. However, in the next 1.1 Ma the S American continent moved, on a track of about 290°, a distance of approximately 150 km. This inferred impact may well have been the most energetic event in the Phanerozoic. Turner *et al.* (1994), using ^{40}Ar/^{39}Ar geochronology, put the age of the erupted basalts between 137 and 127 Ma, which covers our impact date of 135.1 Ma. However, this movement was restricted to S America. The track of E Africa (Figure 6.8b), which is shown on a much more detailed scale than that of Figure 6.7a, merely shows Africa moving serenely northward, without any detectable deviation of direction or rate of movement. This lack of deviation, we suggest, is an indication of the massive inertia, at this time, of Africa and appended continents and subcontinents, which included Australia, Antarctica, Madagascar and India. This constant northward movement helps one understand why the inferred Afar impact track exhibits such small deviation and change in speed of track.

In passing, it may be noted that the Paraña eruptive rocks contain material that originated as molten continental lithosphere. It is difficult to explain how such melt material was generated, if it is assumed that the Paraña eruptive rocks were generated by a plume.

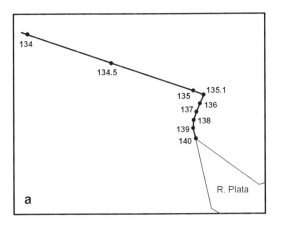

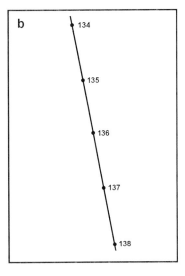

Figure 6.8 (a) Change in track of S America which we take to result from the impact that gave rise to the Paraña CPB. (b) Track of E Africa for the same period (see text).

(e) Antarctic I

The earlier of the two Antarctic PBs, as Heimann *et al.* (1994) indicated, comprises the Ferrar Group (consisting of the Kirkpatrick Basalt and Ferrar Dolerite) which crop out along 3000 km of the Transantarctic Mountains, the formation of which are considered to be related to the break-up of Gondwana. These authors point out that, although a wide range of dates (from 90 to 193 Ma) have been cited, more recent work, using $^{40}Ar/^{39}Ar$ geochronology, shows that the eruptive activity along 1200 km of the Transantarctic Mountains occurred within a short interval of less than 1 Ma, within the time-range of 176.6 +/– 1.8 Ma.

We constructed the track of an easily identified feature on the Atlas 3.3 version of the coast of Antarctica for the periods 170–186 Ma (Figure 6.9a), 179–182 Ma (Figure 6.9b), with a more detailed track for the period 180–180.5 Ma (Figure 6.9c). It will be seen that there was a sharp change in direction of track, at 180.3 Ma, of 15° and that the rate of plate motion slowed by approximately 25 per cent.

It can be inferred that the extreme upper age cited by Heimann *et al.* (1994) for the eruptive event is 178.4 Ma. We would infer from the tracks obtained by using the Atlas 3.3 program that there was a major impact at 180.3 Ma. We suggest that these data are within the error bars inherent in the dating technologies currently used.

(f) Antarctic II

The second Antarctic event, relating to extensive magmatism, was established by Brewer *et al.*, (1992) who also used Ar-Ar geochronology methods. It occurs within the Dronning Maud Land Province and has been shown to have an age of 172.4 +/– 2.1 Ma. These data

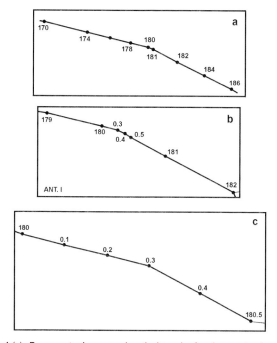

Figure 6.9 (a), (b) and (c) Progressively more detailed tracks for Antarctica I event.

were obtained from specimens collected in the transition region between the Dronning Maud Land and Ferrar provinces in the Theron Mountains.

We present another track for the period 175–166 Ma in Figure 6.10, from which we infer that a major impact occurred at 169.45 Ma, which gave rise to a change in track of 22° and an increase in rate of movement of 78 per cent.

In this instance, the youngest possible rock-date is 170.1 Ma, while the impact, as derived from the Atlas program, is cited at 169.45 Ma. We suggest that the 'cart before the horse' relationship can be attributed to small differences inherent in date determinations.

(g) Karoo, S Africa

The igneous activity associated with the Karoo appears to be complex and of long duration, extending as it does from about 193 to 178 Ma. Here we shall only consider the initiation of this event at (about) 193 Ma.

The track shown in Figure 6.11 is related to the Atlas 3.3 representation of the bay containing 'Lourenco Marques' (Maputo). It will be seen that, from 200 to 194.85 Ma the track runs almost exactly northward. At 194.85 Ma, however, it turns through an angle of 100°, so that it continues slightly S of W, but the rate of movement remains virtually unchanged.

As with other events, we would take this track to be evidence of a major impact. The discrepancy of the dates obtained from radiometric dating and from the track of 1.85 Ma indicates that, as one would expect, the impact initiated the change in direction, and that the erupted rock took a significant time to reach the surface.

The date of the impact is also in reasonable agreement with the initiation of the detachment of the Madagascar–India block from Africa.

(h) Siberian flood basalt

Located in N Siberia, this plateau basalt body dated at 248 Ma is of vast extent, but, unfortunately, has been relatively little reported in the English language. The track of Siberia for the period from 260 to 240 Ma is 'anchored' at the head of the Tazovska Guba (TG in

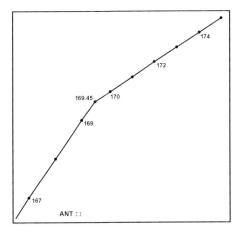

Figure 6.10 Track for Antarctica II event.

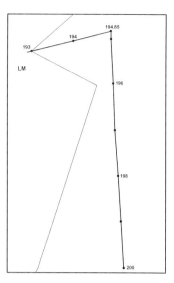

Figure 6.11 Track for the period of the Karoo 'impact'.

Figure 6.12), which runs into the larger Obskya Guba, bounded to the west by the Yamal Peninsula, which in turn faces Novaya Zemlya across the Kara Sea.

It will be seen that from 260 to 250 Ma, the track moves in a generally NE direction. At 250 Ma, however, it suddenly changes track through an angle of 70° and shows a six-fold increase in rate of plate movement. As we have seen, such a track is diagnostic evidence of a major impact. Moreover, one may infer that the energy of the impact event was considerable.

Conclusion: all continental flood basalts were initiated by a major impact.

(The dates of the periods of eruption and dates of track change (and so of inferred impact time) are given in Table 6.2. As noted above, one would expect that a major impact event would occur at a little before the earliest datable erupted rocks which could be sampled. Only two of the eight events listed in Table 6.2 do not show such a relationship and this can probably be attributed to short-comings of earlier dating systems used in the Atlas system.)

6.4 Oceanic basalts – the Ontong-Java event

In the Cretaceous, marine eruptive activity of outstanding violence took place. This activity is exemplified by an event which, more than any other, has induced many scientists to infer that the resulting eruptive body can best be attributed to a *super-plume*. This enormous eruptive feature, the Ontong-Java OPB, was originally dated at 122.4 +/– 0.8 Ma. The earliest biostratigraphic age of basal sediments is 117.5 Ma. In addition, from the seismic data it is clear that the erupted rocks exhibit abundant evidence of erosion. It can, therefore, be inferred that for a period possibly as long as 5 Ma, the upper levels of the erupted rocks were probably sub-aerial. A second phase of igneous activity in the area took place at about 90 Ma.

The setting of this huge igneous province and its relationship to the surrounding basins and plateaus are indicated in Figure 6.13. In this paper (to which the reader is directed for details) these authors present a concise exposition of the various evidence relating to dating of the intrusion and the surrounding area.

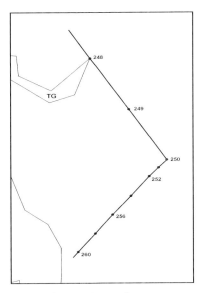

Figure 6.12 Track for the period of the Siberian 'impact'.

Figure 6.13 Map of Ontong-Java PB showing adjacent basin and seamount areas and magnetic lineations on nearby sea-floor areas (after Gladczenko, Coffin and Eldholm).

The disposition of marine magnetic anomalies in the deep-sea areas adjacent to the Ontong-Java plateau (Figure 6.13), range from about M10 to M29, which, using the datings presented by Harland *et al.* (1982), range from 165 to 132 Ma. Hence, one may infer that the ocean floor, into and onto which the flood basalts were emplaced, has ages which ranged from 10 to 43 Ma. From this age-range, one can infer that the Ontong-Java PB was emplaced at no great distance from a spreading-ridge. Indeed, Gladczenko *et al.* (1997) indicate the importance of this nearby spreading-ridge and how it may have influenced the emplacement of the erupting molten rock in Figure 6.14.

It will be seen from this figure that these authors envisage the melt rock to be delivered to the base of the oceanic lithosphere by a plume, which then spread laterally to cause enhanced eruption at the spreading-ridge, as well as being erupted onto the ocean-floor through a series of dykes. As presented in this figure, the reader may possibly infer that the dykes are parallel to the spreading-ridge. However, as we have seen earlier, the stresses induced by the gravity-glide mechanism are more likely to give rise to the development of dykes which trend parallel to the direction of plate motion.

These authors do not discuss the possible mode of origin of the plume and plume-head shown in Figure 6.14. They show, in plan, the likely position of the plume centre and, in section, the dimensions of the plume-head and its subsequent evolution as it reaches the

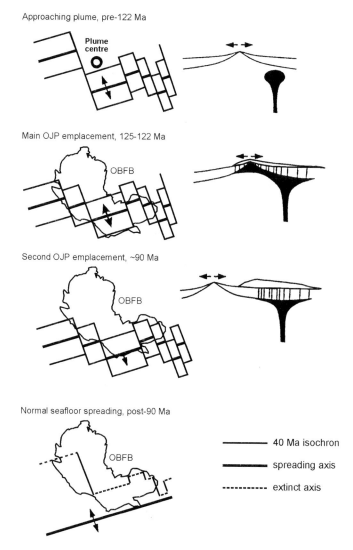

Figure 6.14 Sections and maps indicating possible stages of development of the of the Ontong-Java PB (after Gladczenko, Coffin and Eldholm).

lower limits of the oceanic lithosphere. However, they do not discuss the delivery to, and the generation and collection of, the molten material beneath the oceanic lithosphere. We suggest (in Chapter 3) that the dimension of the plume-head, in Figure 6.14, may be mis-leadingly small. Thus, we have noted that the thickness of 40 Ma old oceanic lithosphere should be 50–60 km. The section of the plume bubble shown in Figure 6.14 represents the plume-head as having an average diameter of about 100 km.

If the molten material that is to be erupted is largely contained within the plume-head, then this 'head' needs to have a diameter in the range 470–540 km. Otherwise it is necessary to postulate a relatively high rate of upward movement in the stem, to supply the required volume of mantle material to the zone of pressure and temperature conditions which are conducive to melting. Also, if we bear in mind McKenzie's warning that only about 30 per cent of such upwelling mantle material is likely to experience melting, then we are faced with the problem of quantifying the cross-sectional area of the stem of the plume and the average velocity of upward flow in the stem, and the driving force that gives rise to what must be a very high rate of flow.

The Ontong-Java plateau basalts are surrounded by other igneous bodies which are of the same general age. For example, the Manihiki plateau, about half the area of, and situated at a distance of less than 3000 km from, the Ontong-Java province, is only about one fifth the estimated volume of the Ontong-Java province. Similarly, the Nauru basin, situated at no great distance from the Ontong-Java province, is only a little smaller in area than the Ontong-Java province, but has an estimated volume only 1/75th of that province. The Pigafetta and East Mariana basins, while having an area of about 27 per cent of the Ontong-Java province, have a volume of only 4 per cent of it. Moreover, the period of emplacement of these two basin flood basalts occurred between 114 and 126 Ma.

This collection of basin and plateau basalts presents evidence of an amazingly energetic eruptive period. Moreover, most of the emplacement activity in this region was completed within 3 Ma. Only the relatively small Pigafetta and E Mariana basins (which are by far the smallest as regards volume of basalt) were active for a period of about 12 Ma.

Here we have two plateau basalts and three basin basalts all emplaced almost simultaneously within relatively close proximity. What mechanisms can be invoked to explain these provinces which are so disparate in their volume and geometry?

The Ontong-Java plateau is adjacent to three basaltic bodies, the East Mariana Basin, the Nauru Basin and the Pigafetta Basin, which are possibly unique. Their mode of emplace-ment represents a problem, for these basin basalts have large areas but contain a relatively small volume of basalt. These three flood basalts have collected in shallow basins, the floors of which are only a little older than the flood basalts themselves. Coffin and Eldholm (1991) suggest that all three of these bodies were formed at the same time as the Ontong-Java plateau.

An objection to discussing these bodies could be made, to the effect that this is a completely separate problem, for which as yet there is no answer. However, the basin basalts developed at the same time as the plateau basalts, and in the same general area. It would be surprising if they were not in some way related.

There is no known traditional explanation regarding the formation of these three coeval basin flood basalts. However, the impact mechanism outlined in the previous paragraphs does permit of an explanation. It will be noted that jetting is an expected feature of such major impact events (Figure 5.29). If such jets were initiated by the impact that, we suggest, gave rise to the Ontong-Java plateau, then, when they fell upon very young ocean lithosphere, at some distance from the ground zero, but within tens of minutes of the impact, the jet would have sufficient mass and momentum to deform the thin lithosphere into a broad shallow basin. Also, the impact of the jets could generate vertical fractures in the lithosphere, which could possibly lead to relatively long-lasting eruptions.

The jet material, still in liquid form, would, of course, rapidly spread out to develop quickly into an extensive, but thin, sheet of basalt on what was the ocean floor. It should be noted that the ocean will have been blown away at this juncture and is not likely to return for some hours, so that initial lateral dispersal of the basalt may be rapid and extensive and thus will be unlikely to exhibit the usual plateau basalt form.

We suggest that this explanation for the development of such basin basalts not only satisfies the known data regarding such structures, but also is further evidence that confirms the suggestion that the Ontong-Java OPB was the result of a major impact.

Clearly, from the analysis by Gladczenko *et al.* (1997), the emplacement of the melt is influenced by a spreading-ridge, displaced by a series of transform faults. We accept that these faults complicate the history of development of the extrusion. However, as will be seen in Figure 6.15a, the track of a marker adjacent to the Ontong-Java PB shows a well-defined change in track close to the date at which erupted rocks were extruded on the ocean floor. Consequently, as an alternative to a plume, we conclude that the Ontong-Java PB was initiated by a major impact.

These authors, cited above, also noted a second phase of eruption at 90 Ma. It will be seen from the small change in track of Bougainville Island (Figure 6.15b) that one can infer a second, relatively minor impact occurred in the Ontong-Java PB at about 93 Ma.

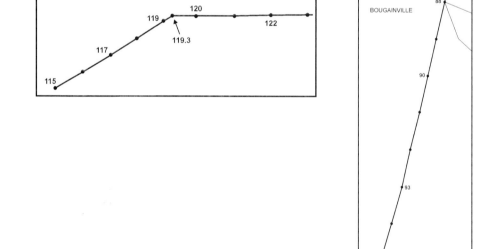

Figure 6.15 (a) Track associated with a major impact (about 120 Ma) that could have given rise to the Ontong-Java PB. (b) Track associated with a less energetic impact at 93 Ma that, we suggest, gave rise to the emplacement of later basalts dated at about 90 Ma.

6.5 Comment

We have seen that all eight continental major flood basalts are associated with tracks that exhibit a change in direction and/or rate of movement of the plate on which the impact has occurred. Dating techniques are being continually improved. The Atlas and similar programmes undergo continual revision. However, the specific reviewed program used here may be somewhat in error when compared with the latest radiometric dating. Moreover, the majority of the track changes, as inferred from the Atlas system occur immediately before the earliest dated extrusive rock. Consequently, we suggest that all (single) impacts precede the eruption of basaltic rocks. However, we note a later extrusion phase at about 90 Ma can be attributed to an impact-induced change in direction of movement for the Ontong-Java plateau.

Two questions immediately come to mind. (1) What mechanism can give rise to such relatively rapid changes in track? (2) Why is the largest oceanic plateau basalt so much larger than all the continental flood basalts? These two questions we shall address in the next two sections.

6.6 Causal mechanism

As we have seen in Chapter 3, the direction and speed of movement of a plate is determined by the initial imbalance between the driving and the retarding forces. The driving force gives rise to acceleration of the plate. Such acceleration induces viscous drag on the LVZ, so that at a specific rate of movement of the plate, the forces come into balance and the plate, thereafter, moves at a constant rate until some major disturbing effect comes into play. From the size, geometry and rates of motion of the plates, one can infer that normal plate motion can only be expected to change very, very slowly. Hence, it is logical to infer that, if a plate experiences sudden acceleration or deceleration and/or change of direction, a major change must have abruptly taken place in one or more of the boundary force conditions. Indeed, in order that the inertia of these immensely heavy plates can be rapidly overcome, one must postulate the action of an immensely powerful mechanism.

6.6.1 Hypothetical impact model

As we have noted, plate movements tend to be reasonably constant for periods measured in millions of years. The relatively even, magnetic striping of the ocean floors is proof of such periods of near-constant speed. To explain how acceleration, deceleration, or sudden change in direction of motion of a plate can develop, we suggest it is necessary to have recourse to the mechanics of the early phase deformation induced by major impacts, discussed in the previous chapter. Consider a simple hypothetical plate as seen, in plan, in Figure 6.16a, in which an impact (IMP) has given rise to a high magnitude stress-wave at ground zero (GZ) which generates a simple crater, which we assume to be 300 km in diameter. The extremely high-magnitude of this initial compressive stress-pulse decays away from GZ and is reduced to a magnitude of 20 kb at a distance of about 10–30 times the radius of the crater (see Figure 5.31). This attenuating stress-wave, induced by the impact, gives rise to a degree of melting and thickening of the original LVZ. As a result, a circle of newly enhanced, low basal resistance at the LVZ layer extends over a very large area. This will result in a dramatic change in the balance of the driving and resistive forces acting in and adjacent to the plate by suddenly introducing a new, basal, low resistive-element over an area of perhaps several millions of square kilometres. It is this reduction of basal resistance of the plate that gives rise to a dramatic change in the stresses in the plate and also to the boundary stresses in adjacent plates which, in turn, gives rise to changes of speed and direction of motion of the impacted plate.

Let us initially assume that our hypothetical target is mainly comprised of oceanic lithosphere which has a thickness of less than 100 km and, prior to impact, is underlain by a LVZ with an average thickness of perhaps 60 km. This LVZ zone normally restricts the rate of plate movement, because of the inherently high coefficient of viscosity of about 10^{19} Pa s. If, as the result of a major impact, a significant proportion of this area can be substantially, and extremely rapidly, heated (or 'unloaded'), then the viscosity of the LVZ within the 1000 km radius about GZ will be very materially reduced.

The degree of melting at a distance from GZ of 1000 km or more, will be mainly determined by the two forms of stress-pulse that are generated by the impact. As we saw in the previous chapter, when a large, fast-moving asteroid or comet makes contact with the Earth's surface, it sets in train a complex succession of events.

Let us consider the first phases of deformation, which relate to the passage of a high-intensity, fast-moving compressive stress-pulse which is initiated at the impact zone. We have seen that the magnitude of this transient stress degenerates away from the point of impact (GZ), and causes a series of effects that may range from vapourisation, melting, plastic deformation and, finally, fracturing (see Figure 5.30 and Figure 5.31). As we shall see, depending upon the size of the impacting body, transient cratering may reach down through the crust to the lower lithosphere or even into the asthenosphere. The spreading of the stress-wave generated in the Earth by the impacting body takes a brief, but finite, time. The compressive stress-wave propagates at a velocity of about 8–10 km s^{-1}, so that the time taken to pass through the various zones of deformation, for a large event, will be about 2–3 minutes.

Adjacent to the crater area, significant melting will take place, and because the transient compressive stress-pulse will exceed its elastic limit, the rock mass will deform by plastic flow (which generates a considerable amount of heat). Hence, significant partial melting of

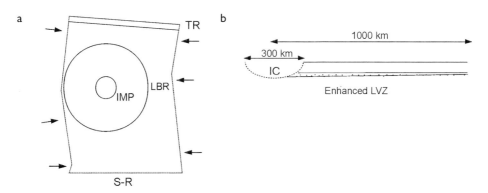

Figure 6.16 (a) Map of a hypothetical oceanic plate hit by a major impact that caused a crater (IC) with a diameter of 300 km. The outer limits of the area of 3×10^6 km^2, in which the impact induced low basal restraint (LBR) are also indicated. It should be noted that the area of LBR is almost certainly extremely conservative. (b) Schematic section through the crater to the edge of the LBR.

the LVZ, in the areas beyond the crater, is likely to occur. Even a small percentage of induced melting in the LVZ will significantly reduce the viscosity of that zone. Moreover, the thickness of the LVZ beyond the limits of the crater is likely to increase as the result of transient pressure-induced melting. Beyond, but still close to the limit of the crater, where the magnitude of the transient stress-pulses are extremely high, the viscosity could be reduced by several orders of magnitude. The basal drag on this specific area of LVZ will be radically, and almost instantaneously, reduced by a comparable magnitude (see Figure 3.13 and relevant equations).

The inner zones shown in Figure 6.16b will be excavated to form a transient crater. However, the plastic and brittle response to the compressive stress-wave can extend to well beyond the crater. Hence, for large impacts, which give rise to, say, a diameter of 300 km, the zone of plastic deformation and fracturing can extend down to a depth well beyond the LVZ, as well as spreading out at the LVZ depth for at least 1000 km beyond GZ. One can infer that plastic deformation of the mantle results in the generation of heat. Consequently, if the plastic deformation extends down to and beyond the LVZ, which is at the PT conditions close to melting, then this deformation can cause further small amounts of melting to take place, thereby lowering the viscosity of the LVZ, at a distance of 1000 km from GZ, possibly by a factor of perhaps 2–3. The overall energy of this stress-pulse event is more than sufficient to supply, instantly, the latent heat to cause melting.

It is known that the development of shear planes in dry rock can cause melting of the fracture surfaces which, in relatively cold rock, give rise to the development of pseudo-tachylyte (Sibson, 1975). The development of such fractures in hot rocks generates a small but further, significant, additional amount of melt, which would further contribute to a reduction of the viscosity of the LVZ. We have seen in Chapter 5 that the stress-pulse caused by a large impacting body, even when it falls as low as 20 kb, is still able to generate heat in the target rock. Depending upon the velocity on impact, this 20 kb stress level may extend from GZ by as much as 10 or more times the 'diameter' of the impacting body.

We have also noted that the compressive stress-pulse may be followed by a weaker, tensile stress-pulse. Nevertheless, although weaker, this tensile pulse may have a considerable effect upon the mantle rock. For example, if the tensile pulse at any distance from GZ has (for a very brief period) an upward component of magnitude greater than 30 kb it is capable of reducing the effects of gravitational loading to zero, down to a depth of about 100 km. In this brief tensile-stress phase, relatively little deformation may take place. The temperature in the LVZ will be little affected. However, as the ambient gravitationally induced rock pressure (P) will be reduced, perhaps to zero, the original PT conditions in the LVZ (where the rock is already at or close to melting) will cause the tensile stress-pulse to further enhances the degree of melting in the LVZ. Moreover, this is not a situation which is quickly reversed. Heat must be conducted away from the LVZ before the original state can be approached. This will be a slow process.

The combined effects of the compressive and tensile pulses are therefore very large, and will result in a significant area of the LVZ experiencing a further, sudden and highly significant reduction in the coefficient of viscosity. Hence, over an area of at least 3,000,000 km^2 this average reduction will completely change the balance of forces in the plate. We estimate that the average viscosity of the LVZ within the zone of 'enhanced LVZ' will be reduced by about 2 orders of magnitude.

In Chapter 3, we noted that the stresses involved in driving lithospheric plates was simply related to the thickness and viscosity of the LVZ (Figure 3.13). For a relatively fast-moving, oceanic plate, the stresses could move the lithosphere over the LVZ at about 7 cm a^{-1}, or, 7 km in 1.0 Ma. If the viscosity of the LVZ is suddenly reduced by an average 2 orders of magnitude over an area of at least 3,000,000 km^2, then, if the effects of inertia are ignored, we suggest that a speed of lithospheric motion of several hundred km per Ma could be attained. This would permit a relatively sudden increase in the speed of readjustment to plate motion in the regions surrounding the major impact, that would give rise to a change in direction and/or rate of plate motion, so that considerable readjustment, regarding direction of plate movement, could be largely completed in a period of between 20,000–50,000 a.

These arguments have been based on the assumption that the 'enhanced LVZ' occupies a very large proportion of the area of the plate. If the enhanced LVZ occupies only a fraction of the plate, the potential rate of change of motion will be correspondingly reduced. The boundary stresses on the plate will also influence the final rate and direction of movement of the plate. It is, of course, difficult to quantify these movement rates precisely. Fortunately, nature has presented an example which indicates that high plate velocities may be engendered by a major impact. We refer the reader to the track associated with the Paraña continental flood basalt, discussed earlier.

It was shown (Figure 6.8a) that in the period 139–135.1 Ma, S America moved northward at a rate of 7.6 km Ma^{-1}. Between 135.1 and 134 Ma, the rate of motion was about 150 km Ma^{-1}. These figures, we suggest, completely support the argument presented above.

We note that the Paraña event relates to movement of a continental area, where, because of the, commonly observed, variable thickness of lithosphere this is likely to provide considerably higher resistance to horizontal movement than that of an oceanic plate. Moreover, there is little evidence of an LVZ layer beneath continental lithospheres. However, where PT conditions below the continental lithosphere are appropriate, a major impact can induce a small percentage of melt such that a layer comparable with an oceanic LVZ may be initiated. Even so, it is likely that basal drag beneath a continent will usually be higher than that for an oceanic plate. Hence, we infer that early rates of adjustment of oceanic plates may be many tens of km per year.

6.6.2 Tomographic evidence

In Chapter 4, we cited Anderson *et al.* (1992), who argued that evidence of hotspots which could be inferred to have given rise to the development of major continental flood basalts should have left an abnormally high temperature in the lower regions of the continental lithosphere. These authors conclude that tomographic evidence of remnant high temperatures was not to be seen beneath the major continental flood basalts. Yet we have invoked melting induced as the result of impacts. Could we not be 'hoist by our own petard'?

In the plume mechanism, the heat is assumed to be brought upward from great depth in the mantle in a cylindrical stream, with a volume of tens of millions of cubic kilometres, which is at a temperature several hundreds of degrees higher than the ambient temperature of the asthenosphere into which it penetrates.

In the *major* impact model, two main forms of melting are envisaged. Firstly, provided the energy of the impacting body is sufficiently high, considerable volumes of the target rock as well as the impacting body will be melted, or even vapourised. This material will be mainly ejected from the crater area, possibly as jets. Only a proportion of the originally, relatively cold target rock will fall back into the crater in a molten form, where, deep in the transient crater, it will encounter rock, the ambient temperature of which is already at or close to its melting PT condition.

The second source of melting occurs where the ambient PT conditions are already near to melting of the LVZ. We have argued that a significant reduction in the vertical pressure, brought about by the passage of a tensile stress-pulse, will cause instantaneous 'pressure release' melting in those volumes of rock where the PT conditions are most critical. This will involve the development of tens of millions of cubic kilometres of melt, but will not require the generation of temperature rises of several hundred degrees above ambient. Therefore, such melting is not likely to give rise to the tomographic evidence that would result from plumes.

6.7 Cratering and the development of oceanic and continental basalts

As we have seen in Chapter 5, the development of a crater and its attendant structures is a complex affair. The surface, or near-surface, features are reasonably well documented. Movements set at some depth below the crater can only be described in general terms. Moreover, precise definition of the limits of such movements and the time necessary for them to reach completion have not yet been definitively ascertained.

Let us now attempt to indicate these limits in general terms, starting with the development of the *transient crater* (Figure 6.17). Such craters have a simple cross-sectional geometry. The profile approximates to a parabola, and the ratio of the depth to diameter of the crater is approximately 1–3.

We have noted that continental flood basalts are much smaller in volume than the largest oceanic plateau basalts. This relationship, coupled with the tremendous quantities of melting associated with the development of the larger OPB, may present traditional theorists with problems. We shall, therefore, begin with a discussion of the effect of impacts leading to the development of the largest of oceanic plateau basalts, namely, the Ontong-Java feature.

The area of this massive eruption is approximately one third of the area of the coterminus USA, and its age was originally estimated to be approximately 119 Ma (Coffin and Eldholm, 1992), while much of the surrounding ocean floor dates from 130–140 Ma. More recently, Gladczenko *et al.* (1997) updated the age of this PB to 122.4 +/– 0.8 Ma.

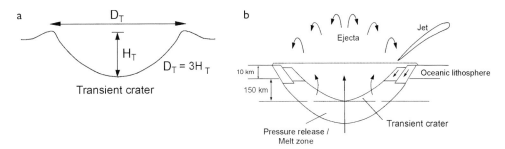

Figure 6.17 (a) Section and relative dimensions of a 'transient crater' in crustal rock. (b) Simple diagramatic representation of the initial geometry of an impact crater in thin oceanic crust.

Because the Ontong-Java PB is only a little younger than the surrounding ocean floor, we can infer that this feature developed in and extruded onto quite young and hence relatively thin oceanic lithosphere.

Parsons and Sclater (1977) modelled the N Pacific Plate and concluded (see Figure 2.5) that for the period 20–30 Ma, the 500°C contour had a depth from 15–20 km. As the 500°C contour will approximate to the lower limit of the strong part of the oceanic lithosphere, the strong layer will have a thickness of about 17.5 km, on average, while the junction between the oceanic lithosphere and the asthenosphere for the period 20–30 Ma will be at about 45–55 km. The oceanic lithosphere, below 17.5 km, rapidly becomes progressively weaker and approaches zero strength (at normal slow geological strain-rates) near the LVZ.

Let us assume that these thicknesses of strong lithosphere are typical of the young oceanic lithosphere, into which a large comet strikes and generates a transient crater 100 km deep and 300 km in diameter (Figure 6.18a).

Upon impact, the lithosphere around GZ instantaneously (or at least in a matter of tens of seconds) goes to zero thickness (i.e. β, in the McKenzie model, is infinite). Here we have the situation that McKenzie (1978) has sought, but *not* in the manner he envisaged. This process is not attained by stretching, but by explosive elimination. In a few tens of seconds the lithosphere, over an assumed diameter of 300 km, is vapourised, melted or comminuted and ejected into the atmosphere and stratosphere.

In the central zone of the crater, the asthenosphere will also be ejected. This part of the asthenosphere that has hitherto been subjected to a confining pressure of about 35 kb at a depth of 100 km, is suddenly subjected only to atmospheric pressures which, instantaneously, result in pressure-release melting in the upward cascade of blocks, blebs and particles that make up the ejecta.

The asthenosphere that remains beneath the transient crater floor also experiences instantaneous pressure-release melting, which is likely to develop to a depth of over 200 km, where it will be subjected to a release of vertical pressure from 35 to 70 kb, and to an ejecting pressure of at least an order of magnitude greater than that experienced in the most energetic of volcanological events. This will cause the crater floor to flow rapidly, even explosively, upward and the crater walls to collapse inward.

The crust and perhaps part of the upper 17.5 km forming the strongest part of the lithosphere will be blown upwards and outwards, possibly to form a recumbent isoclinal fold, the upper limb of which will be inverted (Figure 6.18a). Within the limits of this inverted limb, the thickness of the 'strong' layer may be doubled, and could, temporarily, stand as much as 8 km above the original base level of the oceanic crust.

In the walls of the crater, the weak asthenosphere below the remaining lithosphere, beyond the transient crater, will be eased sideways into the initial crater. The lithosphere adjacent to the crater walls loses its lateral cohesion and basal support and, being more dense than the asthenosphere, will tend to slide towards the crater and sink into the molten upward and sideways flowing asthenosphere, which is now rapidly rising towards the surface.

Soon, some, or even most, of the ejecta begins to return to the crater. That proportion of the ejecta that originated as 'cold' lithosphere and maintained its unmolten form will fall back into the melt and, because it is more dense, will sink. The returning blebs of molten material, whether they originated in the lithosphere or the asthenosphere, will return to the upwelling flood of molten material from the asthenosphere and will meld with it to enhance the upward flow of magma.

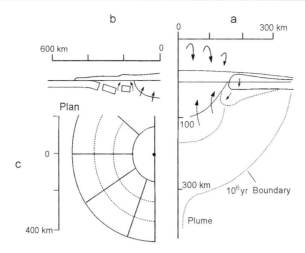

Figure 6.18 (a) Initial development of a transient crater in thin oceanic lithosphere. (b) Simplistic representation of the effects of a major impact. (b1) Section showing the development of an impact crater and the modes of deformation that may operate for 1 Ma. (b2) Section showing break-up and foundering of oceanic lithosphere, thereby allowing molten asthenosphere to reach surface. (c) Plan view showing radiating vertical tension fractures and arcuate failure lines which permit cantilever sections of oceanic lithosphere to break away and founder.

The molten material will occupy a slightly larger volume than the original unmolten material, so that the crater will eventually fill beyond its original level, and extrusive lava flows will begin to develop and extend over undeformed ocean floor. Soon, the former limits of the crater are hidden, or destroyed.

In addition (see Figure 6.18b(2)), melt extrudes up vertical fractures which develop between the sinking blocks, so that the 'igneous body' rapidly extends well beyond the crater's original bounds.

On impact, the pressure-wave blows away the ocean in a tsunami. However, the ocean returns in a matter of hours and rapidly re-establishes itself, so that the lava flows, being emplaced at and after this time, are rapidly cooled and develop into steep-faced plateau basalts.

White and McKenzie (1989) suggest that, in their β-model, the proportion of melt that is produced by sudden pressure-release is about 30 per cent of total volume experiencing pressure-release. One may infer that these authors tacitly assume that the melt gradually collects together and rises to the surface as magma. In the impact model proposed above, the situation is different. There is no retaining layer in the upper level of the plate, keeping the 'lid' on the molten material. In this model, the lid is blown away so that even if only 30 per cent of the volume of rock affected by pressure-release results in molten material, this will give rise to a weak solid/melt slurry that is unconfined at the interface, as defined by the transient crater; so that the slurry of solid/melt may be ejected, without substantial differentiation of the melt. The melt that is generated beneath the walls and floor of the transient crater migrates rapidly inwards and upward. In so doing, it causes the ambient pressure in the lower, and laterally more distant, areas to experience a smaller rate of flow. Progressively slower and slower migration of melt material will work its way towards the surface. Eventually, and it may take a million years, the inward and upward migration of molten material slows and eventually ceases. An indication of the mode of this evolution is shown in Figure 6.18b. If the migration zone extends as indicated in this diagram, the volume of erupted rock will be of the order of 35×10^6 km^3.

A secondary mode of eruption will also come into play. This is related to the radial, vertical tensile fractures which develop in the oceanic lithosphere around the impact. These features, which are discussed in Chapter 5, where they were formed as the result of the Snowball 500 ton explosion, are also represented in Figure 6.18(b2) and (b3). These vertical fractures can be attributed to the tensile stresses generated normal to a vertical line passing through GZ. The tensile stresses only occur when the stress intensity of the compressive pulse falls to a level where an elastic response takes over. This response probably occurs when the radiating stress-pulse (P_r) has a magnitude up to 100–1000 kb. (We discuss the mode of development of these vertical fractures later in this chapter, where we will argue that transient open fissures can develop down to depths of about 140 km.)

The lithosphere is relatively cold and of higher density than the asthenosphere. Consequently, when the radiating fractures are generated, these planes form paths of easy movement for asthenospheric migration. The cold lithosphere, near the crater boundary, will begin to sink, so that a series of cantilevers develop. The tensile stresses that form at the upper surface of the cantilever can cause circumferential as well as radiating fractures, so that quite large areas of lithosphere will break away and founder, to be flooded by a corresponding volume of asthenosphere rising to the surface, where it forms part of the erupted mass.

Because the thickness of the lithosphere is about 50 km, it will take time for the temperature in the sinking body to equilibrate with that in the asthenosphere, so that it eventually ceases to sink. This mechanism, which helps the upward migration of the asthenosphere towards the surface, can give rise to a further volume of extruded asthenosphere which may be comparable with that generated by the transient crater.

Hence, we suggest that the Ontong-Java PB, and other similar features, result from impacts in extremely thin oceanic lithospheres and are caused by only modestly large impacting bodies, which give rise to a transient crater of perhaps 300 km diameter.

We have invoked the assumption that our model, which represents the Ontong-Java PB, results from a major impact event. From what we have already learned, it is to be expected that a record of this event will be manifest in the track of a known point (an island currently adjacent to the Mariana Trench) in the general vicinity of the erupted body. Such a track is shown in Figure 6.15a, where it will be seen that, from 123.0 to 119.3 Ma, the oceanic lithosphere moved westward (270°) at a uniform speed. From 119.3 to 115 Ma, the track runs at about 240° at a slightly slower rate. We can infer from this track that a major impact occurred at a modest distance from the island marker at 119.3 Ma. This figure is close to, but slightly older than, the age (as cited in the earlier literature) of samples taken from the Ontong-Java PB. The more up-to-date age puts the emplacement of the Ontong-Java PB at 121.6–123.2 Ma, which is over 2.6 Ma earlier than the inferred impact. However, the oldest biostratigraphic ages of basal sediments recorded on the Ontong-Java PB is 117.5 Ma. We suggest that the precise timing of the impact and, possibly, the initiation and development of the Ontong-Java PB has not yet been exactly established.

Coffin and Eldholm note that, in addition to the main development of the plateau basalts at about 120 Ma, there was also a later eruption at about 90 Ma; and this too, they suggest, resulted from the emplacement of a minor plume.

We have argued above that the main Ontong-Java PB was the result of a major cometary impact. Rather than assume that the later eruption was the result of a plume, it was reasonable to ascertain whether or not the 90 Ma eruption could have been induced by a relatively minor impact. Accordingly, we plotted the track of the NW tip of Bougainville Island (Figure 6.15b). This island lies immediately to the south of the Ontong-Java PB. It will be seen from Figure 6.15b that from 96 to 93 Ma the tip of Bougainville Island defined a linear track, at a constant velocity. At 93 Ma, however, there was a sudden 3° change in direction of track, which continued at least to 88 Ma. There was no detectable change in the velocity of motion.

We suggest that this behaviour is consistent with the results of a relatively minor impact which probably gave rise to a crater, possibly a little smaller than 100 km in diameter. Such a crater could require a significant period during which basalt was erupted at a relatively slow rate, compared with that associated with the earlier major eruptive event.

We therefore conclude that the evidence supports the thesis that both periods of eruption of the Ontong-Java PBs are the result of impacts. We further suggest that the existence of three basin flood basalts, referred to earlier, which developed in the vicinity of the Ontong-Java FB, lends further support to our argument.

As regards the size of the earlier impacting body, a large meteor, with a diameter approaching 40 km, would experience relatively modest retardation, even if it struck in deep ocean. It is sufficient to note here that the model we proposed for the generation of the Ontong-Java event not only required that the strike took place close to a spreading-ridge, but that it also occurred where the depth of ocean was relatively shallow. That is, the depth was only about 1–2 km. This is confirmed, to some extent, by the observation that the upper surface of the Ontong-Java PB has been eroded, indicating that the basalts once stood above sea-level.

6.7.1 Effect of lithospheric thickness on oceanic and continental PBs

It can readily be inferred from the arguments presented above that large volumes of melt can best develop in oceanic lithosphere if the impact occurs near the ridge, where the oceanic lithosphere is thin and where the depth of water is relatively shallow, so that the transient crater can extend well down into the asthenosphere.

Several of the large oceanic plateau basalts, such as the Ontong-Java, Shatsky and Minihiki PBs, are all surrounded by oceanic lithosphere only a little older than the various cited extrusive bodies, so it can be inferred that they formed when the ocean lithosphere on which they are set was relatively young, and therefore thin.

Mature oceanic lithosphere will approach 100 km in thickness, and so approximates more closely to the average thickness of continental lithosphere. It can readily be inferred from Figure 6.19a that it requires a transient crater diameter of at least 300 km to reach and breach the asthenosphere with a thickness of 100 km. The volume of melt produced by such an impact will be small, relative to that generated by the model shown in Figure 6.18. Also, it will be noted that the upwelling melt will mainly be restricted to the axial regions of the crater, so that the intrusion will be more likely to resemble diapiric intrusion (Figure 6.19b) than the broad upwelling from the base and the walls of the crater represented in Figure 6.19a.

One may infer from Figure 6.19, that impacts of somewhat smaller energy, which produce transient craters that do not completely penetrate the lithosphere, whether it be oceanic or continental, may result in a hotspot which produces a more modest volcanic activity at the

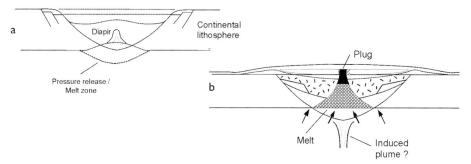

Figure 6.19 (a) Section showing transient crater with a diameter of 300 km which has developed in continental lithosphere with a thickness of 100 km. The volume of melt will be significantly less than that represented in Figure 6.18. Moreover, the melt produced will rise diapirically to produce relatively modest eruption of basaltic material. (b) The impact could induce the generation of a plume. As the plate moves away over the impact site a remnant hotspot may exist for many tens of millions of years. The black central area represents a 'plug' of the type inferred to exist off-shore of Bombay (after Negi *et al.*, 1993).

surface, but which may possibly induce the development of a plume of modest vertical extent. Alternatively, as was noted in the previous chapter, larger impacts may generate antipodal hotspots.

However, even if an impact causes a transient crater of, say, 400 km diameter, so that the asthenosphere is exposed to a depth of about 33 km (assuming the continental lithospheric thickness to be 100 km), although a considerable volume of pressure release melt will be engendered, it will not give rise to the same degree of extrusion that would be associated with an oceanic event with comparable dimensions of crater diameter and lithospheric thickness. Overall, the density of continental lithosphere is significantly less than that of the melt formed in the asthenosphere, so, as indicated in Figure 6.19a, the walls will not sink in the way that those of oceanic craters are able to do. The uppermost crater walls will be deflected sideways and upward, thereby rendering upward migration of the melt; a slower process than in an oceanic event. The development of the recumbent flat-lying fold increases the crater diameter at the surface, as the shock wave associated with the impact causes fold-back of the upper continental units to define the crater rim.

Within the following tens of minutes, ejecta thrown out of the transient crater will begin to return. This fall-back of material will bring the crater floor up to a level of about 3–5 km below the rim. The returning ejecta, especially that made up of crustal material, will not readily sink into the upward moving melt and so will tend to constrict its flow in the upper levels of the erstwhile transient crater.

Neither the OPBs nor the CFBs leave any direct evidence of cratering. This means that the extruded melt must supply not only the estimated volume of basalt flows that extend over wide areas of countryside, but must also in-fill the crater, which is left after ejecta falls back to Earth, but which does not necessarily fall into the crater itself. As regards major oceanic impact events, the returning lithospheric ejecta, as we have noted, tends to be heavier than the uprising melt and, accordingly, sinks into it. Also, because the density of cold blocks of oceanic lithosphere beyond the crater rim, is greater than that of the asthenosphere, it will tend to sink around the limits of the 'hole' caused by the impact. Hence, the melt will be able to rise to relatively high levels, so that again no obvious evidence will remain of the crater on the ocean floor.

It is interesting to note that, when describing some ocean plateau basalts, authors (e.g. Burke *et al.*, 1988) note that the 'crust' may be twice as thick as normal. Why this extra thickness should exist, and how it developed, rarely receives comment, except that it is postulated that it may represent 'some form of compensation for the excess height of the plateau basalt above the surrounding sea-floor, which often amounts to some 4 km'. However, it will be noted that the crater represented in Figure 6.18b is surrounded by a flap of overturned rock. Oceanic lithosphere (except for the crustal region with its sills, dykes and lava flows) is likely to be strong and uniform. This layered, crustal region will be relatively weak and readily deformed into an overturned 'flap'. Hence, we suggest that the thickening of the 'crust' in some OPBs could be regarded as further diagnostic evidence that large oceanic basaltic bodies, where such anomalous thicknesses of crust exist, originated by meteoritic impact.

A series of tensile fractures will almost certainly radiate from the crater, along which molten material in the asthenosphere may migrate upward; thereby 'lubricating' and keeping open the walls of the fractures. These fractures may facilitate the inward movement towards GZ of large wedges of lithosphere in response to the stored stresses within the plate, as well as the result of the boundary forces on the plate. As a consequence, the whole plate, or at least a large area within the plate, will be wholly, or largely, de-stressed. In the main, within a few thousand years, the horizontal stresses in the solid lithosphere, away from the crater, will be those which are generated by gravitational loading. Only if the impact occurred a long way from the spreading-ridge, will the horizontal stresses, induced by the gravity-glide forces, become rejuvenated and (coupled with the vast reduction in the basal drag over an area of several millions of square kilometres) give rise to an increase in the rate of plate motion.

In the situation represented in Figure 6.16, where the de-stressed area occupies such a large proportion of the plate, the effects of the lateral boundary stresses may have a dominant effect and push the plate significantly off its original line of movement. Indeed, the lateral boundary stress could change the line of movement by 90° (or even give rise to a change in track of 120° or more). Moreover, because of the very considerable reduction in basal resistance in the 3×10^6 km^2 area of the plate, the whole process of readjustment in speed and direction of motion of the plate is likely to occur in a period of the order of 10,000 years.

As we have noted earlier, when dealing with the generation of continental flood basalts, one is concerned with an even more difficult task of interpretation. We have noted that continental lithospheric units are usually thicker than oceanic lithospheres and, moreover, that they usually exhibit varying thickness. Most often, they are about 100 km thick, but may sometimes be as thin as 50 km or reach to depths of 200 km or more. Moreover, they do not usually exhibit evidence of an underlying LVZ. However, lateral movement does take place and this is probably dominated by differential motion on an induced shear zone, which may have a viscosity that is somewhat smaller than that of the normal asthenosphere.

Thus, the influence of a major impact on a continent is more difficult to assess. However, it is likely that a continental flood basalt comparable with the Deccan Traps could require an impact crater of at least 300 km diameter. Based on geophysical evidence, Negi *et al.* (1993) concluded that the impact had a GZ at some small distance off the coast of Bombay; from which, one may infer that the impact probably occurred close to the junction between oceanic and continental lithosphere. Consequently, the effects of such an event on, or at, the edge of a continental mass are likely to be reasonably comparable with those of an impact in 80 km thick oceanic lithosphere.

6.8 The splitting of continents

One may readily infer from the known geology that more than one mechanism must be involved in the break-up of continents. For example, from movement patterns of the San Andreas and associated faults of this plate boundary complex, it can be inferred that slices of continent can be cut away from the main continental unit by means of major strike-slip faults. Similarly, the strike-slip Baykal structure of Central Asia also appears to have the potential to split that continent.

The break-up of much of Gondwana, however, appears to have been initiated by tensile failure. In particular, the moving apart of Africa and S America, resulting in the formation of the S Atlantic, appears akin to extension. But it has been argued earlier that the development of tensile stresses, induced by traditional plate tectonic mechanisms, requires special pleading. Moreover, the spreading of the S Atlantic results from the production of basalt at the spreading-ridge. How then can such a break-up by tension be brought about?

It will be noted in Figure 6.20 that features which developed as the result of the Snowball 500 ton explosive experiments included peripheral, radiating tensile fractures, which extended outward from the rim of the crater for a distance at least equal to the diameter of the crater. In the post-explosion survey, it was established that these vertical fractures extended to and beyond the deepest level of excavation. Indeed, it was inferred that in the Snowball 500 ton TNT explosion, the fractures extended to some distance beneath the water-table, downward through the sand-layer to a probable depth of at least 20 m (Jones, pers. comm.). The diameter of the Snowball crater was 87 m, so that the ratio of the depth of the radiating fractures to the diameter of the crater is approximately one quarter.

In the Suffield tests, one may infer that, because the sediments were only submitted to gravitational loading, the pre-explosion, horizontal stresses at any depth would be approximately equal. However, as we have seen, the horizontal stresses in most plates are likely to be biaxial, with the axis of greatest principal stress oriented in the direction of absolute plate motion. One may infer, therefore, that the radiating vertical fractures, which will develop as the result of a major impact, may not be of equal length, but will reach their greatest development in the direction of absolute plate motion (Figure 6.21a). It will be seen that some dykes (A, B and C) associated with the Central Nevada Circular Structures (CNCF), could well be associated with radiating fractures and that their lengths are comparable with, or in excess of, the inferred crater diameter (Figure 6.21b).

Features on the Moon lend further support. Working from the basic map of the Copernicus crater and surrounding area, Shotts (1968) inferred from the linear alignment of groups of

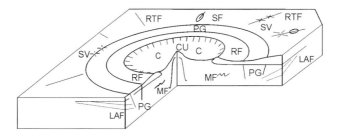

Figure 6.20 Block diagram showing relationship of vertical, radiating tension fractures to other features produced by the Snowball 500 ton explosion. CU = central uplift; C = crater; SV = sand volcano; RTF = radial tensile fracture; SF = shear fracture; PG = peripheral graben; MF = multiple folding; LAF = low angle fault; RF = recumbent fault.

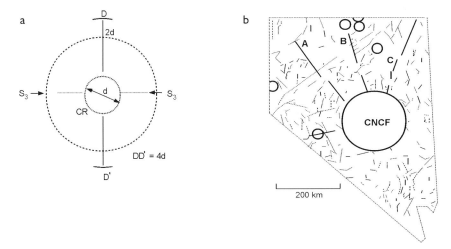

Figure 6.21 (a) Extent of radiating, vertical tensile fractures likely to be associated with a major impact when the target area is under biaxial compression. (b) Dykes associated with the Central Nevada Circular Feature (CNFC), several of which (A, B and C) have been emphasised and could be classified as vertical radiating structures.

three or more small craters that vertical fractures radiated from the Copernicus crater and extended for several crater radii beyond the lip of that major crater. Shotts suggested that the alignment of these minor craters was the result of 'exhudation of volatiles'. Jones (1995) inferred that radiating dykes, which connected and fed the small craters, represented an immediate post-impact event.

6.8.1 Opening of the S Atlantic

A 'traditional' interpretation of how the development of the Paraña (or Tristan) plume, combined with that of the smaller St. Helena plume, gave rise to the splitting of S America from Africa and the opening of the S Atlantic (Wilson, 1992) is shown in Figure 6.22. We have already argued that a plume-head, with a diameter of about 2000 km and a maximum elevation of about 2 km cannot generate the magnitude of lateral stress required to split open a super-continent. A much more powerful mechanism is required.

In detail, the break-away of S America from Africa was almost certainly a complex process which is not yet fully understood. It has never been our intention to attempt to explain the many historic events involved in plate motions. We are primarily concerned in this text with the various driving mechanisms involved. Consequently, here we shall initially restrict our comments to the opening of the southernmost portion of the S Atlantic, to the south of the Paraña-Entendeka flood basalts which, as indicated in Figure 6.22, was situated some 800–900 km north of the current estuary of the Rio Plata.

6.8.2 Original relationships of landmasses adjacent to the 'S Atlantic'

Before we discuss the opening of the S Atlantic, we must first indicate the apparently problematical relationship of Graham Land (or Antarctic Peninsula) to Tierra del Fuego and its hinterland. This problem is clearly seen in Figure 6.23a, where the landmasses of S America

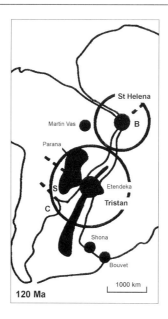

St Helena

Martin Vas   B

Parana

S   Etendeka

C   Tristan

Shona

Bouvet

1000 km

120 Ma

Figure 6.22 Traditional reconstruction of Africa and S America during the initial stages of opening of the S Atlantic, supposedly generated by plumes (after Wilson, 1992).

(SA) and that of the northern portion of Graham Land (GL) are superimposed one upon another. This is an impossible situation. The compilers of the Atlas program were, of course, aware of this problem, but the data that permitted a solution to it did not, at that time, exist. So, wisely, they left the decision making, as regards this impossible situation, to interested workers and the accumulation of relevant data. In the early literature of this area, Sir John Barrow (1830), followed by Arctowski (1895) and Suess (1909), suggested that 'the Andes are seen again in Graham Land'. Such coincident viewpoints are interesting, but not direct evidence, and, moreover, are difficult to maintain in the light of the existence of the Tierra del Fuego Orocline and the relationship of S America and Graham Land to the Scotia Arc.

We shall discuss this problem further in the following chapter. It is sufficient here to note that it can be shown, from paleomagnetic data, that the orocline came into existence somewhere between 80 and 65 Ma.

Moreover, Antarctica is nowadays considered to be made up of a number of platelets, of which Graham Land is one. Lawvers (1985) indicated that this platelet did not take up its current position with respect to the rest of Antarctica until 119 Ma ago. Hence, we suggest that the overlap anomaly in Figure 6.23a is resolved. An uncurved southern tip of S America and Graham Land were adjacent (Figure 6.23b). Indeed, there is no reason why one should not infer that Graham Land was a continuation of S America. One cannot but applaud the intuitive conclusions reached by the early explorers in this area.

We have noted that major impacts are associated with major changes in track direction and/or changes in rates of plate motion. The most spectacular example of this type of behaviour that we have so far encountered, is represented in Figure 6.8a. This figure represents the movement of the point where the Rio Plata widens, over a period from 140 to 134 Ma. From 140 to 135.1 Ma the point followed a smooth, northward curve, at a near constant

a

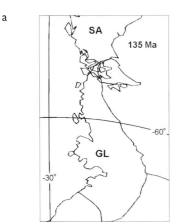

b

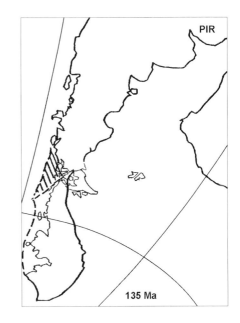

Figure 6.23 (a) Impossible superimposition of S America and Graham Land at 135 Ma as represented by the Atlas program. (b) Showing the straightened version of S America, prior to the development of the Patagonian Orocline and the juxtaposition of Graham Land which, 135 Ma ago, was not connected to Antarctica.

rate of about 6.0 km Ma^{-1}. At 135.1 Ma there was a sharp change in track to about 300°, accompanied by a prodigious increase in rate of about 150 km in 1.0 Ma.

What happened to Africa? The track in Figure 6.8b defines the movement of a cape near 'Mozambique' at a much more detailed scale than is shown in Figure 6.8a. Hence, the answer to the question posed is 'nothing of any consequence!'. At this time Gondwana constituted a coherent mass which included Africa, Madagascar, India, most of Antarctica and Australia. Also, to the east of this landmass, there existed oceanic lithosphere, which for thousands of kilometres was devoid of subduction zones which trended parallel to the African coastline. Such a huge continental and oceanic mass is not easily deflected from its path. As can be inferred from Figure 6.8b, it continued on its northward migration without detectable deviation of speed or direction.

As regards that portion of Gondwana that was to become S America, this extended for only about 2000 km to the E of the impact point, at which distance the continent was, and still is, thrusting over a subducting oceanic lithosphere. Subduction zones offer the least resistance to plate movements. Hence, there was only one direction in which proto-S America could move. However, this amazingly high rate of movement of 150 km Ma^{-1} could not possibly be sustained.

The rate slowed as S America was forced to move westward, and to over-ride the down-going oceanic lithosphere of the appropriate Pacific plates. The induced increase of strain-rates in the continental rocks of what was to become S America, led to increased stress intensities which could not be supported by the continental lithosphere. It can be inferred that the Andes, at this time, must have been subjected to considerable uplift and deformation, and for at least a million years, portions of this mountain chain may have attained altitudes approaching that of the Himalaya.

There is no known current, conventional, mechanism accredited with driving plates, that could have caused such an abrupt change in track direction and, possibly unsurpassed rate of plate movement.

On the basis of the arguments put forward in this chapter, it is more than reasonable to assume that 135.1 Ma ago, Gondwana was struck by a very large impacting body that gave rise to the Paraña Flood Basalt Complex, which is related to the much smaller Etendeka flood basalts adjacent to the W African coast. From the reaction of Gondwana, it is reasonable to infer that the diameter of crater that resulted from this impact event must have been large and may possibly have been over 500 km. (It will be recalled that Oberbeck *et al.* (1993) indicated that three craters with a diameter of 750 km would be expected to develop in the last 2000 Ma (Figure 5.4).) As will be seen in Chapter 8, we argue that the numbers of events of specific dimensions cited by these authors are underestimates by a factor which probably exceeds 2. Thus, it may be inferred that at least one very large crater occurred in the last 250 Ma.

Let us now consider how development of radiating, vertical fractures results from an interaction of the very brief, near-hemispherical pressure wave induced by the impact, which is superimposed upon an existing, probably orthogonal, stress trajectory that existed in the lithosphere prior to the impact. A rigorous analytical treatment of the interaction of these stress fields is well beyond the scope of this book. However, a simplistic assessment of the interaction of these stress fields, for a very brief period, is instructive.

We shall initially consider the stresses that could develop in an oceanic plate, where the density of rock in the lithosphere is reasonably uniform and a little greater than that of the asthenosphere. Firstly, we reduce the problem to a two-dimensional vertical plane situated beyond the impact crater (Figure 6.24a). This vertical plane we take to contain the *in situ* principal plate stresses prior to impact, with the least horizontal stress assumed to act normal to the plane of the section. The trajectory of the stress-pulse is represented in this figure as dashed lines. Further, we take this plane (Figure 6.24a) to give rise to an outwardly moving spherical surface, in which tensile failure will take place. In this section, all the stresses are shown as compressive, so how may tensile failure take place? This can be inferred from Figure 6.24b, which shows in plan view the stress-pulse which causes transient, radial displacement of a given arcuate marker from A to B. This, in turn, gives rise to a lateral tensile strain, and hence tensile stress develops. This elastic behaviour only develops such tensile stresses when the stress wave reaches the region beyond the crater rim, where the rocks behave in an elastic manner.

Let us further assume that the onset of elastic deformation caused by the 'front' of the stress-pulse takes place at 300 km from GZ and, within milliseconds, pushes back the initial marker circle (A) to the outer marker (B). This displacement induces tensile strains in the outer, displaced marker (B). This is a process akin to 'blowing up a balloon'.

If we assume that the marker (A) is displaced by 3 km, one can readily calculate the circumferential stretching that is induced in marker (B). Thus, the circumference of the initial marker at a distance of 300 km from GZ is C_a = 1886 km, while the circumference of the outer marker is C_b = 1905 km, so that a circumferential stretching of 19 km, i.e. a tensile strain of −0.010 has taken place. If we take the Young's modulus (E) of the upper lithosphere to be 5×10^6 bar, the stretching will induce a tensile stress of −50 kb.

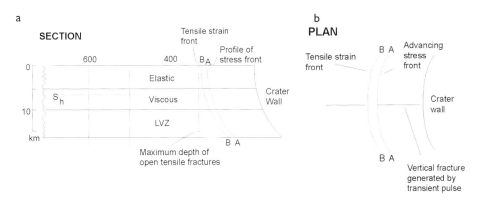

Figure 6.24a Showing diagramatic section in the plane of S_1S_2 outside the crater and the profile of an elastic stress-pulse. The depth to which open tensile fractures may develop is also shown. (b)Plan view of pulse stress shown in (a).

Let us now consider the initial tectonic stresses that exist in the plate immediately prior to impact. If we assume that a basaltic lithosphere has a density of 3.2 gr cm^{-3}, then the vertical stress induced increases by 320 bars for every kilometer of depth. Hence the vertical stress at 50 km is 16 kb and at 100 km and 150 km the vertical gravitational stress will be 32kb and 48 kb respectively.

As we have seen in Chapter 2, this vertical stress, the result of gravitational loading, induces a horizontal compressive elastic strain in the upper lithosphere such that a horizontal stress (S_h) is generated by the relationship

$$S_h = S_v/(m-1)$$

where m is Poisson's number, which, for basaltic rock, has a value of about 4.0. Hence at a depth of 50 km, a horizontal compressive stress of 5.3 kb is generated.

Plate stresses, in the direction of movement (which is in the direction indicated in Figure 6.24) are likely to be about 8 kb and these will induce a further horizontal component of stress of about 2 kb acting normal to the plane of the section shown in Figure 6.24a.

At depths significantly deeper than 50 km, the gravity-induced stresses become 'hydrostatic', i.e. they equal the vertical stress in all directions. Thus, at a depth of 150 km, the lithospheric (or possibly the asthenospheric) pressure will be about 50 kb.

The magnitude of the tensile stress ($-P_t$) generated by the compressive radial stress (P_r) is given by the relationship

$$-P_t \approx (v/(1-v)).P_r \tag{6.1}$$

where v is Poisson's Ratio (i.e. the reciprocal of m, Poisson's number). From seismic signals and experimental rock mechanics, the value of m for basic rock is commonly about 4.00. Hence, it follows for these specific values that $-P_t = -P_r/3$. These stresses would, for a very brief period, be superimposed on the ambient compressive stresses which exist in the mantle immediately below the oceanic lithosphere. The tensile stress-pulse would induce a relatively small quantity of pressure-release melt, thereby introducing a 'hydrofrac' situation where the transient $-P_t$ stresses will be augmented by the fluid pressure of the pressure-release melt fluid.

We have estimated that a tensile stress of about 50 kb is sufficient to induce tensile fracture at a depth of just over 150 km in oceanic asthenosphere. Also, it is interesting to note that if 12 radiating fissures were to develop, they could each attain a transient, maximum width at the surface of about 1.5 km.

If we assume that the crater diameter generated by this event was 500 km, the duration of the compressive wave would not be more than about 100 seconds. The development of vertical fissures will only come about when the stress-pulse is reduced to elastic behaviour. Thus, one is concerned with fissures that are open for only a very brief period.

The events which, we suggest, ensue in this period are indicated in Figure 6.25. These tensile fissures extend laterally (for at least 500 km) and downward, until they reach their maximum depth of possibly 150 km.

It is important to note that the estimates of stress given above are based on a constant density of rock which is consistent with oceanic lithosphere. However, we are here concerned with the generation of fractures in continental lithosphere and the underlying asthenosphere. Continental rocks are less dense than oceanic lithospheric rock, so that the vertical and horizontal stresses in continental environments will be less than those that develop in oceanic lithosphere, for a corresponding depth. Hence, for the magnitude of transient stress-pulse assumed, open fissures could develop to a deeper level in continental lithosphere than they would in oceanic lithosphere. However, to introduce a safety factor into our calculations we will assume that an open fissure is unlikely to develop in a continental lithosphere to a level greater than 150 km.

At this point, it is pertinent to cite the work of Petford et al. (1993). These authors carried out a theoretical analysis of the rate of intrusion of granitoid melt, with a viscosity of as much as 10^6 Pa s. They assumed that the dykes were 2–7 m wide, and that the density contrast between the magma and the country rock was 200 kg m^{-3}. These authors concluded that such a dyke would be emplaced in a 30 km thick crustal unit in about one month.

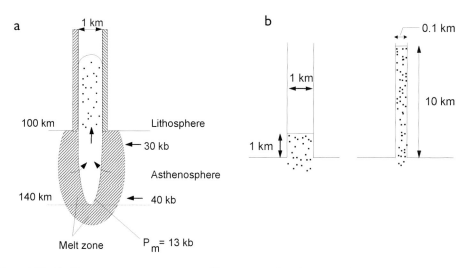

Figure 6.25 (a) Diagrammatic representation of fissure extending down into the asthenosphere, thereby generating a considerable cross-sectional area of pressure-release melting which begins to be ejected up the fissure. (b) As the walls of the fissure begin to close, this requires the melt to be forced rapidly upward.

Let us convert these conclusions to our model. Firstly, the viscosity of molten basaltic magma is about 4 orders of magnitude smaller than granitoid melt, at 10^2 Pa s. Hence, with a modest driving pressure, commensurate with that of the granitoid melt, basalt magma would be emplaced in a 30 km high column in a little less than 10 minutes.

However, we are concerned with a fissure which, briefly, may be over 1 km wide, so that, initially, no work would be involved in pushing back the fissure walls. Also, as we may infer from Figure 6.25a, the generation of such an open fissure suddenly transforms the ambient stress conditions in the asthenosphere, so that adjacent to the open fracture the 'containing' pressure is reduced to zero. Thus, at the walls of the fissure in the asthenosphere, which had been at a hydrostatic stress of 28–40 kb, is suddenly to (a somewhat enhanced) atmospheric pressure. Instantaneous pressure-release melting ensues in the asthenosphere adjacent to the open fracture. With the huge ambient hydrostatic pressure initially existing close to the fissure, the melt (and solid) slurry implodes into the fissure and moves upward into the void space in the overlying continental lithosphere. The driving pressure of the molten asthenosphere adjacent to the open fracture will initially be about 40 kb, an order of magnitude greater than an extremely vigorous volcanic eruption. Hence, the rate of ejection from the asthenosphere and injection into the fissure will be extremely rapid. If the fissure in the continental lithosphere remains fully open for only a few seconds, the slurry may be injected to a height of several (possibly tens) of kilometres into the lower layer of the continental lithosphere. As the pressure pulse passes outward, away from the crater, and decays in magnitude, the walls of the fissure begin to close (Figure 6.25b). However, the pressure driving the melt/slurry out of the asthenosphere is still being injected into the fissure, which now is being held open, as a dyke, by the hydraulic pressure of this inflowing material from below.

Here one has the problem of trying to squeeze toothpaste back into the tube, exacerbated by the fact that the tube is still experiencing compression. The melt/slurry can only move upward. Let us assume that the fissure was originally 1 km wide at the bottom of the continental lithosphere. Also, let us further make the extremely conservative assumption that in the first second in which the fracture opens, the fracture (above the asthenospheric/lithospheric interface) is injected by the melt-slurry to the modest height of 1 km above the lithospheric/asthenospheric boundary. If, thereafter, the walls of the dyke try to close, the dyke may becomes reduced in width to (say) 100 m. However, this lateral squeezing of the dyke must ensure even more rapid injection of the molten basalt to a height of 10 km. Closing of the tensile fracture will result in continued rapid injection to a height of 100 km or more into the continental lithosphere.

This situation is indicated in Figure 6.26, together with the potential static, vertical and horizontal over-pressures that will develop in the injected 'magma'. This static relationship is over and above the injection pressure of the 'slurry'. It can be inferred that the high horizontal over-pressure in the magma may blast its way through to the surface. Alternatively, it may generate a magma chamber in the upper levels of the continental crust. Such an intrusion will start extending the walls of the magma chamber laterally, thereby generating extension of the superficial layers in the crust to form a graben.

The development of this dyke into a ridge is indicated schematically in Figure 6.27. The ease with which the walls of the dyke may be pushed apart will depend upon the shear resistance at the LVZ. We have noted that a major impact may reduce the resistance to shear of the LVZ by about 2 orders of magnitude, to a radius of 1000 km or more, in oceanic lithosphere. If we take the radius of such a zone to occur below a continent through which

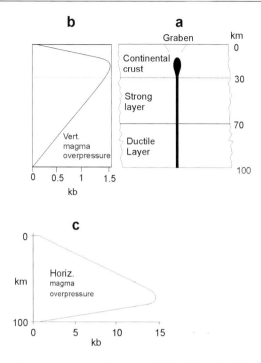

Figure 6.26 (a) The near-instantaneous emplacement of this dyke in the lithosphere is indicated, which subsequently gives rise to the development of a high level magma chamber, and a graben. (b) and (c) Indicate the static vertical and horizontal magmatic stresses, per unit extent in the direction of radiation of the dyke, owing to the mass of the molten material in the dyke. In the early stages of emplacement these static stresses will be greatly overshadowed by the dynamic pressure generated by melting in the asthenosphere. However, as these dynamic pressures decay, the magnitude of the static stresses will be sufficient to push back the walls of the dyke.

the dyke cuts, to have a comparable dimension, and the average lateral magmatic stress in the dyke to be 8 kb, the walls of the dyke would be rapidly pushed apart to a degree represented in Figure 6.27a. (It can readily be shown, by drawing circles about GZ, based on the size of crater, and extent of zone of easy slip, that the width of the easy-glide zone goes from about 2000+ km, 250 km from GZ, to 850 km, 600 km from GZ.)

This initial surge of energy would gradually decline, so that the dyke would evolve in stages (a–d), as indicated in Figure 6.27b, with the emplacement of uprising magma and diapirs cooling and 'underplating' the margins of the continental lithosphere.

By the time the ridge is well established, with the widening of the base of the initial dyke to 250–500 km (Figure 6.27c) in the first million years after impact, the continental lithosphere would be separated by an uplifted and developing spreading-ridge. At this stage, in addition to the hydrostatic pressure of the asthenosphere acting upon the lithosphere, gravity-glide of the continental lithosphere would generate a further 8 kb of lateral pressure, as indicated in Figure 6.27d. With only a small viscous retardation on the LVZ (over the short distance of about 250–500 km), to either side of GZ, this combination of asthenospheric pressure and gravity-glide would, in favourable conditions, give rise to the splitting of a continental mass. However, before we can apply this to the break-away of S. America from Africa, we must first consider the paleogeography of that particular area of Gondwana.

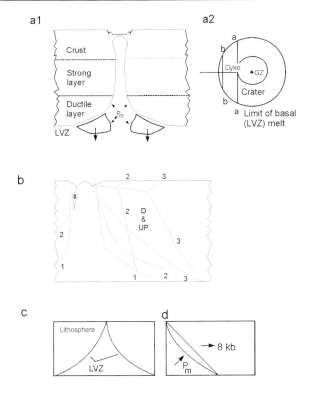

Figure 6.27 (a¹) Shows the early stage of the pushing aside of the dyke walls to generate the initiation of a spreading-ridge. The width of the initial dyke is determined by the ease with which the continental lithosphere can be pushed back. This is largely determined by the extent to which the 'viscosity' of the LVZ has been reduced by the impact. (a²) It will be seen that the section of aa and bb, the low viscosity LVZ about the impact site are extensive, so that the dyke will be wide. (b) Shows three stages (1, 2 and 3) in the early development of the spreading-ridge. The generation of the sloping lithospheric/asthenospheric interface is in part the result of high lateral asthenospheric pressure on the weaker layers of the lithosphere and the addition of rising asthenosphere diapirs (D) giving rise to under-plating (UP). (c) By the time the base of the asthenosphere has parted to about 200 km, the athenosphere will be comparable to the early conditions shown here. (d) As an approximation this diagram indicates the horizontal stress that can be generated as the result of gravity-glide, to which must be added the influence of the asthenospheric pressure (P_a).

It is reasonable to infer that one or more such radiating fractures must have been formed by the Paraña impact. We suggest that 135 Ma ago, a major and dominant fracture developed which trended at about 200°. This fracture, shown in Figure 6.28a, rapidly evolved into a young and initially narrow, but extremely energetic, spreading-ridge, which propagated to the ESE, until it encountered the Andean/'Antarctic-Peninsula' subduction zone. The resistance to movement of S America, as the result of the generation of this young spreading-ridge, would be roughly proportional to the length of the continental lithosphere (a, b and c in Figure 6.28a) in the direction of plate motion, approximately normal to the embryonic spreading-ridge. Hence, although the initiation of the 135 Ma phase of break-up of Gondwana was initiated by an impact deep within that super-continent, the opening of the S Atlantic progressed from the ESE to the WNW.

By this same mechanism, the portion of S America cut off by the new spreading-ridge would be driven to the SE, where it eventually made contact with Antarctica, to become the Antarctic Peninsula. However, such a collision between the two continental units requires one to assume that there was either a subduction zone between them, or else they made contact along a major strike-slip fault. This southerly spreading-ridge has been represented as a straight line. However, its development may well have been influenced by pre-existing weakness in Gondwana.

As may be inferred from the track shown in Figure 6.28b, this development and widening of the new spreading-ridge must have occurred within 1.1 Ma (from 135.1 to 134 Ma) so that from 134 to 133 Ma, the energy of the impact 'ran out of stream'. Thereafter the pace of spreading slowed significantly, with a significant change in the direction of movement of the S American plate.

A comparable major fracture must, almost certainly, have been initiated, running NE to NNE from the Paraña impact site. However, the track of this fracture is more clearly defined by the junction of the coastlines between S America and Africa (Figure 6.29).

This north-easterly major fracture, which would also be struggling to become an embryonic spreading-ridge, did not have the energy to penetrate far into the massive N African area of Gondwana. A simplified diagram of the mechanistic situation is indicated by the insert diagram in Figure 6.29. This represents a (horizontal) cantilever, which, along its 3000 km length, is acted upon by about 8 kb of magmatic pressure, which, of course, generates a tremendous force on the S American Plate. The bending of the (horizontal) cantilever generates tensile stresses along a vertical plane at right angles to the 3000 km long (NE–

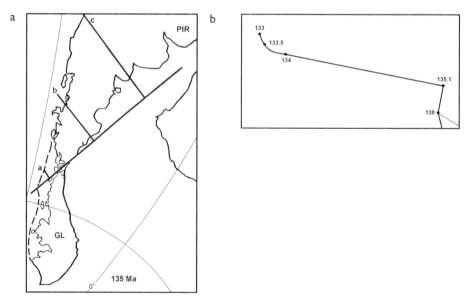

Figure 6.28 (a) Showing how the S Atlantic, although initiated to the north by the Paraña impact, 'opened' from the south. (b) Detail of track from 138–133 Ma, showing the sharp change in direction and rate of movement of the developing S American Plate: the initial amazing rate of movement from 135.1 to 134 Ma and the rapid slowing and change of direction from 134 to 133 Ma.

SW) fracture. These tensile stresses eventually become sufficiently large to cause vertical failure of the lithosphere, more or less at right angles to the embryonic spreading-ridge. Once this vertical fracture develops it becomes infilled with melt from below. (Indeed, it is probable that the St Helena hotspot developed at this time.) S America was then able to detach itself along this and other similarly formed fractures by strike-slip movement.

This conceptual model is supported by simple mechanical analysis and experimental experience. Hence, we suggest that the proposed model provides a viable explanation for the parting of S America from the rest of Gondwana and the opening of the S Atlantic – which the plume theory fails to provide.

6.9 Summary and conclusion

From a simple model, we predicted that plates should, as a result of major impacts, exhibit sudden changes in speed and/or changes in direction of motion.

We plotted the tracks for periods of several millions of years around the impact dates of three *certain* impact events, for which the diameter of the crater was at least 100 km. To this, we added the track of a 'near certain' impact event: together with the track of an impact for which the point of impact was not known with any certainty, but whose date of impact was closely constrained by stratigraphic evidence.

All five tracks showed the characteristics that our model predicted.

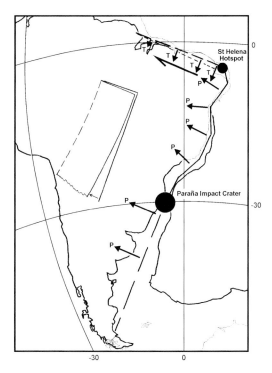

Figure 6.29 Opening of the northern part of the S Atlantic and the mode of splitting from Africa as the result of vertical fractures induced by the spreading-ridge pressure on the 'cantilever' represented by the inset diagram.

We then applied this concept to the eight continental flood basalts. All eight showed the same relationship between the age of the erupted basalts and associated change in track.

The Ontong-Java PB was similarly treated and it was shown that it also exhibited two tracks with a change in direction of movement at about 120 and 93 Ma which was compatible with the date of two phases of emplacement of basalts.

Thus, we have constructed tracks using the Atlas 3.3 program for five certain, or highly probable impact events, eight major CFBs and one ocean flood basalt, that have been formed within the last 250 Ma. The ages of emplacement of the various CFBs have been ascertained by various dating techniques. In all 14 instances we have shown that the tracks either exhibit a sharp change in direction, and/or a sudden change in speed of motion of the plate. In each of these instances, the age of the impact, or the age of the emplacement of the basalts was in reasonable agreement with the age of the track change.

We suggest that a score of 14 out of 14 is impressive. To quote that famous fictional detective, Hercule Poirot: 'If a thing happens once – it happens! If it happens twice – it is a coincidence! If it happens three times – it is a clue!!!' (We have cited a fair number of such 'happenings' in this chapter and will adduce more in those that follow.) This we suggest is more than a clue, it is extremely strong evidence that a *principle* is being revealed.

Chapter 7

Arcuate subduction features

7.1 Introduction

In Chapter 3, where we discussed subduction and the initiation and development of trenches, we only considered the development of such features in a two-dimensional plane, defined by a vertical line and the direction of plate motion. It was shown that trenches are most easily initiated near the junction between oceanic and continental lithospheres. Once such trenches are formed, their subsequent migration will usually be restricted and controlled by continental migration. However, quite a number of these trenches, which are closely related to the oceanic/continental junction, exhibit a well-defined arcuate form (Figure 7.1). In this chapter, therefore, we shall first briefly describe the geometry of such large, curved subduction features, which often have radii of curvature of over 2000 km, and shall outline the mechanisms which have been proposed to explain the development of these features.

However, we are not primarily concerned here with the mode of arc development, which may exhibit large or irregular radii of curvature; but shall concentrate on those examples in which the arcs form almost perfect semicircles and have radii of less than about 600 km. These features, especially ones which occur in an oceanic context, are few in number and, notwithstanding their intrinsic interest, have received little comment. These trenches and/ or associated island arcs, which are often quite remote from continental masses, are either

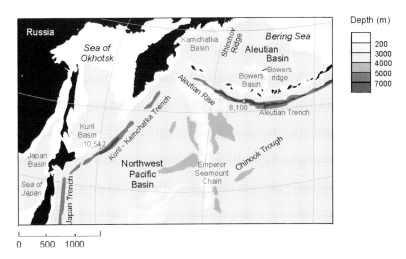

Figure 7.1 Trenches adjacent to the Asiatic Plate with very large radii of curvature.

almost perfectly semicircular in plan, or they exhibit a portion of a circular arc, which may be as small as 60°. Only five such features which exhibit an almost exact circular arc, with a spread from 60° to 180°, are identified. These features cannot readily be explained in terms of normal plate tectonics mechanisms. To date, in the few papers that deal with their development, these features have been discussed in terms of kinematics, i.e. how (without reference to force and mass) the positions of the various elements changed with time. As far as we are aware, no mechanistically based theory has been put forward to explain the initiation of these relatively small arcs.

The initiation of each of these five structures is shown to be associated with an event which gives rise to an abrupt change in direction of plate movement and/or a change in rate of plate motion. We have already seen that such changes in track, which are associated with major continental flood basalts (CFBs) and oceanic plateau basalts (OPBs), may be attributed to major impacts. Hence, based on the evidence of the change in direction in track and speed, and their well-defined arcuate morphology, we present a model for their initiation and development that requires them to be the result of major meteoric impacts.

However, we shall first briefly discuss the conventional explanations regarding the development of the larger arcuate features.

7.2 'Conventional' development of arcuate oceanic trenches

It is now understood that oceanic lithosphere may eventually become subducted at a trench, and that a belt of volcanoes forms above the site where the subducted oceanic plate reaches a depth of about 150 km (Figure 7.2). The simple geometry represented in this section can develop, whether the over-riding lithosphere is (a) continental, (b) oceanic or (c) transitional between (a) and (b).

The position of the subducting slab is inferred from geophysical data. However, these data do not generally permit one to infer the orientation of the movement-vector of the subducting plate. For this, one must know the direction of plate motion, which may often be oblique to the local trend of the trench.

Oceanic ridges, continents and trenches (with associated volcanic arcs) all migrate. The development of island arcs has been discussed in terms of slab-pull. However, without doubting the secondary and minor influence of slab-pull, it has been argued (Chapter 3) that the initiation of a subduction zone, using only the slab-pull concept, is impossible. Indeed, it was concluded that the most obvious site for subduction is where thick piles of sediments have accumulated onto the oceanic lithosphere, adjacent to the continental

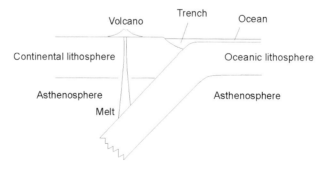

Figure 7.2 Schematic section through a hypothetical trench, subducting plate and attendant volcano.

margin. The load of these sediments causes deflection of the oceanic lithosphere, which is already subjected to high lateral compressive stresses, as a consequence of the gravity-driven sliding of the elastic part of the lithosphere down the slope of the LVZ, away from the spreading-ridge.

Because of the compressive stresses that occur in the upper lithosphere, where the gravitational loading of the sediments presses it down to form a syncline, coupled with the high lateral stresses, the lithosphere fails by thrusting and thereby may initiate the subduction of the oceanic lithosphere.

The initiation of a trench would occur at the weakest zone, where the sedimentary pile is at its thickest and causing the greatest flexure of the lithosphere, and where the lateral compressive stresses from the oceanic plate are high and of optimum orientation, e.g. at point A in the hypothetical situation shown in Figure 7.3. Once initiated, the trench will migrate laterally, following a gentle curve (to the right of A) of the continental/oceanic interface, or it may continue off-shore and traverse transitional, or even oceanic, crust. Alternatively, if the continental/oceanic interface makes a high angle to the direction of plate motion, the trench will die out, and the interface may become a strike-slip fault (Figure 7.3).

The rigidity of the ocean plates, normal to the direction of plate movement, is such that it is difficult to envisage how a trench or arc, once it has developed, may experience changes in curvature, except as the result of strike-slip movement on vertical or near-vertical fracture planes which cut and displace the trench.

Such a mechanism of lateral development and possible modification of a trench by strike-slip motion on faults may be invoked to explain the curved trenches which define the Pacific/Asian interface in the vicinity of Japan (Figure 7.1).

How ocean-plate/continent interface can completely dominate the siting and orientation of the off-shore trench, is shown by the feature which follows the line of S America. In the north, where Colombia, Ecuador and Peru bulge into the Pacific, the trench is curved to accommodate this bulge. For the length of Chile, from Arica in the north, to about 35°S, the trench is as straight as the coast. South of this specified latitude, the Pacific plates are currently at such a marked angle to the trend of the coast that subduction is apparently inhibited.

Thus, if the trench is formed where oceanic lithosphere is being subducted beneath continental lithosphere, whether a trench and associated volcanic activity is linear or curved can often be explained in terms of conventional plate tectonic mechanisms. However, these

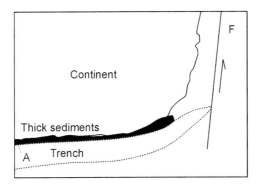

Figure 7.3 Schematic plan showing relationship of a trench to a continental plate.

simple relationships do not cover all curved subduction trenches that are adjacent to continental masses.

Some arcuate trenches, such as the Aleutian, Keril and Java Trenches, form relatively smooth, curved features with a radius of curvature of several thousand kilometres. Yamaoka (1988) has suggested that such features result from the buckling of the lithosphere, as the result of *uniform* horizontal compression of a spherical shell. It has been already noted that plates are usually in *biaxial* compression. Hence, such buckles should not have a circular, but an elliptical, plan. However, a curved section of an ellipse normal to the shortest horizontal axis would give rise to an arcuate form normal to the least axis, which would approximate to a circular segment. Hence, when such arcs are broad, buckling may well have given rise to these features. As the concepts expressed by Yamaoka are an extension of the simple two-dimensional buckling mechanism outlined in Chapter 3, we suggest that such a curved trench, with a very large radius of curvature, is only likely to be initiated in the vicinity of continental lithosphere.

7.3 Mid-ocean trenches which form arcs of a circle

There are certain arcs which should attract special attention, because they usually satisfy one or more of four attributes: (1) they define an almost exact circular arc; (2) they are relatively small features, with a radius of 250–625 km; (3) subduction takes place at the outward bow of the arc, i.e. the arc points *into* the direction of plate motion; and (4) they usually exhibit 'ocean/ocean' lithospheric contact.

This latter point is very important for, as may be inferred from our earlier comments, we find that it is extremely difficult to envisage how such a subduction zone, especially one that is curved and of limited extent, can be initiated within an oceanic plate by conventional mechanisms, normally associated with plate tectonics.

A list of the arcs which are almost exactly circular is given in Table 7.1, where the radius of curvature and the angle subtended by the limits of the arc and the inferred centre point are also given, as is the type of contact.

The largest of these features, the Mariana Trench, is shown in Figure 7.4. Only the Scotia and the Banda Arcs closely approximate to a semicircle. The arc formed by the Leeward and Windward Islands and the Mariana Trench sweeps through 150° and 120° respectively, while the Amirante Arc forms less than a quarter circle.

These arcuate features are so obvious that they must prompt the observer to speculate upon a possible mechanism which caused them. Yet relatively little is to be found in the literature regarding their probable origins and mode of development.

A statement, attributed to Tuzo Wilson, was that any inclined plane within the Earth will intersect the surface along an arc of a circle. This statement, which is, of course, true,

Table 7.1 Circular arcs.

Feature	Radius of curvature	Plan angle	Type of contact
1 Mariana Trench	625 km	150°	?/Ocean
2 Caribbean Arc	550 km	150°	Ocean/Ocean
3 Scotia Arc	300 km	180°	Ocean/Ocean
4 Amirante Trough	300 km	60°	Ocean/Ocean
5 Banda Arc	225 km	180°	Sea/Sea?

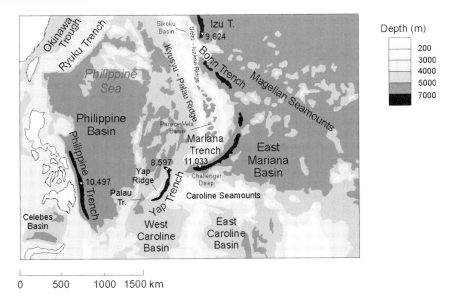

Figure 7.4 Map of the Mariana Arc.

appears to have stifled further thought on the matter. But to what plane within the Earth could the statement refer? Could it be the plane of the down-going slab? Some slabs dip at about 45°, others dip almost vertically, while others have only gentle dips. Moreover, the dip of the slab is not an inherently fundamental characteristic, but is determined by external factors such as the absolute motion of the trench relative to the down-going slab.

We can suggest only one plane which is inherent to all down-going slabs, namely the *original thrust* which permitted initiation of the slab. As we have seen in Chapter 3, the dip of the initiating thrust will be approximately 25°. If this were taken to be on a single plane throughout a segment of the Earth, one could readily estimate the radius of curvature of trench which is consistent with this planar constraint. Thus, assuming the Earth is a perfect sphere (Figure 7.5), it follows that the inclination of the plane d at the surface is q. (The dashed line indicates the inclination of many down-going slabs.) It can be inferred that d is the radius of the arcuate feature which will be generated at the surface. Clearly, $d/R = \sin q$, where R is the radius of the Earth (6366 km). If we take q = 25°, then $\sin q = 0.423$ and $d = 2693$ km. It is clear from Table 7.1 that all the radii of curvature cited are very much smaller than the limiting value set by Figure 7.5 of 2693 km. Hence, it is clear that the radii of curvature of the arcs cited in this table cannot be explained in terms of the statement made by Tuzo Wilson.

Turcotte and Schubert (1982) suggest that arcuate structures can be qualitatively understood by the 'ping-pong ball analogy'. They point out that when such a ball is indented, the indented portion will have the same radius (but opposite sign) as the original surface of the sphere, thereby forming a perfect mirror image basin to the original spherical surface (Figure 7.6). They use this concept to explain the dip of the subduction plate of the Aleutian Trench. They do not, however, explain how the inferred basin would develop to a depth of 764 km. Indeed, no mechanism is implied by their analogy. These authors are aware of the shortcomings of this concept and cite situations for which the model is not valid.

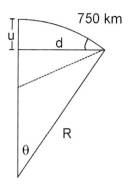

Figure 7.5 Minimum radius of curvature of trench that can be defined by a plane dipping at 45°.

Figure 7.6 The 'ping-pong ball' mechanism.

Let us, therefore, discuss the development of the Scotia Arc in some detail to demonstrate how such arcuate features can develop.

7.4 The Scotia arc

The configuration of the Scotia Arc and Trough (the latter is sometimes termed the S Sandwich Trench) and the interrelationship of the various elements of the region are dramatic and intriguing (Figure 7.7). The trench itself is a well-defined, near semicircular, feature which occurs from about 55°–62°S at 25°W, and forms the eastern boundary of the relatively small Scotia Plate. This plate is bounded to the north and south by strike-slip faults; the northern fault connects the northern limit of the trench to the tip of Tierra del Fuego, while the other inferred boundary fault runs from the southern limit of the trench almost to the eastern end of the South Shetland Islands. The final boundary to this small plate is defined by the Shackleton Fracture Zone.

It will be seen (Figure 7.7) that to the west of the Scotia Plate lies a deep abyssal plain at a depth of over 5000 m. To the east of the Scotia Trench is the S Atlantic, which, in this area, reaches a depth of 4–5000 m. The Scotia plate itself contains extensive, and also smaller, areas, where the depth of ocean ranges from 3000 m to as little as 1000 m. The structure of the Scotia Plate (Figure 7.7) gives some indication of why the ocean floor is relatively shallow. This diagram shows relatively short lengths of floor of different ages, indicating that ridge-like spreading has taken place in both a mainly east–west, and also a north–south, direction.

As regards the initiation and development of the Scotia plate and Arc, the problems that must be faced are (1) *when* and (2) *why* this feature developed. As a starting point, we take the position of the Antarctic Peninsula relative to the southern portion of S America, prior to the initiation of the southern-most, S Atlantic, 135 Ma ago (presented in Chapter 6,

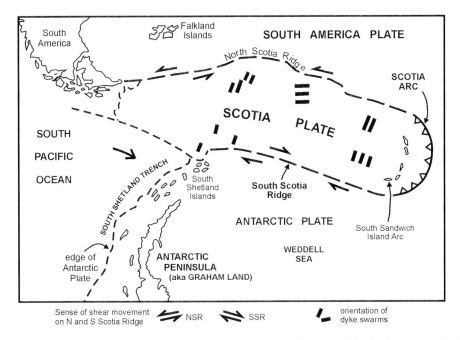

Figure 7.7 Schematic map showing the Scotia Arc and associated features (after Barker *et al.*, 1984).

Figure 6.23, and repeated here as Figure 7.8a). Following this event, what was to become the Antarctic Peninsula drifted to the south, but probably continued to overlap the 'straight' tip of S. America for several tens of millions of years.

It will be recalled that Sir John Barrow (1830–31), followed by Arctowski (1895) and Suess (1909), suggested that 'the Andes are seen again in Graham Land'. Two attempts to explain the development of the Scotia Arc have been predicated upon the *original* linear alignment of the Andes and Graham Land (i.e. the northern tip of the Antarctic Peninsula). The first of these was by Hawkes (1962) (Figure 7.9a), and a similar initial configuration and subsequent *migration* of elements in the chain was presented by Dalziel and Elliot (1973), (Figure 7.9b). The outstanding feature of these constructions is that it was assumed that the swing of the Andes from N–S to E–W at Cape Horn did not initially exist, and, moreover, that the Andes were aligned with an uncurved Graham Land.

However, few recent workers have had the temerity to straighten out the southernmost extent of the Andes. Indeed, Dalziel changed his mind (as a co-author) and supported a construction presented by Barker *et al.* (1984), which comprises a different configuration of elements, but shows them as changing from a cuspate point which, these authors suggested, existed 30 Ma ago (Figure 7.10). Although this reconstruction, by Barker *et al.*, shows more detailed geological information than that presented by Hawkes, the approach is intrinsically the same; namely, they both indicate the inferred track of various elements of the Scotia Arc from its inception (Hawkes) or over the last 30 Ma (Barker).

The pictorial development of the elements which 'frame' the Scotia Arc (Figures 7.9 and 7.10) is equivalent to the *Bewegungsbild*, or *movement picture*, that plays an important role in the thinking of some structural geologists. However, it must be emphasised that such move-

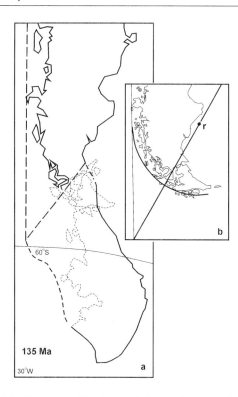

Figure 7.8 (a) Overlap of the Patagonian Orocline and the northern tip of Graham Land, as shown by the Atlas 3.3 Program 135 Ma ago showing the *impossible* superposition of these two land masses. (b) The geometrical construction used to straighten the orocline.

ment patterns, desirable though they may sometimes be, are NOT mechanistic explanations regarding the geological structures or the development of the Scotia Arc. Indeed, as far as we are aware, no mechanistic model exists, which permits one to understand how the Scotia Arc developed.

The fine lines in Figure 7.8a (derived by the Atlas 3.3 program) show the northern tip of Graham Land, at 135 Ma, superimposed upon Patagonia. The straightened Andes and its relationship with the micro-plate of Graham Land are shown by thicker lines in Figure 7.8a. How the geometrical straightening of Patagonia in Figure 7.8a was derived is shown in Figure 7.8b.

It is important to note that Lawvers *et al.* (1985) argue that, at this time (i.e. 135 Ma ago), Graham Land was not attached to Antarctica (Figure 7.11); indeed they state that Graham Land did not become attached to Antarctica until 119 Ma.

The question that now poses itself is, at what time did the development of the Patagonian Orocline and hence, the curvature of the southern part of the Andes, take place?

Following Carey's (1955) suggestion regarding the development of the Patagonian Orocline, Dalziel and Elliot (1973) carried out a preliminary palaeomagnetic study of the area, which indicated that a degree of such bending had taken place. This was confirmed by Burns *et al.* (1980) who stated that bending in the range 40° to 27° had occurred between 80 Ma and 21 Ma ago.

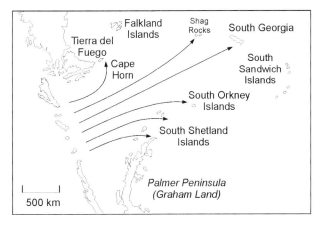

Figure 7.9a Development of the Scotia Arc (after Hawkes, 1962).

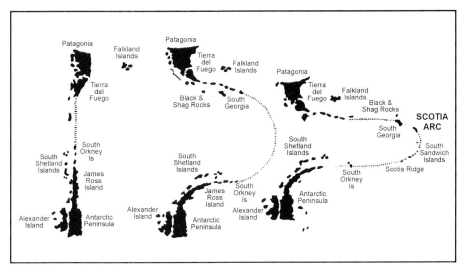

Figure 7.9b Development of the Scotia Arc (after Dalziel and Elliot, 1973).

Cunningham *et al.* (1991) who presented the results of a more recent study, stated that neither of the earlier studies (by Dalziel *et al.* and Burns *et al.*) provided data that pass modern confidence 'filters' for data reliability. The new data presented by Cunningham *et al.* enabled them to conclude that this orocline is the result of approximately 90° of anticlockwise rotation of the crust, and that this has taken place since the mid- to late-Cretaceous (i.e. about 100–65 Ma ago).

If the early, intuitive interpretation regarding the original 'straightness' of the Andes has proved to be correct, then it is not unreasonable to question whether Graham Land too was originally straight. In fact, Grunow *et al.* (1987) have shown, from palaeomagnetic data, that Graham Land has rotated 30° in a clockwise direction since about 190 Ma. ago. However, it is interesting to note that Grunow *et al.*, in a construction which purports to represent the distribution of S America and the Antarctic Peninsula, for 90 Ma, show the S American

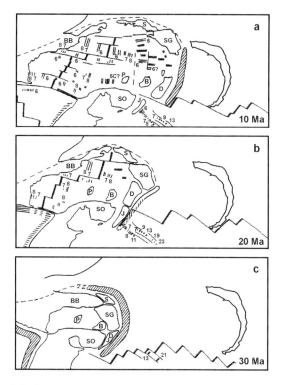

Figure 7.10 Initiation of the Scotia Arc from a cuspate form at 30 Ma (after Barker *et al.*, 1984).

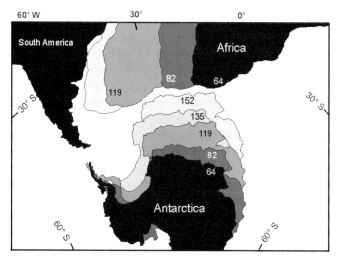

Figure 7.11 The motions of Africa and Antarctica relative to S America from the late Jurassic to the early Tertiary. The current Antarctic Peninsula did not become attached to the rest of Antarctica until 119 Ma ago.

Orocline at its full development. We would reiterate Cunningham's statement that, as regards this particular point, Grunow's construction for the 90 Ma relationship is incorrect.

Consequently, we repeat the conviction of the early explorers in this region, namely that the 'Andes are seen again in Graham Land' and suggest that the most obvious conclusion is that, 140 Ma ago, the 'Antarctic Peninsula' was, in all probability, physically attached to S America as a direct continuation of the southernmost 'straightened' Andes.

It was further noted (as argued in Chapter 6) that the S Atlantic Ocean was initiated at its southernmost tip of (straightened) S America at around 135 Ma. Hence, it is suggested, that the 'Antarctica Peninsula' was separated from S America at this time.

From 135 Ma to about 65 Ma, Graham Land migrated southward toward Antarctica. The precise position of specific reference points on the S American and the 'Antarctic Peninsula' can only be given within a circle of error at the 95 per cent level of certainty, where the diameter of the circle is at least 100 km. Consequently, one can only state that, at some time between 80 Ma and 65 Ma, the northern tip of what by now was truly the Antarctic Peninsula, may have cleared or had only a modest overlap with, the still-straight portion of S America. This position is not too dissimilar from that assumed by Hawkes (1962) and Dalziel and Elliot (1973), shown in Figures 7.9a and b, and 7.10 a, b and c.

7.4.1 Mechanisms

Let us now consider the mechanisms which could give rise to a curved trench. Three questions must be answered: (1) how or why did the plate in the S Pacific break through on a relatively narrow front? (2) why is the subduction trough almost exactly semicircular? and (3) how or why did the subduction sense become reversed and develop in the Atlantic lithosphere?

A linear trench can be given apparent curvature (dashed lines in Figure 7.12) if the trench is cut by a sequence of transform faults which exhibit a consistent amount of strike-slip displacement. To obtain a greater angle of arcuate development, it would be necessary to have a mirror image of the faults shown in Figure 7.12. The different sense of shear along the transform faults could, presumably, be attributed to variations in the speed of the ocean plate between the transforms, where the lowest rate of movement is in the centre. However, such structures would be readily observed. Such hypothetical structures could develop if an area of well-rounded, plateau basalt or ridge on the Pacific side failed to subduct beneath S America (cf. the Nazca-Sala-y-Gomez Ridge). However, the force of the Pacific lithosphere would still be opposed and balanced by that of the Atlantic lithosphere, so that there is no specific reason for breakthrough to take place.

The *conventional* mechanisms discussed in Chapter 3 do not provide a satisfactory explanation of how a breakthrough could have taken place. The circular form of the trench can only be derived by special pleading, and the question of how the subduction reversal took place must be taken on trust. However, such questions can readily be answered in terms of what was set out in Chapter 5, where aspects of *impact tectonics* were discussed.

A schematic representation of the ridge-push forces (P_{1-4}), which are likely to have obtained when the tip of the Antarctic Peninsula was adjacent to, or slightly overlapped, the southern tip of S America, is given in Figure 7.13. We suggest that the geometry of the Scotia Arc itself is evidence regarding the probable mechanism by which it formed, and are aware of only one mechanism capable of developing this geometry on such a scale. That mechanism is a *major impact* (Figure 7.14).

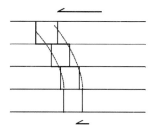

Figure 7.12 Schematic representation of how a linear trench can be changed to a pseudo-arc by shear along transform faults.

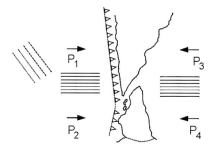

Figure 7.13 The balance of forces to the east and west of S America and Graham Land prior to the initiation of the Scotia Arc.

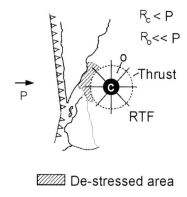

Figure 7.14 Sketch map showing location of a major meteoritic impact relative to S America and Graham Land, together with the circumferential thrust(s) and radiating tensile fractures.

At the time we first considered this problem, the only faults with a circular plan which we remembered to be associated with a major impact were the normal fractures which defined the peripheral graben (Figure 7.15). Because we are concerned with a situation in which the maximum principal stresses in the ocean lithosphere some 100 km from the ridge are horizontal and compressive, it follows that we were faced with an analytical problem.

It can be shown (Price and Cosgrove, 1990) that, for normal faults with an angle of dip of 60° or more, it is usually impossible for a horizontal maximum, principal stress to cause re-shear, in the reverse mode, on such faults. Hence, the stresses will behave as though the normal faults did not exist. What was required by our hypothesis was the existence of arcuate thrusts which crop out around the crater. We went back to the data supplied by Jones (1995). Originally, we were unaware of the existence of such thrusts (see Price, 1975). In this paper I presented a sketch of a major impact, described verbally by Jones, which shows concentric normal faults and radiating vertical fractures, but no thrusts. It will be seen in Chapter 5 that such arcuate thrusts did, in fact, develop in the Snowball explosive experiment. A hypothetical model had been erected which required a certain element to exist. When that element was in fact found to exist, we were suitably encouraged.

It was shown in Chapter 5 that these thrusts crop out at a distance from ground zero (GZ) at a radius of about twice that of the crater-rim, depending upon the velocity of the impacting body. The radius of curvature of the Scotia Trench is approximately 300 km, so that, using the Snowball 500 ton explosive experiment as a guide, the impact crater could have had a diameter of 300 km.

The approximate location of the impact is, we suggest, as shown in Figure 7.14 and in greater detail in Figure 7.16. It is clear from this diagram that subduction took place using only one half of the thrust system that developed. Hence, the question arises as to why the subduction took place on the eastern part of the thrust array, rather than the west (or indeed any other direction).

The section through the suggested impact structure, shown in Figure 7.15, is parallel to the direction of S American plate motion. If we assume that this plate moved at a rate of about 2 cm/yr, then the rocks cut by the thrusts at the western end of the section are about 15 Ma older than the rocks at the eastern end. This means that they are older, colder and

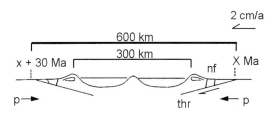

Figure 7.15 E–W section through crater and thrust of impact structure shown in Figure 7.14.

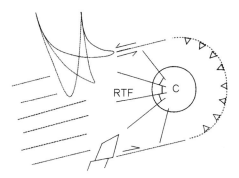

Figure 7.16 Envisaged mode of breakthrough of Antarctic/Pacific plate into the S Atlantic, as the result of a major impact, with map of inferred impact showing stages in the development of the orocline and the movement of wedges into the crater area and the initiation of the Scotia Arc.

more dense, and therefore more likely to be subducted. However, this tendency will be a very small one, and could easily be overcome by a stronger, contrary influence.

It is further maintained that the position of impact was adjacent to S American and Graham Land landmasses, and that this point of impact was vital in determining the direction of subduction. If we assume that the impact GZ was in the Atlantic Ocean, about 200–400 km to the east of the southern tip of S America, then GZ would have been about 2000 km from the S Atlantic spreading-ridge, but very close to S America. The elastic strain, which develops in the strong layer of this oceanic lithosphere of the S American plate as the result of ridge push, may easily approach an average value of about 0.5 per cent (see Chapters 2 and 3). Hence, the elastic strain stored in the oceanic plate to the east of GZ could be released shortly after impact, and a down-going slab, with an initial increment of movement of 10–30 km, could be initiated on the eastern side of the impact structure.

However, if, as indicated in Figures 7.14 and 7.16, the impact GZ is situated at about 350 km east from the tip of S America, the amount of strain release within the S. American plate to the west of the structure would result in under-thrusting of only a 100 m, or so. Thus, the side of the impact structures on which the down-going slab is initiated is, at first, controlled wholly by the amount of elastic strain energy stored in the oceanic element of the plate.

Let us now consider the changes in the balance of forces which take place as the result of the impact. In this area, as we have seen in Chapter 5, the impact causes massive changes in the rock properties in the crater area (c, in Figure 7.16). The changes near GZ involved considerable melting of the target rock, and certainly resulted in considerable minor 'Grady–Kipp' fracturing (Figure 5.30).

The changes in the material properties in the rocks in the outer area, beyond the crater rim, also involved a degree of melting of the LVZ. In addition, this outer area was greatly weakened by fractures, namely the arcuate thrusts and also the radiating vertical tensile fractures, which probably have extended outwards for hundreds of kilometres beyond the crater rim and may even have extended into (and possibly even through) the 'straight' continental area of S America. As we saw in Chapter 5, the rock-mass in the central area (c) is in part excavated, and what remains is, initially, relatively weak and lends little support to the 'outer area'. Consequently, inward migration by shear movements on the major, vertical, radiating tensile fractures can readily develop.

The elastic strain to the east of the structure is released by the initiation of a down-going slab. The de-stressing of the hachured area (Figure 7.14) to the west of the impact causes little deformation of the impact structure. The de-stressing, on the Atlantic side, however, leaves the force P of the Pacific/Antarctic plate almost completely unopposed. Breakthrough of a portion of the Pacific/Antarctic plate into the Atlantic element of the S American plate is, initially, unopposed and inevitable.

It may be thought that such a breakthrough of the Pacific/Antarctic plate would continue the down-going trend of this plate beneath the S Atlantic. However, this caveat applies to the slow processes normally associated with general plate motions. With the impact model proposed, we are dealing with far quicker rates of lithospheric movement than are usually associated with plate tectonics. It is suggested that the models described by Price and Audley-Charles (1983, 1987) are applicable here. They showed how gravity-glide from the ridge caused a buckle geometry of the lithosphere at the trench (Figure 7.17). The strains induced in the buckle can, in such circumstances, greatly exceed the elastic limit of the lithosphere (see Chapter 3). Normal faults will develop at the upper levels of the anticlinal lithosphere

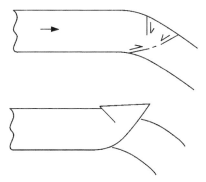

Figure 7.17 Break-up and separation from the horizontal plate of a down-going slab along a through-going fracture (after Price and Audley-Charles).

and thrusts develop on its lower side, beneath the neutral surface. At relatively fast strain-rates, one or more such thrusts and normal faults would tend to extend, and occasionally join up, to form a through-going fracture, which would facilitate over-riding of the down-going portion of the slab. In the over-riding element of the Pacific/Antarctic, slab-accelerated motion results from the loss of resistance caused by the separation along this thrust plane. In turn, this sudden release would enable stored strain-energy to give rise to a relatively rapid (unbending) strain-rate in the curved lithosphere above the thrust plane of the down-going slab. This promotes break-away along an upwardly convex surface of the through-going fracture.

This breakthrough results in the relatively rapid uplift of the upper portion of the hitherto down-going slab. Such recovery, and detachment from the down-going slab, within a period of about 10^{4-5} years, would return the Pacific/Antarctic ocean floor (in this locality, opposite the impact site) to the horizontal.

This part of the lithosphere would have a ski-like, upward-curving front, which would remove any tendency for further down-warping of the Pacific element of the Pacific/Antarctic plate as it broke through to the Atlantic.

The uplift and straightening of this part of the lithosphere at the front end of the Pacific/Antarctic plate would increase the horizontal extent of that portion of the plate, opposite the impact de-stressed released zone, by about 10 km. Hence, this extension, together with the elastic recovery of the compressive strains in that segment of the Pacific/Antarctic floor element, would result in a very rapid initial incursion of this element of the Pacific plate into the landmass of S America and part of the Antarctic Peninsula by 20–50 km. This incursion could have been exacerbated by movement along the radial tension fractures induced by the impact.

This eastward movement of the plate in the Pacific would cause major fracturing of the continental rocks, thereby forming minor elements from which the islands, which now flank the current Scotia plate, are derived. The mode of breakthrough which is envisaged is indicated in Figure 7.16. The relatively weak central area of the impact (c) is deformed by wedges of outer and stronger rock, which are defined by the radial tension fractures (RTF) that are a feature of the Snowball explosion experiment.

The breakthrough from the Pacific side into the Atlantic would probably have taken place within 100,000 years of the impact event (i.e. almost immediately in terms of normal plate motions), as the result of elastic rebound of the Pacific/Antarctic ocean floor and enhanced by the isostatic recovery of that plate at the breakthrough area.

The forces at the appropriate ridges would gradually build up the compressive stresses in the de-stressed portions of the floors of both the Pacific/Antarctic and S Atlantic. This requires regeneration of the original elastic strains induced in the oceanic plates by ridge-slope gravity-glide. The rate of stress build-up would initially be relatively fast, because the horizontal stresses would be reduced to that which can be induced by gravitational loading. Continued build-up of the stresses would become progressively slower, as the elastic stresses induced by gravity-glide became larger and reached about 0.5 per cent. The time required to 'replace' the original elastic strains will depend upon the length of the oceanic plate from ridge to impact, and upon the average rate of plate movement. Thereafter, there would be a slower but inexorable incursion of the Pacific/Antarctic oceanic lithosphere into the Atlantic, at the rate of the relative motions of the two opposing plates. This incursion would decelerate and eventually cease, when the resistive elements along the transform faults, which border the Antarctic Plate incursion, approached equality with the driving forces of the East Pacific Ridge.

7.4.2 Subduction along a semicircular trough

It can be inferred that a strong, elastic, oceanic lithosphere would experience great difficulty in subducting down a semicircular trench. Such subduction is possible only because the ocean plate contains a number and variety of fractures. The important fractures in this context are the major and minor strike-slip, parallel shear fractures in the S Atlantic, which trend parallel to the E–W boundaries of the Scotia Plate. These fractures will permit the ocean lithosphere to be subducted in sections (Figure 7.18). In the central portion of the arc, the subduction is relatively easy, but becomes progressively more difficult at the northern and southern portions of the arc, where the flexure is required to develop at an acute angle to the transform fractures. Figure 7.19 shows the displacement of the magnetic anomalies along strike-slip, shear faults. The preponderance of seismic activity related to the subduction of the arc coincides with the areas where the arc joins the transform faults, where, it is inferred, the stresses are highest.

Grunow (1993) states that the 'Northern and southern mid-Cretaceous (palaeomagnetic) poles of Graham Land are very much alike, suggesting that the "S" shape of the Antarctic Peninsula is not due to oroclinal bending since (approximately) 110 Ma'. In his paper, the

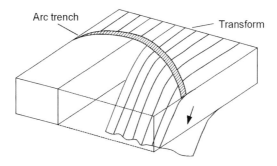

Figure 7.18 Block diagram representing subduction of S Atlantic lithosphere beneath the Scotia Plate.

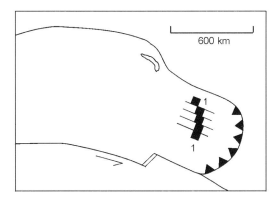

Figure 7.19 Plan of Scotia Arc and break-up of internal band of basalt by vertical shear fractures.

description of the various sites from which the samples were taken and the difficulties of obtaining good rock samples are emphasised. The spread of results inherent in the palaeomagnetic data obtained from these sites could easily hide a modest amount of 'bending' to have taken place since 110 Ma. However, this is not incompatible with his remark, except for the northernmost part of N Graham Land and the S Shetland Islands. Indeed, the 'curvature' may be more apparent than real, in that it may well be the result of differential shear displacements on strike-slip faults rather than bending.

7.4.3 Timing and evidence of the impact event

The question now arises – precisely when did the impact take place? As we have seen, Cunningham *et al.* (1991) indicated that the rotation could have occurred sometime between the middle and end of the Cretaceous (i.e. 100 Ma to approximately 65 Ma), while Burns *et al.* (1980) suggested a range in which rotation occurred of approximately 80–20 Ma. Hence, the most likely period for rotation to have taken place is between 80 and 65 Ma. Let us attempt to date the event more precisely by noting the movement of an easily identified point on the computer map-simulation of Patagonia (Figure 7.20a). This point is the Atlas representation of the southern tip of Wellington Island, which is off the west coast, at the northern limit of the bending of the Andes.

It will be seen in Figure 7.20a, which represents the track of the tip of this reference point between 70 and 60 Ma, that there was a sharp change in direction of track at about 65.5 Ma. It can also be inferred that between 70 and 67 Ma the speed of motion was significantly slower than from 67 to 65.5 Ma. A detail of this track is given in Figure 7.20b, where it will be seen that the actual change in direction took place at 65.45 Ma, and that the increase in the rate of motion between 70 and 67 Ma actually continued to point A which represents 66.5 Ma.

It is suggested that the change in speed at 66.5 Ma and the change in direction at 65.45 Ma are signals that two distinct impact events are recorded. The 66.5 Ma event may possibly be a record of the distant Chicxulub event in Yucatan, while the 65.45 Ma event most probably relates directly to the impact that gave rise to the Scotia Arc.

In the light of this evidence, we suggest that the impact model presented is a plausible and likely series of mechanistic and geometrical events that gave rise to the Scotia Arc.

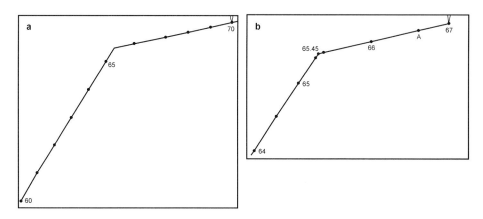

Figure 7.20 (a) Track of a point on the west coast of S America between 70 and 60 Ma. (b) Detail of track between 67 and 64 Ma.

The case for an impact-linked mechanism, which has been made for the Scotia Arc, is the only mechanism, of which we are aware, that can account for the development of a near semicircular trough. We hold, therefore, that it is not unreasonable to infer, a priori, that the other cited arcuate troughs may also be the result of cometary impacts.

It was noted in Chapter 5 that, because of the intense action of the oceanic waters which results from a major marine impact, the reworking of the oceanic sediments will largely, if not completely, obliterate evidence of ejecta produced by the impact. However, in place of this potentially lost evidence, we suggest that the change in plate motion and/or the change in plate speed is valid evidence that can be adduced to support the impact theory.

Slow changes in speed and direction of movement of a plate can, of course, result from normal plate tectonic causes, such as collision of continental units or the development of a spreading-ridge. However, *rapid* changes of plate speed or direction of movement are an important, even a diagnostic, feature of major impacts which result not only in CFBs and OFBs, but also give rise to arcuate subduction troughs. Let us now consider the other arcuate features listed in Table 7.1, to ascertain whether they too can be linked to major impacts.

7.5 The Caribbean (or Lesser Antilles) Arc

The Caribbean area is set between the continents of N and S America and, in the west, is currently separated from the Pacific by the platelets of Central America. The Caribbean did not start to form in the east until the development of Tethys began to separate Africa from the 'New World'. The current geology, as represented in the tectonic map of the area prepared on a scale of 1 : 2,500,000 by Case and Holcombe (1980), is clearly complex and is, as one may infer from this map and articles in the literature, to some extent conjectural and not well understood.

The general geography of the Caribbean Sea, its main islands and adjacent landmasses, is shown schematically in Figure 7.21, together with the structural elements of the region. There are belts of deformed sediments completely surrounding the Caribbean Plate. Active external subduction is occurring at the *Middle American Trench* on the Pacific sea-board, and outside the *Lesser Antilles Arc* where the Atlantic lithosphere is being subducted. Intern-

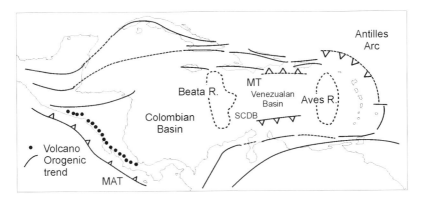

Figure 7.21 Major tectonic-geomorphic elements of the Caribbean region, showing structural trends and zones of seismic activity.

ally, subduction is taking place at the *Muertos Trough* (MT) to the north, and at the northern limits of the *South Caribbean Deformed Belt* (SCDB) and *North Panama Deformed Belt*. Intense seismic and volcanic activity occur at both E and W ends of the region, while zones, or small areas, of seismic activity also occur along the northern and southern margin of the region.

The *Colombian and Venezuelan Basins* are two main structural elements of the Caribbean, where thicknesses of sediments have accumulated, which range from a few hundred metres to several kilometres of relatively undeformed pelagic sediments and turbidites. In these two basins, the age of the sediments is usually younger than 90 Ma. In addition, there are three areas (the *Nicaraguan Rise, Beata* and *Aves Ridges*) whose early sediments, which are moderately deformed at intermediate to shallow depths of water, usually date from between 90 and 73 Ma.

It is important to note that, because magnetic anomalies over the Caribbean are of low amplitude, it is not easy to identify sea-floor spreading patterns. Indeed, Burke (1988) points out that, for the Caribbean ocean floor proper, there is, as yet, no wholly persuasive identification of magnetic anomalies.

There is debate as regards the extent of the arcuate feature which is so noticeable at the eastern limit of the Caribbean. Some refer to a variety of small segments of the Greater and Lesser Antilles (Pindell and Barrett, 1988). Burke (1988), however, suggests that the whole of the Lesser Antilles, and even part of the Greater Antilles, constitute one huge feature related to the penetration of the Pacific into the Caribbean (Figure 7.22). Just how a near semicircular arc could have developed from this intrusion of Pacific units between N and S America requires special pleading.

A reconstruction of the evolution of the western part of the Caribbean area based on the Atlas 3.3 program is shown in Figure 7.23. It will be seen that, since about 60 Ma, the migration of the component elements has been mainly from south to north. From this, one may infer that the Beata Ridge probably marks the most eastward incursion of the Pacific Plate. The main body of this ridge is over 1000 km away from the northern portion of the Antilles Island Arc.

The structural incursion along the north coast of S America extended much further to the east (Figure 7.24). However, as may be inferred from Figure 7.25, this structural incursion disrupted an arcuate system which was already in place. It will be seen from this latter figure that the Lesser Antilles islands are not currently a perfect arc, in that they comprise three sets.

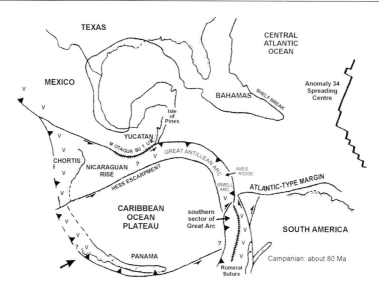

Figure 7.22 Schematic representation of the Caribbean 80 Ma ago as the result of penetration of the Pacific plate (after Burke *et al.*).

Backtracks relative to Mexico

A   Yucatan   160 - 140 Ma   SE to NW : 140 - 0 Ma. No movement
B   Nicaragua   160 - 61 Ma   Mainly W to E : 60 - 0 Ma. S to N
C   Panama   160 - 0 Ma   Continental movement : U-turn 60 Ma
D   Jamaica   160 - 61 Ma   W to E : 60 - 0 Ma   S to N

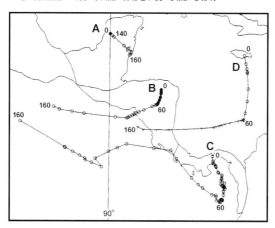

Figure 7.23 Using the 'backtrack' of platelets forming the Central American Isthmus. In the last 60 Ma movement in tracks C and D has been northward.

The development of one main arc and two minor ones is attributed by Bouysse and Westercamp (1990) to modifications of the original arc, as the result of collision of two, three or even four elongated ridges in the Atlantic. These authors refer to an ancient arc, older than 50 Ma, and recent arcs aged less than about 25 Ma, with an intervening hiatus of volcanic activity that lasted about 10 Ma. As regards the age of initiation of the ancient arc,

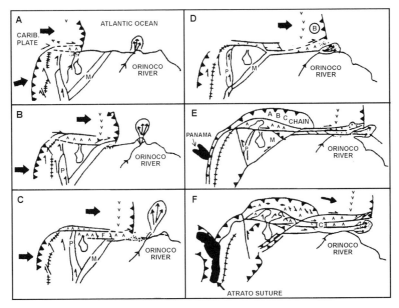

Figure 7.24 Inferred movement of the Caribbean-South America plate boundary zone (PBZ) from early Paleocene (60 Ma) to the present day.

they are understandably vague, and suggest that it could have been active from the early Cretaceous. It is, of course the initiation and development of the ancient arc which most concerns us here. Let us now define the limits of what we consider the original Caribbean Arc to be. It will be seen from Figure 7.25 that the only well-developed trench in the eastern Caribbean occurs north of Puerto Rico, and that this trench only curves around the northern isles of the Great Antilles as far as Antigua and Barbuda. However, because of the seismicity and volcanic activity that occur in the vicinity of this island arc, it is obvious that it is associated with an active, down-going slab that exists at least as far south as Grenada. These islands define an almost perfect circular arc through 110°, the radius R of which is 400 km, while the radius of the trench is 550–600 km. The geometry of these features, albeit modified by later events, is consistent with what one would expect to be generated by a major impact.

7.5.1 The impact event

At what date did the impact occur? We can reasonably put an upper limit on impact at about 120 Ma, which represents, in this area, the age of 'old', normal Caribbean floor.

We have argued that impact events cause noticeable changes in the track, usually regarding direction and rate of plate movement. The track of the easternmost cape of Puerto Rico is shown in Figure 7.26. The track between 76 and 64 Ma is shown in Figure 7.26a. It will be seen that there was a modest change of direction and pace at about 66.5 Ma. This, we have already seen, relates to the Chicxulub impact event.

The other, earlier event, occurred at about 73.1 Ma, where there was a much more marked, sudden change in the direction of motion, and a 30 per cent reduction in rate of motion following the impact (Figure 7.26b). This event is taken to be the impact that gave rise to the Antilles arcuate features. One may infer from the greater angular change and the greater

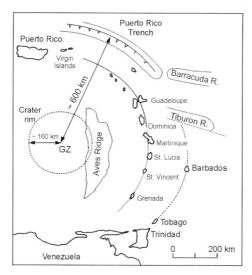

Figure 7.25 The Caribbean Arc revealing complexities of the subduction zone which developed some 40–50 Ma ago as the result of the influence of a series of approximately E–W ridges. It also shows morphological regions within the island arc, with indications of the minimum age of the sediments together with the inferred position of the impact crater.

increase in rate of motion, that this event was probably more energetic than the Chicxulub impact, from which it is not greatly separated in either distance or time.

The Aves Ridge is unique, in that Coffin and Eldholm (1992) are in doubt as to this feature's origin. These authors suggest that this submarine ridge may be either a 'drowned' continental flood basalt, or an ocean basin flood basalt. It has already been argued that ocean basin flood basalts can result from the jetting of melt generated by a major impact. It is interesting to note, therefore, that corroboration of the impact date comes from the palaeontological evidence, which indicates that the Aves Ridge is younger than about 73 Ma.

We have already noted that the depth of water was moderate to shallow. In shallow water, the peak impact-stress of a major, fast moving, cometary body, would be capable of generating a significant volume of melt. Hence, jetting would be a distinct possibility. Alternatively, in somewhat deeper water, a large impacting body could cause an anticlinal, 'down-range', bulge of oceanic lithosphere, especially if the lithosphere was relatively thin. Fractures would develop in such a structure, thereby permitting the intrusion of asthenospheric material to enter the ridge and even to develop extrusions on the sea-floor.

7.6 The Banda Arc

The Banda Arc is a relatively small structure. Nevertheless, it has a diameter of about 450 km, so that it is approaching Alpine in scale. This arc is a puzzle surrounded by an enigma, in that it has received relatively little structural study in the far from well-understood complexities of the larger area which comprises the East Indies (Figure 7.27).

Harris (1989) concludes that 'the Banda orogen is an arcuate mountain system comparable in scale and structural style to other arcuate orogenic belts, such as the Alpine, Carpathian, Aegean, and Caribbean. These, and other examples of arcs, form loops where a mountain chain changes in strike by up to 180°. Whether the orogenic processes that give rise to these

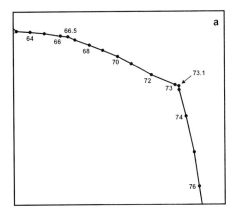
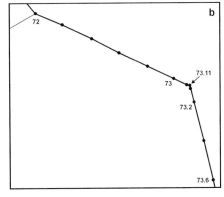

Figure 7.26 (a) Track of the easternmost cape of Puerto Rico in the period 64 to 76 Ma. (b) Detail of track shown for the period 72–73.6 Ma.

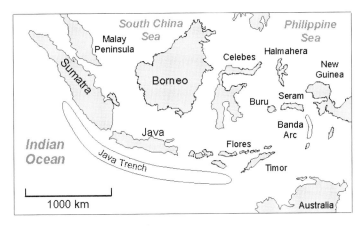

Figure 7.27 General sketch-map of the East Indies, indicating the setting of the Banda Arc.

features are the result of the relative motions of larger bounding plates, or the produce of body forces which act independent of plate kinematics is controversial' (Wezel, 1988; Dewey, 1988). Thus, conventional mechanisms as applied to the Banda Arc are vague.

The fundamental reasons for the disposition of the main units within this Indonesian complex are understood in terms of the main plate movements at the southern and eastern limits of the area. The magnificent chain of islands that extends from Sumatra eastward to Timor and beyond, with an attendant trench, is obviously the consequence of the Indo-Australian plate, which moves generally northward and is subducted beneath the islands and shallow seas that form the southern East Indies. Australia has already made contact with the eastern area at Timor, and has caused the development of the mountain range that forms the spine of New Guinea. This continental unit is now probably at the stage which India had reached some 30–40 Ma ago.

The view one takes of the Banda Arc depends upon the scale of the map one studies. When viewed on the grand scale (Figure 7.27), the Banda Arc looks as though it may be a natural continuation of the main island chain that starts with Sumatra. Alternatively, the curvature of the Banda Arc could possibly be attributed to the influence of the E–W compression of the Pacific plate. However, when viewed at a more detailed level (which is permitted by the map of the SW quadrant of the Plate-Tectonic Map of the Circum-Pacific Region, Doutch, 1981, a modified detail of which is given in Figure 7.28), it can be inferred that the amount of slip along the Molucca-Sorong fault has been insufficient to account for the degree of curvature. Moreover, one would expect that, were the curvature to be attributed to the influence of this major strike-slip fault complex, then there would be smaller, parallel or sub-parallel strike-slip faults cutting and displacing the Banda Arc; thereby providing a stepped, apparent curvature of the arc. As will be seen from Figure 7.28, such faults are conspicuous by their absence.

It is suggested here that the Java Trench extends only as far east as Sumba, and that the triangular Savu Basin separates Timor and Sumba from the islands of Flores to Wetar. It may also be noted that the volcanic islands of the Banda Arc are minuscule compared with the islands of the Sumatra-Timor chain. Consequently, one is left with the impression that the Banda Arc is a more recent, coincidental, appendage to the main island chain, and that its origin is not linked with the development of the main island chain.

It will be seen that the Banda Arc is different from both the Scotia and the Caribbean structures, in that it exhibits no trench, at least, where one would expect one to be. An obvious circular arc of the Banda structure is the dashed line in Figure 7.28, which marks the subsurface limit of continental crust. It will be noted that the circular arc AA′ approximates very closely to the inferred subsurface limit. The contours of the subducted slab are emphasised by the figures which indicate the depth of the slab in hundreds of kilometres. It can be inferred that the line AA′ approximates to the submarine outcrop of the upper

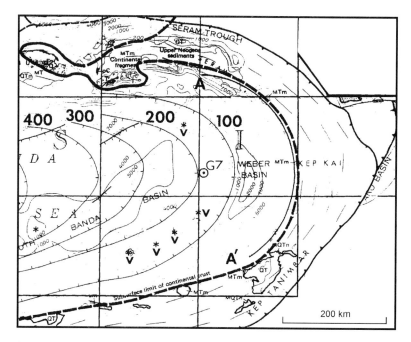

Figure 7.28 Modified detail of the SW quadrant of the Circum-Pacific Plate-Tectonic Map, showing structural elements about the Banda Arc (courtesy of Institute of American Association Petroleum Geology).

surface of the subducted slab. At depth, the contours of the slab become progressively less circular, thereby defining an E-W line of symmetry at about 5° 45'S. The arc of volcanoes occurs above the 150 km contour of the slab, which, in the south and west, appears to elide with the Weter-Flores chain of islands.

In normal plate-tectonic situations, where the lithosphere is under sufficient compression to give rise to subduction, one would expect the break to occur at the weakest point or line, which may occur at or near the junction between continental and oceanic lithosphere. This gives rise to the situation where the oceanic lithosphere becomes the down-going slab and disappears beneath the continent. This is what happened when the Indian Ocean plate encountered the Asian plate along the Sumatra-Flores island chain. The section shown in Figure 7.29 represents the reverse situation. It may be inferred, therefore, that a break occurred within the oceanic lithosphere. The eastern side subducted and carried continental cover westward, until these continental rocks encountered the trough, leading one to infer that subduction of the continental rocks proved too difficult, so that an *underthrust* developed and an embryo trough (the Seram-Uru Trough) formed behind the wedge of continental rocks.

7.6.1 Mode of origin

Conventional interpretations, based on plate movements, put forward to explain the development of the Banda Arc are not rigorously constrained. Therefore, let us try the alternative explanation and techniques, which we have already applied in this chapter, to explain the development of such a circular arc.

We take our reference point on an island (Kep Aru) adjacent to the Banda Arc, south of Western New Guinea. The *track* data for this island are given in Figure 7.30. For the period 40–0 Ma there is a marked change in the direction of track at about 34.5 Ma. It will be seen from the large-scale track (Figure 7.30b) that the change in direction actually took place at 34.3 Ma. Not only is there a kink in the track at 34.3 Ma, but there is also an increase in plate speed after this date. After 34.3 Ma (Figure 7.30a), the track follows a gentle curve and runs smoothly to the 5 Ma point, when there is an abrupt change in direction which continues to the present day.

Therefore, it is suggested that, from the geometry of the various features of the Banda Arc, coupled with the evidence of the *track* data outline above, this arc may reasonably be identified as the result of a meteoritic impact. The diameter of the line AA', which is taken to be the subsurface location of what was, originally, a trough (that formed at the site of the arcuate thrust associated with the impact), is approximately 450 km. Consequently, one would expect that the crater that developed as the result of this impact may have had a diameter of about 200 km. It may also be noted in passing that the date of this event is within 0.5 Ma of the Popagai impact event in N Siberia.

7.7 The Amirante Arc

The Amirante Trench, in the Indian Ocean, is an arcuate feature centred on the Seychelles. We have seen that an argument can be made for the Deccan Traps to be attributed to a major impact event. It is suggested that an even stronger case can be made for the Amirante Trench being initiated by another simultaneous, or possibly somewhat earlier, impacting body.

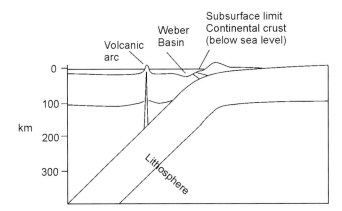

Figure 7.29 Section showing inferred structural relationships of the Banda Arc.

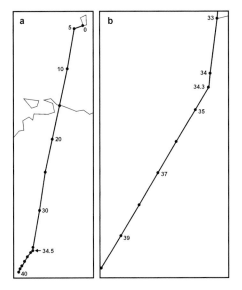

Figure 7.30 (a) Track of Kep Aru, an island south of New Guinea and to the east of the Banda Arc, for the period from 40 to 0 Ma. (b) Detail of track in (a) for the period 40 to 33 Ma.

White and McKenzie (1989) maintain that the Deccan Traps and the basalts of the Seychelles, Mascerene, Saya and Nazareth Banks resulted from the same plume (Figure 7.31a). We have indicated that the contention that the Deccan Traps can be attributed to a plume can be criticised; and shall now look at the evidence regarding the dating of the basalts in the areas of the Deccan Traps and the Seychelles.

The pertinent data for the Seychelles, and also a summary of the dates for the Deccan Traps, have been compiled by Devey and Stephens (1992). The disposition of the various morphological areas of the Seychelles is given in Figure 7.31b, which also indicates the boreholes and seismic data line from which the data they discuss in their text are derived. Much of their paper is given to a comparison of the geochemistry of the two areas. The conclusion they reach is that the basaltic rocks in the two areas are *similar*. The two extrusive areas were not too widely separated at the time of their development. Most of the extruded

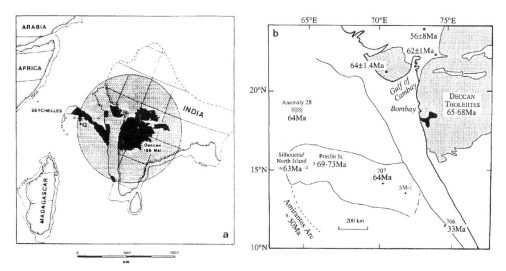

Figure 7.31 (a) Deccan/Seychelles eruptive province at approximately 65 Ma. The circle represents the assumed extent of a plume. The black areas indicate the extent of the volcanism and the contemporaneous off-shore basaltic volcanism (after White and McKenzie, 1989. © American Geophysical Union). (b) Tectonic situation at 65 Ma and ages of volcanic activity around the time of the Deccan magmatism (after Devey and Stephens, 1992).

material derived from the lower lithosphere or asthenosphere and we argue that the extrusions resulted from the same trigger mechanism. It is not surprising, therefore, that the basalts of the two areas are similar. However, of the areas in the Seychelles, it is the basalts of the Saya de Malha Bank which are most similar to those of the Deccan Traps. A reconstruction of the Seychelles area and its relationship to the Deccan area 64 Ma ago is shown in Figure 7.32. Devey and Stephens (1992) tentatively suggest that the rocks in the vicinity of SM-1 came from the Deccan flood basalt province. Their close physical relationship with the 'mainland' basalts of India lends support to this suggestion. So, to this extent, at least, the contention by White and McKenzie (1989) finds some support.

Although acid igneous rocks crop out in the Seychelle Islands, the majority of the erupted rocks forming the Bank, shown in Figure 7.32, are mainly submarine basalts. The rapid cooling of lava as it makes contact with the ocean tends to inhibit the size of vents and, therefore, the rate of extrusion is also inhibited. Consequently, one may infer that much of the material was intruded, thus causing the ridge to grow vertically rather than giving rise to extensive horizontal flows. These intrusions would keep the interior of the ridge relatively hot, so that eruptive activity would continue longer (perhaps 10 Ma) in the Seychelles area, rather than the 1.0 Ma or so in the Deccan province.

There are two aspects of Figure 7.32 which require comment. The first of these relates to the Fortuna Bank as a whole, which forms a significant submarine feature that is 50 km across. As a result of geophysical modelling, it has been suggested that the crust is 20 km thick at the Fortuna Bank, while, in the surrounding areas, the crustal thickness averages 10 km (Girling, 1992). Its magnetic and gravitational signature indicates an anomalous feature, which has been interpreted by Girling as a major volcanic centre.

The second feature is the Amirante Arc (Figures 7.31b and 7.32). This trench defines a circular arc which extends through an angle of about 60°. It is significant that the centre of

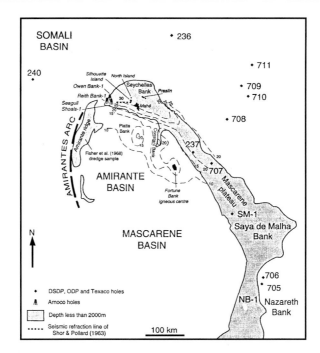

Figure 7.32 Geological setting of the Seychelles region, showing various volcanic features discussed in the text (after Devey and Stephens, 1992).

radius of the arc coincides with the *igneous centre* of the Fortuna Bank (shown in black in Figure 7.32). This feature may well represent the type of 'plug' discovered in the vicinity of Bombay (Negi *et al.*, 1993). With this black unit as centre, the radius of the Amirante Arc is approximately 300 km.

The age of the trench is not well defined. Fisher *et al.* (1968) dated samples dredged from the Amirante Bank at 82 +/– 16 Ma (which gives a minimum age of 66 Ma).

7.7.1 Tracks

It has been noted earlier that the change in speed of the Indian sub-continent at about 67.23 Ma, marked the event which gave rise to the Deccan province (Figure 6.6a and b). It has also been argued in this chapter that, because of sharp changes in direction and/or rates of movement, other arcuate trenches can be interpreted as being associated with a major impact event.

Details of the displacement of the Reunion hotspot relative to Madagascar are shown in Figure 7.33. The change in the relative direction of track was 120° anticlockwise, and there was also a small increase in speed. As this change in direction occurred 67.68 Ma ago, we suggest that the Amirante trench formed as the result of an impact at this time, which gave rise to failure of the lithosphere along an arcuate circumferential thrust zone, which extends from GZ for a distance equal to at least twice the radius of the impact crater. The Amirante Arc is situated 300 km from what can be inferred as GZ. Hence, by using the Snowball 500 ton explosive experiment as a guide, it can be inferred that the diameter of the impact crater was not more than about 300 km.

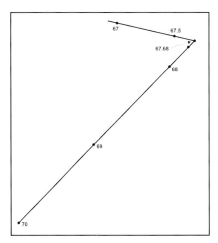

Figure 7.33 Backtrack of Reunion Island, showing abrupt change in direction and speed of motion at 67.68 Ma.

The age (inferred from Atlas 3.3) at which the track of the Seychelles changed direction is 67.68 Ma, while that which, we suggest, initiated the Deccan Traps was 67.24 Ma. The difference in age of these two events is approximately 0.5 Ma. This discrepancy could be a measure of the intrinsic uncertainties of the Atlas program. Alternatively, the two events may really have been separated by about 0.5 Ma. This point will be discussed later in this chapter, when we deal with the impact events that occurred near the K/T boundary.

7.8 The Mariana Trench

We have noted that the west Pacific rim exhibits a variety of arcuate subduction zones with very large radii of curvature, which are adjacent to the Asiatic landmass. In this section, we shall only be concerned with the development of the Mariana arcuate feature. However, before we discuss this feature, it is apposite to comment briefly on the *tracks* of a point in Figure 7.34 which relates to the movement of a point, currently at the most eastern point of the Mariana Trench.

The current positions of this reference point is indicated by 0. The Mariana track has been plotted back for 300 Ma. In the period 0–195 Ma, the movement of this track runs a little north of east, while from 195–300 Ma, the tracks run a little east of north. The Mariana trench is the largest of the five arcuate structures that we discuss in this chapter. As will be seen from Figure 7.4, the arc has a radius of curvature of about 625 km and extends through an angle of about 150°. This structure is adjacent to some of the oldest known, oceanic crust and is adjacent to the East Mariana Basin which contains seamounts and widespread eruptive material.

The extreme tips of the arc change their curvature and, to the north, turn to become almost continuous with the Agasawara Trench, while to the south it almost reaches the northernmost tip of the Yap Trench. To the west, it abuts upon an area in which the sea-floor contains a series of ridges and basins (Figure 7.4), which are adjacent to the Philippines.

From the amount of oceanic lithosphere that has been subducted, we can infer that this arc is a mature feature. Thus, if the oceanic lithosphere has been subducted to a depth of

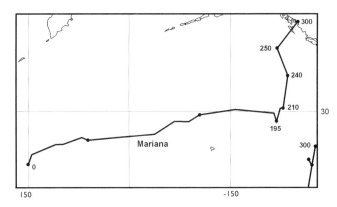

Figure 7.34 Tracks of points currently adjacent to the Mariana Trough from the present day (0) back to 300 Ma.

600 km down a 45° dipping unit, the length of lithosphere subducted is at least 850 km. If we further assume that the rate of subduction has averaged 8.6 cm a^{-1}, then it can be inferred that the arc has been in existence for at least 100 Ma. Also, as we have already noted, the trench is adjacent to some of the oldest oceanic lithosphere currently existing. Hence the age of the oceanic lithosphere, in which the Mariana structure developed, may date back at least as far as 200 Ma and could possibly extend as far back as 240 Ma.

If, in the light of the previous examples cited in this chapter, we assume that the Mariana Trench was the result of an impact, then, clearly, we are looking for an extremely large event, possibly the largest oceanic impact event to have occurred in the latter half of the Phanerozoic.

The overall *track* for the Mariana reference point from 0–300 Ma is shown in Figure 7.34 and details of this track are shown in Figure 7.35. For the period from 280 to 160 Ma, the track is shown in Figure 7.35a. It can be inferred from this portion of it that several track discontinuities, which we attribute to major impact events, occurred in this period, at 250, 238, 220, 208, 195 and 175 Ma respectively. The oldest is the 250 Ma event, which we have already related to the initiation of the Siberian plateau basalt. We shall discuss the other inferred impacts in the following chapter.

(In passing, the reader should note that the scale of the tracks shown in Figure 7.35 changes from example to example. The relative scaling can be obtained by comparing the length of track for 1 Ma, or 5 Ma intervals in the different diagrams. Thus, the scales in Figure 7.35a and Figure 7.35b are approximately the same. The scale in Figure 7.35c is 3 times greater than in Figure 7.35b, in Figure 7.35d the scale is 6 times greater than in Figure 7.35c, and the scale in Figure 7.35e is approximately equal to that of Figure 7.35c.)

It will be noted that, in the period 220–207.9 Ma, the oceanic lithosphere moved only a few tens of kilometres in 13 Ma, which, compared with the relatively rapid movement before 220 Ma and after 207 Ma, was a period of near 'standstill'. The sudden reduction in speed at 220 Ma and the increase at 207.9 Ma can best be understood in terms of the effects of two separate impact events. (See the inset map in Figure 7.35a for details.)

The sudden 145° reversal of track at 195 Ma and the major change in track, both in direction and increase in speed at 175 Ma, are both features which compare with others we have seen earlier which we have attributed to major impacts. (Both these features are shown again in Figure 7.35b.) As will be seen in this figure, minor changes in direction and speed of track also occur at 160 Ma and at 156 Ma. In Figure 7.35c, other relatively minor changes of track occur at 135, 128 and 124 Ma respectively. A detail of one such change is shown in Figure 7.35d at 119.2 Ma and finally, in Figure 7.35e, a single change occurs at 110 Ma.

Of the various 'impact events' cited since 195 Ma, only the event at 119.2 Ma and 175 Ma gave rise to a change in direction of plate movement of 30° or more. The smaller of these angles is associated with the 119.2 Ma event, which we have already indicated to be the impact event that gave rise to the initiation of the Ontong-Java plateau basalt.

The major impact event which initiated the change in track at 175 Ma, not only caused a significant change in direction of more than 30° but also resulted in a five-fold increase in speed. As can be seen in Figure 7.35b, this rate of movement was maintained for at least the next 25 Ma. Hence, we suggest that it is reasonable to attribute the Mariana impact event to the 175 Ma track change.

However, we have put the probable earliest date of this plate at about 240 Ma. The 195 and 175 Ma impacts would therefore have developed at no great distance from a newly created spreading-ridge. The 625 km radius of the thrust, for a really major impact, must not extend beyond the spreading-ridge. Any arcuate feature associated with a major impact, with a distance from GZ to the ridge significantly less than 600 km, would have been truncated by the spreading-ridge, with the eastern-most part of the arc being carried away to the east.

If it is assumed that at 175 Ma, the point of impact could have been about 650 km from the spreading-ridge, providing the spreading rate was close to 10 km Ma^{-1}, with a radius of curvature of the thrust arc set at 625 km, the thrust arc would not be truncated.

It will be noted from Figure 7.34, that the Pacific oceanic lithosphere extended for several thousands of kilometres to the west of the impact point where, we suggest, the arcuate thrusts were generated. This oceanic lithosphere moved in a generally westward direction,

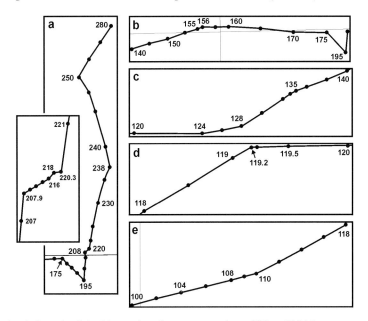

Figure 7.35 (a–e) Details of the Mariana's reference point from 280 to 100 Ma.

carrying the arcuate fracture zone within the plate. The oceanic lithosphere to the west of the 175 Ma impact must have been subducted, probably along a southerly extension of the Japan Trench. During this westward journey, the Mariana thrust structure probably experienced little further differential movement. Only when the *circular* feature encountered the Japan Trench did the rear arcuate aspect of the impact structure develop to dominate the subduction in that collision area.

7.9 Impact size and minor subduction arcs

It will be inferred from the arguments presented in this and the previous chapter that CFBs and OPBs have been studied, among other things, with the object of determining their date of emplacement. It has been established that these igneous features are usually emplaced within about a million years, though minor activity may continue for a further ten or so million years. Thus, the time of initiation and main development of these large basaltic features are established within the limits of error of the dating technique used.

When the sudden changes in track, whether they be in speed and or direction, have been established, it is often a relatively simple task to correlate the specific track change with a specific OFB. The magnitude of the effect of the inferred impact event can, in part, be estimated from the changes that the tracks reveal. However, we can also infer that the magnitude of change in these tracks is influenced and constrained by the type and size of plate, and also, possibly, the position of the impact in relation to the plate boundaries. As we have seen, prodigious changes may take place on one side of an event, while on the other, virtually nothing is recorded (e.g. the considerable change of track in S America compared with the lack of change of track of Africa following the 135 Ma Paraña event). Hence, although we can correlate the time of a track change with the initiation of emplacement of a CFB, the size of the impact event can only be inferred from the volume of extruded rock that results, coupled with an assessment of the thickness of the lithosphere in which the impact took place.

As regards arcuate subduction features, the reverse situation exists. These features are mainly 'on-going' structures and have attracted relatively little investigation regarding their initiation. Dating of the volcanic activity associated with the subduction usually tells one little, if anything, of the date of initiation of the subduction event. The Amirante Arc appears to be no longer active, so is, of course, an exception to this statement.

The track changes, together with other evidence, are sufficiently specific to identify the time of impact of the Antilles, Amirante, Scotia and Banda features. However, the Mariana arcuate feature is somewhat ambiguous as regards the specific track change to which it can be attributed, but we conclude that it was probably the event that occurred at 175 Ma.

We have noted that arcuate thrusts developed as the result of the Snowball 500 ton TNT explosive experiments, and these were shown to crop out at the surface as far as twice the crater radius from GZ. Therefore, it is suggested that such thrusts brought about by a major oceanic impact could result in the development of an arcuate subduction zone. There is, however, the problem regarding the size of the impacting body that could possibly give rise to such arcuate features.

The radius of curvature of the Mariana Trench is about 600 km. Hence, if this natural feature also exhibited a relationship between crater diameter and outcrop diameter of the most distant thrust from GZ comparable with the 'Snowball' explosive experiment, it is necessary to postulate that the impacting body gave rise to a crater radius of about 300 km.

The Lesser Antilles Arc in the Caribbean requires a crater of almost comparable size. We have also noted that the impact that gave rise to the Paraña could well have been associated with a crater of comparable diameter.

It can reasonably be postulated that one such 500 km diameter crater could possibly have developed in the latter half of the Phanerozoic; but to suggest that there are three such events may stretch the credulity of even the most open-minded reader. Let us, therefore, consider mechanisms of impact a little further.

The extent to which thrusts develop from GZ will be determined by the rate of attenuation of the pulse stress as it propagates outward from GZ. The outermost thrust occurs where the magnitude of the pulse stress falls to a value of differential stress, below which it is no longer able to cause the rock to fail in shear. Let us, therefore, consider the stress conditions which can give rise to arcuate thrust fault development.

7.9.1 Rock failure in terms of pulse and ambient stresses

For strong, dry, unweathered, igneous basic rocks the relationship between principal stresses at failure is given, with reasonable accuracy, by the relationship:

$$S_1 = S_0 + K.S_3 \tag{7.1}$$

where S_1 and S_3 are the greatest and least principal stresses at failure, S_0 is the uniaxial strength of the rock and K is a constant determined by the angle of sliding friction of the rock (a), such that:

$$K = (1 + \sin a)/(1 - \sin a) \tag{7.2}$$

For a value of a = 37° the value of K = 4.0, and the angle which the shear plane makes with the axis of maximum principal stress is 26.75° (Price, 1966).

It can be inferred, therefore, that the development of arcuate thrusts is controlled by three main factors: (1) the material constants of the rock (i.e. the value of S_0 and K; (2) the magnitude of the stress pulse; and (3) the orientation and magnitude of the ambient stresses in the lithosphere at a particular time, immediately prior to the arrival of the stress pulse.

The Snowball 500 ton TNT explosion generated several parallel shear planes which dip at a relatively low angle, inward toward GZ. These fracture planes, which cropped out at a maximum distance from GZ of approximately twice the radius of the crater rim, were exposed in deep radial trenches at least as far back as directly beneath the crater rim. From the orientation of the fracture planes and the thrust movements sometimes seen on these planes, there can be little doubt that these fractures originated as thrusts. (In those trenches which showed normal fault movement on these planes, this inward movement can be explained by a relatively small degree of *rebound* 'slumping' of the sediments towards the crater.)

Beyond the crater, several radiating vertical tensile fractures developed. As we have seen in Chapter 5, the stress conditions necessary to generate such vertical tensile fractures are significantly different from those required for the generation of thrusts.

For vertical tensile fractures, the least principal stress (S_3) must act circumferentially, while for thrusts, it is the intermediate principal stress (S_2) that must act circumferentially, while the least principal stress (S_3) must be near vertical. The maximum principal stress (S_1) in both these instances acts radially (Figure 7.36).

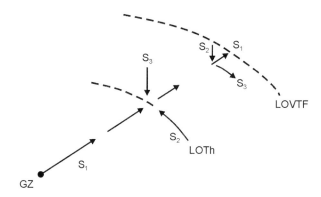

Figure 7.36 Schematic representation of stress orientations S_1, S_2 and S_3 at different distances from GZ. LOTh = limit of thrusts and LOVTF = limit of vertical tensile fractures.

It has been argued in Chapter 6 that, for the necessary stress conditions which can give rise to the location of (LO) and generation of vertical tensile fractures (LOVTF), one must invoke the action of the circumferential extension that is caused by the propagation of the hemispherical, compressive, pressure-pulse induced by the explosion or impact. Once one or more major vertical, radiating, tensile fractures is/are initiated by these circumferential tensile stresses, then, provided they reach down into the asthenosphere, where pressure-release melting takes place, rapid migration of the fluid melt will widen and extend these tensile fractures by the hydraulic fracture mechanism (Price and Cosgrove, 1990).

The development of these vertical fractures will, therefore, give rise to a positive increase in the magnitude of the circumferential stresses, which immediately changes from the least principal stress (S_3) to the intermediate principal stress (S_2) (LOTh in Figure 7.36). These stress conditions are maintained throughout the brief period that remains of the propagating stress pulse P. It is suggested that it is during this brief period that the pulse stress, superimposed upon the ambient stress field, gives rise to the circumferential thrust planes.

As we are concerned here with the generation of thrusts that can be induced in oceanic lithosphere by such a pressure pulse (P_p), it is now apposite to recall the magnitude of the probable ambient stresses that exist in such rocks, down to a depth of about 30 km. Below this level, except in very old, thick, ocean lithosphere, a thrust, or a number of such fractures that penetrate to a depth of 30 km, will so weaken the lithosphere that the lower, more ductile lithosphere will, sooner or later, yield and, as the result of ductile shear, permit an extension of the lower part of the thrust to cut completely through to the asthenosphere. Because of the weakness of the lower portion of the oceanic lithosphere, the lower zone will behave as a weak plastic material. The upper, brittle thrust fault will gradually degenerate, with depth, into a shear zone which is likely to increase gradually in dip until it reaches an angle of 45°.

As we saw in Chapter 2, the ambient stress in elastic lithosphere is made up of a series of elements. The first of these is the vertical stress (S_z) that is generated by the weight of a 'column' of rock in oceanic lithosphere (Table 7.2a), where the vertical stress is given by S_z = d.g.z (where d is the density of the rock, g is the gravitational constant and z is the depth in the appropriate units). At a depth of 30 km, the vertical stress in oceanic lithosphere will be approximately 9.9 kb. The values of S_z from 0 to 30 km are listed in Table 7.2a, and from 0 to 20 km in Table 7.2b.

Table 7.2a

			Ambient stress		Pulse (20 kb) + ambient			Comment
Z(km)	S_z (kb)	$S_h = S_z/(m-1)$ (kb)	S_x (+6) (kb)	S_y (+1.5) (kb)	S_1 (S_x + 20) (kb)	S_2 (S_y + 5) (kb)	$S_3 = S_z$ (kb)	
0	0	0			20.00	5.00	0	Thrust
5	1.65	0.55	6.55	2.05	26.55	7.05	1.65	..
10	3.30	1.10	7.10	2.60	27.10	7.60	3.30	..
15	4.95	1.65	7.65	3.15	27.65	8.15	4.95	..
20	6.60	2.20	8.20	3.70	28.20	8.70	6.60	Near failure
25	8.25	2.74	8.74	4.24	28.74	9.24	8.25	
30	9.90	3.30	9.30	4.80	29.90	9.80	9.90	Switch of S_z to S_2

Table 7.2b

			Ambient stress		Pulse (20 kb) + ambient			Comment
Z(km)	S_z (kb)	$S_h = S_z/(m-1)$ (kb)	S_x (+6) (kb)	S_y (+1.5) (kb)	S_1 (S_x + 20) (kb)	S_2 (S_y + 5) (kb)	$S_3 = S_z$ (kb)	
0	0	0			20.00	5.00	0	Thrust
5	1.65	0.55	6.55	2.05	22.05	11.55	1.65	..
10	3.30	1.10	7.10	2.60	22.60	12.10	3.30	..
15	4.95	1.65	7.65	3.15	23.15	12.65	4.95	..
20	6.60	2.20	8.20	3.70	23.85	13.20	6.60	Near failure

The second element contributing to the ambient stress is the lateral stress generated by the vertical stress. This increment of horizontal stress is determined by the equation cited in Chapter 2, namely $S_h = S_z/(m-1)$, where m is Poisson's number which, for the rocks in question, has a value of 4.0. Hence, the horizontal stress that results is one third the vertical stress at any specific depth (i.e. at 30 km, $S_h = S_z/3 = 3.33$ kb).

These two elements, in combination, determine the ambient stress in oceanic lithosphere *at rest*, were the lithosphere not influenced by any other mechanism. However, as has been argued in Chapter 3, oceanic lithosphere is normally acted upon by a gravity-glide mechanism which is driving plate motion. The average magnitude of this glide-stress depends upon the boundary retarding forces and will also vary with the distance from the spreading-ridge.

For convenience, let us assume that this gravity-glide stress (S_{gg}) has an average value of 6 kb. This stress will give rise to horizontal stress at right angles to S_{gg} of a magnitude of 6 kb/ $m = 6$ kb/4 = 1.5 kb, so that, at a depth of 30 km, the value of horizontal stress will be $S_2 = 3.33 + 6 = 9.33$ kb and $S_3 = 3.33 + 1.5 = 4.85$ kb. (These gravity-glide induced stress magnitudes are not valid at shallow depths, for they are capable of generating thrusts near the surface.) Let us now assume (for Table 7.2a and Table 7.2b), that the magnitude of the compressive stress generated near the end of the compressive, transient pulse-stress is 20 kb. Then, as can be inferred from Table 7.2a, the pulse stress of 20 kb when added to the larger horizontal principal stress becomes 28.20 kb (at a depth of 20 km), with the other horizontal principal stress of 8.70 kb. The vertical stress (S_3) is 6.60 kb. The stress required to cause failure by thrusting is $S_1 = 2.0 + (4 \times 6.6) = 28.6$ kb. It will be noted that the maximum horizontal stress at this depth is only 28.20 kb, so that failure will not take place for these assumed conditions. The limiting depth at which thrusts will form for the assumed conditions is actually 19.75 km.

Similarly, for the stress conditions listed in Table 7.2b, it can readily be inferred that the maximum depth at which thrust faults will develop is between 15 and 20 km. The precise level is 16.25 km

Thus, the limiting depths at which a thrust can develop varies with position relative to the ambient stress axes. When the stress pulse is oriented parallel to the direction of the ambient maximum horizontal stress, this limiting stress is set at 19.75 km. However, when the stress pulse is parallel to the minimum, ambient, horizontal stress, the maximum depth at which a thrust can develop is set at 16.25 km.

One can infer from these figures that if a circular pulse stress of 20 kb is generated at a specific distance from GZ (say 200 km) then, from Figure 7.37, it follows that the thrust plane would crop out at distances from 238.5 to 232.5 km from GZ. That is, a circular stress pulse will give rise to an outcrop of the most distant thrust, which forms an ellipse rather than a circular arc. However, this deviation from an exact circle is only a few per cent of the average radius of the thrust structure. Within the accuracy of determining the geometry of a major arc, it is suggested that such a minor deviation is not likely to be detected.

This analysis is, of course, only an approximation. Clearly, for the values of depth, pulse stress and Poisson's ratio chosen, we have a critical condition. Had a pulse stress of 22 kb been stipulated, or a depth of 25 km, or a value of Poisson's ratio of m = 3.8, any of these changes would have resulted in a stress situation in which circumferential thrusts would have developed at somewhat different depths.

We have also assumed that linear-elasticity theory holds, and that the elastic moduli remains constant throughout the 25 km thick layer, ignoring the fact that the components of compressive stresses, caused by the gravity-glide mechanism or the pulse stress, does not leave the thickness of the 25 km layer unchanged. However, despite these various approximations, we can reasonably conclude that a pulse stress with a magnitude of about 20 kb will be sufficient to initiate near-circular circumferential thrusts down to a depth of about 20 km. What we now need to specify is at what distance from GZ such thrusts can be initiated by a 20 kb pulse stress.

Let us initially assume that the impacting body is a solid spherical meteorite. We have noted that the rate of attenuation of the far-field stress pulse for such a body is determined by the impact velocity (Figures 5.29 and 5.30). For example, Ahrens and O'Keefe (1977) have shown that the magnitude of the attenuating pulse falls to a value of 20 kb at a distance from GZ of 15 times the diameter of the impacting body, when the impact velocity is 5 km s^{-1}. In addition, we have noted (Figure 5.37) that the ratio of the diameter of a 100 km crater to that of the impacting body, striking at a velocity of 5 km s^{-1} is 6 : 1 (i.e. the impacting body has a diameter of 16.7 km).

Hence, we may infer that the stress pulse falls to a value of 20 kb at a distance of 15 × 16.7 km (i.e. 250.5 km) from GZ. At this distance from GZ, the 20 kb pulse will be able to generate a thrust plane, which crops out at the surface at a further distance of over 30 km (Figure 7.37). Consequently, the diameter of the circumferential thrust will be 250.5 km plus 30 km, giving a total of about 280 km. Thus, for this specific model, the ratio of the radius of curvature of the peripheral thrust to that of the radius of the crater will be 50 : 140, or almost 1 : 3, rather than the 1 : 2 ratio attained by the Snowball explosion. (One may infer that the Snowball experiment was a model simulating an event with a higher impact velocity than 5 km s^{-1}.)

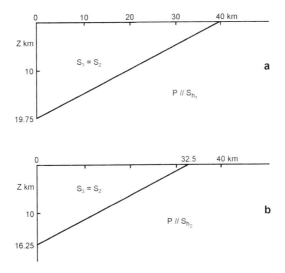

Figure 7.37 (a) Distance of outcrop of thrust plane from position where the pulse stress of 20 kb is oriented parallel to the greatest, horizontal 'ambient' stress. (b) As in (a), except that the pulse stress is parallel to the least, horizontal 'ambient' stress.

If the ratio of 1 : 3 for the radii of the impact craters to the corresponding peripheral thrusts were to hold for all the arcuate trenches mentioned above, the impact crater diameter would range from 400 km diameter for the Mariana Trench down to as little as 150 km for the Banda Arc.

In reaching these conclusions, however, we have assumed that the impacting body was a solid boloid, rather than a comet. As we have seen, a 100 km crater, caused by a solid, stony meteorite, with 5 km s^{-1} impact velocity, would require this body to have a diameter of 16.7 km. Such an impacting body is likely to hit a continental target at more than 15 km s^{-1}. An ocean splash-down, where the water is 5 km deep, would cause only a relatively small degree of retardation in velocity of the meteoritic impacting body.

We have also noted that it is now generally accepted that major impact events (i.e. giving rise to craters with diameters greater than 100 km) result from comets. It has been indicated that a simple, solid near-spherical asteroid can be likened to a high-velocity rifle bullet, whereas a comet is more likely to approximate to the effects of a 'super' shotgun. We have also noted that a single body that splashes down into the ocean must have a diameter that exceeds 1/15th to 1/20th the depth of water to the ocean floor, if it is to have the capability of generating even a small crater.

7.10 Impact of comet-bodies in deep ocean

Readers who have seen films, shot by underwater cameras, in which a person, usually a man, is seen swimming at a depth of as little as 1.0 m beneath the surface, will have observed that he is completely safe from being wounded by bullets fired at him. The bullets very rapidly decelerate in the water and, within a depth of 0.5 m or less, begin to sink at their terminal velocity. The actual depth of penetration of a bullet depends upon its calibre, shape and muzzle velocity, the distance between rifle and target, and the angle at which it enters the water. Hence, the problem is a complex one. Nevertheless, such scenes are testimony to the general accuracy of the predictions of Ahrens and O'Keefe (1977) regarding the minimum diameter of impacting bodies, which are required to cause even small craters on the ocean floor. However, as we have seen in Chapter 6, to assume that a comet which splashes down into the ocean can be likened to a shotgun charge, is a model analogue which is certainly too simple.

7.10.1 Comets and comet-derived bodies

As we have noted in Chapter 1, the physical properties of comets are not at all well known (Lewis, 1996). Halley's comet in its most recent passage was the only one to have been studied by a 'fly-by' mission. From the studies of gases and dust emitted by comets as they pass close to the Sun, it is inferred that this particular comet contains fine-grain dirt and ice in approximately equal mass. However, it is thought that the density of this material varies from comet to comet. Some appear to contain at least 90 per cent of void space. Others are, possibly, firmly compacted and have a density of 1.5–2.0. These may be comprised of compressed permafrost with dirt and ice.

Comets break up even under relatively low gravitational disruptive forces. However, this disintegration may represent the parting of loosely bonded units of much stronger, dense 'permafrost' blocks of dust and ice (and possibly solid rock). When they enter the Earth's atmosphere, small units of cometary material are easily destroyed, so that only the more dense and stronger, larger, permafrost blocks are likely to penetrate nearer than 50 km from the Earth's surface.

As we have noted earlier, it is now held by those studying impact events that any impact crater with a diameter of more than 100 km is almost certainly attributable to meteor or meteor-derived bodies. Let us, therefore, assume that the conceptual model shown in Figure 7.38 and discussed briefly in Chapter 6, represents a large comet or comet-derived impacting body, which, in space, has a spherical form, with a diameter of several tens of kilometres. If much, or all, of the comet's carapace has been removed in multiple passages past the Sun, the remaining core could be composed of relatively strong chunks of permafrost which are only loosely bonded one to another. Provided the average dimension of the permafrost chunks is over 250 m in diameter, even such chunks that become disrupted from the original body would survive passage through the atmosphere. Though reduced in size, such bodies, on impact, could still measure about 150 to 200 m across.

The largest (non-giant size) comets, or comet-derived bodies, have a diameter of about 40 km (Taylor, 1992). Such a large cometary body could contain as many as 2,000,000 chunks of permafrost with an average diameter of 250 m. Even if the number of bodies of this size are as few as 200,000, this is a more complex situation than is modelled by a shotgun charge.

Let us first consider a relatively small comet with an initial diameter of 5 km and also assume that this comet splashes down in deep ocean. Then, even if the leading large permafrost bodies of this comet are reduced to a diameter of 150–200 m, they will still penetrate the water to a depth of 1.5–2.0 km before their speed can be reduced to near their terminal velocity in water. However, this situation would not arise. The earliest 'wave' of these 150–200 m diameter blocks to hit the ocean's surface would disrupt the water by their individual shock waves, which would be so close to one another that they would quickly combine to push aside the upper levels of the ocean. In so doing, the first arrivals would experience a rapid reduction in their velocity, and would also begin to spread laterally one from another, as the oceanic water was either blown away in a radial direction, or burst through between

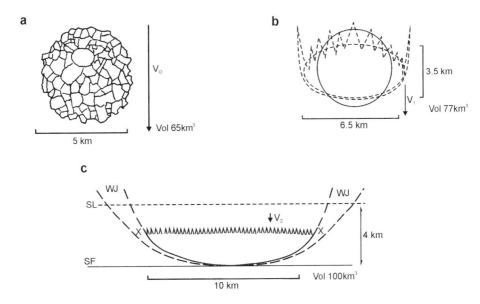

Figure 7.38 Changes of dimensions of a hypothetical comet as discussed in text.

these first arrivals as jets and fountains of water. Thus, the first arrivals would create a wide but shallow 'crater' in the water of the ocean.

The second wave of cometary blocks would not encounter the surface limits of the 'crater' of water until they were perhaps 0.5–1.0 km below the original oceanic surface, and would therefore maintain the impact velocity they would have had, had they fallen on a continental mass. Hence, they would rapidly catch up with the blocks of the first wave. The first wave of blocks, which would have a significantly reduced velocity, would now be struck in the rear by the blocks of the second wave. The energy of the components of the second wave would ensure that the rate of progress of the first wave would continue, albeit at a reduced velocity. Individual blocks of the second wave would tend to seal the water jets as they found, by chance, the breaks in the forefront of the impacting bodies.

This process would be continued by successive waves of impacting permafrost blocks. The later waves would catch up with the earlier ones, and thereby prevent total loss of momentum of the frontal blocks. The blocks of later 'waves' would compact and fuse together with the earlier arrivals, forcing most or all water to escape radially from in front of the impacting body. At this stage, this deformed cometary body would become a reasonably compacted, oblate spheroid, with an impact velocity of about 5 km s^{-1}, even if the water was originally 4–5 km deep. That is, the impact on the ocean floor would be about 30–40 per cent of the velocity at which a similar comet would impact upon a continental unit.

As the density of the component elements of such a cometary body could be about half to a quarter that of an asteroid for a given volume, the energy of a comet-derived body would be about half to a quarter that of an asteroid, of corresponding volume which impacted at a comparable velocity (say 5 km s^{-1}).

One can infer that the dimensions of the impacting body will also have changed. It would have greatly decreased in volume as the smaller elements were melted as it passed through the atmosphere. Also, the initial spherical form would have become an oblate spheroid (Figure 7.38). In that figure, we start with an assumed initial sphere with a diameter of 5 km which, as it approached the surface of the ocean, changed to a diameter of 6.5 km and a vertical thickness of 3.5 km respectively. When the body made contact with a 4 km deep ocean floor, its diameter was assumed to have increased to 10 km and its thickness reduced to 2 km. These dimensions relate to only a modestly large comet. The largest such body may have an initial diameter of 40 km, so that the dimensional changes that would take place in this large cometary body at the base of the atmosphere could result in a diameter of 52 km and a maximum thickness of about 28 km.

If we assume that, as the result of passing through the Earth's atmosphere, the blocks are reduced from 250 m to 200 m in diameter, the volume of the major component parts is reduced by about 50 per cent. It can be inferred, therefore, that the void-space between the component parts of this icy body may be large.

When the ice-body enters the ocean, it will spread even more dramatically, so that its diameter could approach 80 km, while the vertical extent of the ice-body may be further reduced to about 16 km. However, this latter figure does not take into account the deceleration of the component elements of the ice-body or its 'porosity', which initially, could be at least as high as 80 per cent.

The momentum of the individual components of the ice-body, once the leading edge hits the 4 km deep ocean floor, is such that it begins to reconstructs itself into a relatively non-porous, pancake-like impactor with a diameter of about 80 km and a thickness of perhaps as little as 2 km. This reconstruction, or recompaction, would be completed in a matter of a second or so.

The example cited above is, of course, based on simplifying assumptions. For example, we have, for convenience, referred to waves of impacts, whereas the process would approximate to a continuous stream of impacting component elements of the main body. Nevertheless, we suggest that the process outlined above demonstrates how pancake types of cometary impactors may result, when large agglomerates of ice-bodies strike the ocean floor. It is suggested that an impact velocity of 5 km s^{-1} for such a body is not unreasonable. The peak-stress will be significantly lower (possibly only 10 per cent) of the 95 GPa (950 kb) cited by Ahrens and O'Keefe (1977) (Figure 7.39). It is assumed that the rate of attenuation will exhibit similar slopes, so that a pulse stress of 20 kb may well extend for a distance of 7 times the radius of the impacting body. The stress level of 20 kb can give rise to a set of thrust planes, the lowest of which will reach the surface at a distance of about 35 km beyond the point at which the transient pulse falls to a value of 20 kb. If the radius of the crater is 100 km, and the attenuation curve of the pulse stress falls to 20 kb at a distance of 7 times the radius, the outer-most thrust would develop at a distance of about 735 km from GZ. This significantly exceeds the radius of the Mariana Trench.

The extrapolation that has been made of the Ahrens and O'Keefe curves could possibly result in an exaggeration of the ratio of the radius of the impact crater to the radius from GZ over which the pulse stress falls to 20 kb. If we assume that the extent of the thrusts from GZ which resulted from the 500 ton TNT Snowball explosion, represents a reasonable scale-model of a major impact of a large comet in deep ocean, then the thrusts will have a radius about twice that of the crater. From this, we can infer that the Mariana Trough would have required an impacting body capable of generating a crater, however rudimentary, of a little over 300 km diameter. As we have inferred in Chapter 5, it is probable that craters of such size developed at least five times in the Phanerozoic. (In the following chapter, evidence will be introduced regarding the flux of meteoritic impacts, which indicates that the number of craters in the Phanerozoic with a 300 km diameter could have been in the range of 7 to 10.)

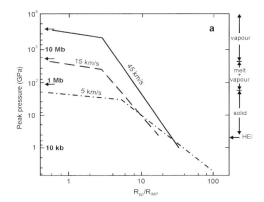

Figure 7.39 Attenuation curves of pulse stress at different velocities of impact, presented earlier, representing the behaviour of a solid, spherical stony meteorite.

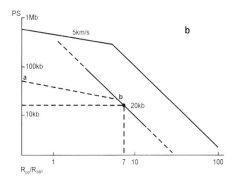

Figure 7.40 Hypothetical analogy of pulse stress attenuation of a major cometary impact (impact velocity 5 km s^{-1}), with an assumed peak impact stress 90 per cent smaller than that of a solid, stony meteorite, but with a similar stress attenuation.

The hypothetical model presented here of the evolution of a comet or comet-derived body as it passes through the Earth's atmosphere and then through the ocean is simplistic and possibly contentious. It has been tacitly assumed that the body in space approximates to a sphere, whereas it may have been deformed by the gravitational fields of the Sun and planets into a somewhat elongated body with the long axis in the direction in which it is travelling. Although the model presented here leaves much to be desired, it must be pointed out that, despite the fact that many asteroids are known to be irregular and wildly non-spherical in shape, they are almost invariably modelled as spheres immediately prior to impact. The limitations of the analytical methods available constrain the theorists to make this simplifying and widely accepted assumption. However, as pointed out by Boslough et al. (1995), within 2 seconds of impact, the shape of the impacting body has little significant effect upon the form and extent of the pressure wave. It is suggested, therefore, that although comet bodies would make impact as something approaching oblate spheroids, after 2 seconds following impact, the extent and attenuation of the stress wave would approximate to that calculated by Ahrens and O'Keefe (1977) (Figure 7.39). Figure 7.40 shows a detail of Figure 7.39, which is based on a 5 km s^{-1} impact speed, but with the peak stress reduced to about 50 kb, from which it can be inferred that a stress magnitude of 20 kb (which is sufficient to induce

Table 7.3

Impact event	Atlas dates (Ma)
Amirante Arc (Figure 7.33)	67.68
Deccan Traps (Figure 6.1)	67.23
Chicxulub (Figure 6.7b)	66.25
Scotia Arc (Figure 7.3)	65.43

thrusting) can be attained at a distance of 7 times the radius of the impact crater. If these figures are reasonably representative, one can infer that a thrust that crops out at 650 km from GZ can be induced by an impact that gives rise to a crater of about 200 km in diameter.

7.11 Comment on the K/T boundary events

In the following chapter, it will be argued that there have been about 30 major impact events in the Upper Phanerozoic. Many of these may be solo impact events, while others, such as those clustered near the K/T boundary, may be single or multiple events.

A number of impact events have been cited that have occurred within a period of less that 2.5 Ma of the K/T boundary, which we take to be 65 Ma ago. Four of these impacts are listed in Table 7.3, all of which were dated by sudden changes in track.

Brief mention has been made of the initiation of the N Atlantic Igneous Province (NAIP), but no convincing track data could be found for this event. Therefore, although radiometric dating of this event also places it close to the K/T boundary, data for the NAIP have not been included in Table 7.3.

The average impact date for the four events in Table 7.3 is 66.87 Ma (+ 0.83 or −1.44 Ma). The range of dates is probably outside the errors likely to be incurred for events near the K/T boundary (Harland *et al.*, 1982). There are three possible ways in which one can interpret these age data.

Firstly, we can assume that all four of these events, as well as the NAIP and possibly other smaller events, were coeval, within the limits of less than a few days. Certainly, such a group of major simultaneous impact events would have had a far greater effect than Chicxulub alone, and would support the viewpoint of those who hold that the dinosaurs were exterminated in a horrendous Armageddon.

Secondly, one may assume that there were two phases of impacts. The first related to the 67 Ma period, which are also geographically linked, in and around the position of western India at that time. The second phase related to 65.4–66.25 Ma.

Thirdly, it is possible that there were four (or even more) separate unrelated events, or perhaps the Indian Ocean group represented one event and the Chicxulub and the Scotia Arc events represented two additional, separate events which were unlinked in time.

Paleontologists who support the view that the dinosaurs were not made extinct in a single event would probably favour the third option. Others may argue that major impacts occur so infrequently that one may reasonably expect that when a major comet strikes Earth, it may well be that the event is similar to the Shoemaker-Levi 9 multiple impact on Jupiter, so that a string of major and minor impacts on Earth could have occurred within a few hours.

An alternative model, based on chaos theory, has been put forward by Shaw (1994), in which he argues that there is a period in which a number of comets collect themselves into a 'reservoir'. This build-up period is then followed by a 'dumping' of the contents of the 'reservoir', several of which are likely to impact upon Earth. We tend to favour this latter model.

The pros and cons of this argument are currently receiving considerable debate. (See for example Special Paper 307 (1996) of the Geological Society of America [Eds Ryder, Fastovsky and Gartner] entitled 'The Cretaceous-Tertiary Event and other Catastrophies in Earth History' for an extensive introduction to the debate.)

7.12 Summary and conclusions

A mechanism has been presented whereby relatively small, arcuate features could have developed. This mechanism is primarily based on the results obtained in the Snowball 500 ton TNT experiment conducted in Suffield, Alberta, Canada (Jones 1994). As we have seen (Figures 5.19 to 5.21), the crater formed by this explosion is underlain by a series of inward-dipping thrusts that cropped-out at a distance from GZ approximately equal to twice the radius of the crater.

A great deal of theoretical analysis will be required before the concepts developed in this chapter can be put on a firm base. However, it is suggested that, at this time, the best evidence one can present is that of the experimental work conducted at Suffield, which provides ample evidence on which to base an explanation of the development of the relatively small, arcuate subduction zones.

Finally, we note that all but one of the cited arcuate features can reasonably be correlated with specific abrupt changes in speed and direction of movements. The preferred dating for the initiation of the feature that eventually developed into the Mariana Trough, can, we suggest, reasonably be taken as occurring 175 Ma ago.

It has already been shown that changes of track can be associated with major oceanic plateau basalts such as the Ontong-Java event. Consequently, we see no inherent difficulty in using the track anomalies as evidence that the cited arcuate oceanic features are the result of major impacts in the latter half of the Phanerozoic.

Chapter 8

Periodicity, regional tracks, impact control and future risks

8.1 Introduction

In the previous chapters relating to meteoritic or cometary impacts, we have been concerned with details of impact mechanisms and the structures which result from them. These relate to specific features, each of which has a track which exhibits a sudden change in direction and/or a change in rate of motion, and include all three known certain, Phanerozoic, impact craters with a diameter of 100 km or more and the tracks of two non-certain impact features. All eight main, continental flood basalts, we argue, can be attributed to major impacts, as can the Ontong-Java plateau basalt and five arcuate subduction trenches.

Thus, we have cited 19 examples, all of which showed a correlation between the abrupt track change and the geological event, regarding the age of the inferred impact (as determined by the Atlas 3.3 program) and the ages of the various geological events obtained by radiometric dating. In addition, we have set out a mechanism that is able, in general terms, to explain how and why such changes in rate and direction of plate motion came about.

This correlation goes well beyond the possibility of coincidence. As noted earlier, we have established a principle, namely that several types of major geological phenomena are linked with, and are indeed generated by, major impacts.

The next obvious step is to accept that sudden changes in rate and/or direction of plate motion are the result of certain impacts. All we can expect at this stage of the investigation is that the age of the impact event can be ascertained. Accordingly, we set up a search for major track anomalies.

To this end, we established the tracks of specific identifiable points in seven regions back in time from the present day to about 260 Ma. (Initially, we imposed this time limitation because of the larger degree of inaccuracy that could be expected as we regress further in time.) By correlating the inferred impact events in the seven different areas, we were able to obtain some indication of the magnitude of the major events. As a result of this exercise we were able to show that 30 major, abrupt changes of track occurred, which we equated to 30 major impact events in the Upper Phanerozoic, some of which related to multiple impacts.

It will be recalled that the average impact flux throughout the last 2000 Ma has been assessed by Oberbeck et al. They indicated that the number of major impacts with a crater diameter in excess of 100 km, which are likely to have occurred on Earth in the Upper Phanerozoic, was about 14. However, from a study of impacts, which occurred on the far side of the Moon in the last 250 Ma, McEwen et al. (1997) concluded that during the Upper Phanerozoic, Earth experienced an impact flux which was approximately double the rate experienced in the Lower Phanerozoic. We therefore followed the tracks in two areas back to the beginning of the Phanerozoic; and established that this change in flux is reflected in a smaller number of track discontinuities, each of which we now take to indicate a major impact. From this exercise, we infer that the impact rate in the Lower Phanerozoic was, indeed, approximately half that which Earth experienced in the Upper Phanerozoic.

As we shall see, the results of this survey were surprising, for they indicated that the stratigraphic record in the Phanerozoic is determined, or otherwise controlled, by the incidence of major impacts.

An assessment of impact activity has also been inferred from the major phases of extinction that can be determined from the geological record. We shall show that the recognition of abrupt track anomalies as evidence of major impacts, and the reasonable assumption that there is a correlation of impacts with extinctions and other events in the geological record, provide almost identical results. This, of course, is highly significant. We suggest that the two different approaches are pointing to a single, irrefutable conclusion as regards the incidence and importance of major impacts throughout the Phanerozoic – and hence, by inference, throughout most of the life-span of Earth.

The effects of small to moderate-size impacts upon Earth, and more specifically the risk to individuals, communities and even the whole of mankind, as well as other relatively large animals, has become a necessary, if bizarre, topical interest. Accordingly, we shall present a brief recapitulation of the evidence, arguments and conclusions which indicate that, during the Phanerozoic, major impacts have been of primary importance in determining major changes in the evolutionary progress and development of the lithospheric areas.

Finally, we make the point that geology is currently a 'backward-looking' science. In the light of the evidence presented in this chapter, it is apparent that major impacts will occur on Earth in the future. A significant body of the scientific community must become 'forward looking' if mankind has any real chance of avoiding the fate of the dinosaurs.

8.2 Periodicity and relationship of cyclical events

In this book, we have adduced a significant amount of evidence that correlates known geological features. We suggest that these correlations are so conclusive that we can now, with little danger of error, assume that every major abrupt change in track, even though it cannot initially be linked to a specific event, must of necessity be attributed to a major impact. For convenience, it has long been tacitly assumed that impacts are random events, which nevertheless result in a near-linear, logarithmic, accumulation curve. However, in the last decade or so, many workers have been studying and correlating a variety of geological phenomena, which include extinctions, continental flood basalts, sea-level lows, ocean-anoxic and black shale events, and abrupt changes in rates of sea-floor spreading. It transpires that these varied events appear to be cyclical.

It was suggested by Kearey and Vine (1990) that current plate tectonic concepts permit one to infer that all major aspects of the Earth's long-term regimes should be related. For example, the global carbonate-silica cycle, which is taken to be driven by plate tectonic processes, gives a direct link between plate motion and the composition of the oceans and the atmosphere and also of the climate. It has also been shown that there have been abrupt changes of sea-floor spreading which may be correlatable to other events. Let us now briefly consider some of these aspects.

8.2.1 Galactic control of periodic events of Earth?

As Napier (1998) has pointed out, claims that the terrestrial record shows a periodicity of approximately 30 Ma have intermittently been made for at least the last 70 years, dating as far back as Arthur Holmes (1927). Such periodicity, it has been claimed, can be seen in climate variations, sea-level changes, mass extinctions, geomagnetic reversals and global volcanic episodes. It was first proposed by Napier and Clube (1979) that the cyclical changes cited above, together with ice-ages and plate tectonic processes, might ultimately be controlled by the galactic environment.

These authors assumed that the Oort cloud of comets was an intermediary, which was acted upon by our Galaxy, and was periodically disturbed during our Solar System's journey around the fringes of our Galaxy. These disturbances, they suggested, dislodge comets from the Oort cloud. Many of these are lost to outer space, but some are thrown inward towards the planetary system, and so gave rise to episodes of bombardment on Earth, which may endure for a few million years.

The Galactic hypothesis, which has been further developed by Clube and Napier (1982, 1984, 1986, 1996) and also by Rampino with various colleagues (1984, 1986, 1992, 1994, 1997), is based on the relatively recent findings that (a) the Oort comet cloud is unstable in the galactic environment and (b) the expected rates of collision between comets and Earth are now estimated to be sufficiently high that such periods of bombardment would result in 'geological trauma' on Earth.

For decades, it was generally considered that the major reservoir of potential Earth-impactors was the asteroid belt. However, beside the probability that asteroids from this belt are likely to give rise to impact craters of less than 20 km in diameter, there is no obvious mechanism which could be invoked that would give rise to a periodicity which has acted over a very long time. For such an extended impact scenario, 'the Earth becomes essentially a uniformitarian stage on which the drama of random impact is occasionally enacted' (Napier, 1998).

If, however, disturbances of the Oort cloud are periodically involved, then a more continuous process can be envisaged, which not only involves the direct effects of occasional cometary impacts, but also the effects of 'stratospheric dusting'. This latter effect can be ascribed to the fact that the mass distribution of comets is 'top heavy'. That is, most of the flux into the inner planetary system comes in the form of rare, giant comets – the disintegration of which leads to multiple impacts. In addition, Napier (1998) argues that when a large comet in an Earth-crossing orbit, with a diameter about, or in excess of, 100 km, breaks up, the finer particles created by this disintegration result in a prolonged 'stratospheric dusting' so that the 'optical depth' of the stratosphere may be increased for millennia (see also Bailey et al., 1994 and Clube et al., 1997).

Although it has been argued that such events may well be random (Alvarez et al., 1980); Napier and Clube (1979) suggested that these events are cyclical and that galactic periodicities must be expected to leave a terrestrial record with a periodicity of about 30 Ma. Other estimates of periodicity which relate to terrestrial impact craters are given in Table 8.1.

This periodicity of about 30 Ma is similar to that of successive vertical oscillations of the Sun as it passes through the plane of our Galaxy. However, as Napier pointed out, in the late 1970s, there was no known viable mechanism which connected our Galaxy to Earth. Nevertheless, it soon became clear that the vertical galactic tide (which is proportional to the density of the ambient material) was the dominant, though slowly fluctuating, external

Table 8.1 (After Napier, 1998.)

Authors	Periodicity (Ma)
Seyfert and Sirkin (1979)	28
Alverez and Muller (1984)	28
Rampino and Stothers (1984)	31
Shoemaker and Wolfe (1986)	31
Yabushita (1994)	16,31

force which acted on the Solar System. This force is considered to be of particular importance in disrupting the tracks of icy bodies in the Oort cloud.

Although the majority of such affected bodies are probably lost from the Solar System, other such bodies are thrown into the inner Solar System to form long-period comets. The flux of such comets reaches a maximum as the Sun passes through the plane of the Galaxy, during its 'vertical' oscillation, relative to the plane of the Galaxy. Thus, Napier (1997) infers that there is a potential mechanism connecting Galaxy and Earth in the form of cyclical episodes of impact bombardment on Earth. He further suggests that the dynamical transfer of comets from the Oort cloud into short-period, Earth-crossing orbits occurs as the result of these bodies interacting with major planets to produce a large population of dark, Halley-type comets.

Seyfert and Sirkin (1979) suggested that there appears to be a periodicity as regards the maxima of impact activity, which enhanced the claim that the 'Holmes cycle', of approximately 30 Ma, was induced by some unspecified external forces (Figure 8.1). Shaw (1994) extended this exercise back to about 500 Ma, i.e. to the Cambrian/Ordovician boundary, but the picture does not become any clearer. These data certainly show sequences of high and low flux, but the periodicity between peaks or between troughs appears to be somewhat variable.

Grieve *et al.* (1988) and Grieve and Pilkington (1996) argue that the accuracy and precision of dating of the crater age data does not permit such periodicity of impact events to be clearly and unambiguously demonstrated. These authors have, of course, a pertinent point.

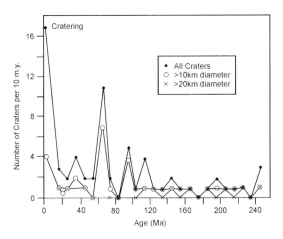

Figure 8.1 Histogram of craters plotted against age and diameter for the Upper Phanerozoic. The various points relate to the centre of 10 Ma bins (after Shaw, 1994; Seyfert and Sirkin, 1979).

Seyfert and Sirkin (1979) also held that these impact epochs could be correlated with a wide range of disturbances, which included the development of flood basalts, sea-level changes, orogenies and mantle plumes. They also attributed sharp changes in the rate and direction of motion of plates, which were even then known to occur, to impacts. Unfortunately, these prescient authors weakened their argument by concluding that the energy of major impacts would be so large that they would destroy direct evidence of the occurrence of such impact events.

Analyses of non-cratering events, which embrace a range of geological phenomena have also been carried out, particularly by Rampino and co-authors (1984, 1986, 1992, 1994, 1997). The ages of specific types of geological results are listed in Table 8.2 and the mode of analysis is briefly outlined below.

The various aspects cited above have been collated by Rampino and Caldeira (1993) who reworked the data of Raup and Sepkoski and other authors, which they gathered on a page-by-page survey of major journals back to 1975. They argue that this collation is possible because, during the decade or so prior to their publication, more accurate methods of dating had been developed. Accurate dating, they correctly aver, is the fundamental key to their analysis. For example, they point out that, as regards the geological time-scales published in the last decade, the stage boundaries within the Tertiary, Cretaceous and Triassic vary between limits of 2 per cent, 3 per cent and 6 per cent respectively. Consequently these researchers, like myself, were mainly constrained to a study of the last 250 Ma, which correlates approximately to the Permian/Triassic boundary, where, even at the age limit set, errors as regards time may be currently as much as 10 Ma.

The data in Table 8.2 were subjected to statistical testing to obtain 'significant' correlations between the data and *time*. The first of these techniques was the *Moving-window analysis*. A 10 Ma moving window, centred every 0.5 Ma, was initially applied to the age data given in Table 8.2. The number of items that fell within the moving window for every 1.0 Ma interval was then obtained and plotted, revealing a succession of peaks (Figure 8.2a). In order to test the effect of window size on the location of these peaks, the authors used moving windows which extended from 2.0 to 15 Ma. They found that the peaks remained stable in position, despite the varying size of the window. The number of geological events weighted by a Gaussian function with a scale length of 5 Ma is shown in Figure 8.2b.

Rampino and Caldeira then computed the Fourier transform of the auto-covariance function of the original un-windowed, time-series data rounded up to the nearest million years, utilising a standard Tukey window with a bandwidth of 4.7 Ma. The highest peak in the Fourier power spectrum occurs at a period of 26.6 Ma (Figure 8.3). The Fourier transform of the time-series permits one to infer that a phase with the most recent maximum of the 26.6 Ma cycle occurred about 8.7 Ma ago. The authors then applied other techniques and concluded that the errors in their results, regarding cyclicity, were unlikely to exceed the 0.01 per cent level.

Rampino and Caldeira further maintain that their preliminary results support a model in which there are rapid 'jumps' in sea-floor spreading and outbreaks of hotspots or flood basalts, which are related to rifting, volcanism, orogenesis, oscillations of sea-level and changes in the composition of the Earth's oceans and atmosphere. This is similar to the 'pulsation tectonics' scenario of Sheridan (1983). Figures 8.4 show (a) power spectrum of 11 extinction events with a highest peak of 25.9 Ma; (b) power spectrum of stratigraphic sequence boundaries, with highest peaks at 28.3 and 60 Ma; and (c) power spectrum of combined extinction events and stratigraphic sequence boundaries, with a highest peak at a period of 26.2 Ma.

Table 8.2 Compiled data of some important geological events during the past 250 Ma.

Date (Ma)	Mass extinction	Anoxic events	Flood basalts	Sea-floor spreading	Evaporite deposits	Sequence boundaries	Orogenic events
0–9	1.6			2	5		0.6, 2.5, 4, 5
10–19	11.2		17	10, 17		16	12.5
20–29							25
30–39	36.6		35		30, 36.6		
40–49				40			40
50–59				53		52	
60–69	66		62, 66	63		60.6	65
70–79				77			
80–89		84			86, 88	80, 87	
90–99	91	91	92	94		97	
100–109							
110–119	113	113	110	112		113	
120–129							
130–139		130, 135		138			
140–149	144		148	144	144	145	
150–159		156			154	155	
160–169		163					
170–179	176		170		173		
180–189							
190–199	193	193	190		196		
200–209		208	200		208		
210–219	216						
220–229					220	220	
230–239					230		
240–249	245	245					
250–259			250		258	250	

The main sources for the data represented in this table were obtained from the references cited below:
(1) *Mass extinction* The data relating to extinction is mainly derived from Sepkoski (1989, 1994).
(2) *Ocean anoxia* Times of open-ocean 'anoxic' or widespread black-shale events on the continental platforms are mainly from a compilation by Leary and Rampino (1990).
(3) *Flood basalts* Estimated dates of eleven events were taken from the compilation of Rampino and Stothers (1988).
(4) *Sea-floor spreading* Anomalous patterns of movement of ocean crust, recognised from changes in trend of linear island/seamount chains, typically related to 'jumps' in the spreading centres have been summarised by Schwan (1980); other data obtained from various sources.
(5) *Evaporite deposits* Data obtained directly from the compilation by Ronov (1983).
(6) *Sequence boundaries* Data after Vail et al. (1977) and Haq et al. (1987).
(7) *Orogenic episodes* Data after Haq and Van Eysinga (1987).

At this time, Rampino and Caldiera postulated that these cyclical effects could be related to either *internal* or *external* causes.

The assertion that these cyclical effects could be controlled by external causes did not go unchallenged. It was suggested that apparent tectonic oscillation could be solely a result of internal core/mantle dynamic processes (Loper and McCartney, 1986; Courtillot and Besse, 1987). Time-dependent numerical models of thermal convection in the Earth's mantle, it was suggested, permit one to infer that, at high Rayleigh numbers, the route to chaotic thermal convection in the mantle may be through periodicity and quasi-periodicity (Loper *et al.*, 1988).

However, ridge 'jumps', it was considered, may be too rapid to be caused by direct plume activity. So it was postulated by Stewart and Rampino (1992) that rapid changes in spreading

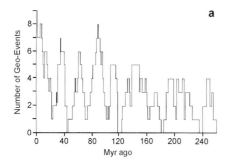
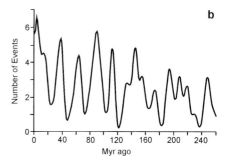

Figure 8.2a Number of geological events listed in a 10 Ma moving window as described in the text. (b) Number of geological events weighted by a Gaussian function with a scale length of 5 Ma (after Rampino *et al.*, 1984, 1986).

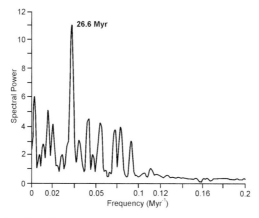

Figure 8.3 Major episodes of geological change, showing spectral power for all data sets described in text directly from the data in Table 8.1. The highest peak is at a period of 26.6 Ma (after Rampino *et al.*, 1986).

patterns may be related to sudden changes within ocean slabs sinking into the mantle. Alternatively, the driving force may be the result of changing configuration of the Earth's plates.

It was further suggested (Pal and Creer, 1986; Muller and Morris, 1986) that impacts were responsible for magnetic reversals as well as the emplacement of plateau basalts. Loper and McCartney (1990) rejected the possibility of Galactic influence, and stated that, as regards the magnetic reversals, 'the models that have been proposed to explain magnetic reversals require the action of several unsubstantiated novel processes'. Moreover, they continued, the 'proposals by Rampino, Stothers, Alt and others that impacts can trigger prompt flood basalts are found to be dynamically untenable'. It is concluded by Loper and McKartney that no viable model has yet been proposed to link impacts with reversals or flood basalt mechanism. *This latter comment, that the generation of flood basalts cannot be linked to impacts is incorrect and unfounded, and is, of course, a conclusion which we completely reject.*

As regards external causes, it was noted that planetesimal impacts potentially represent the most energetic events that can perturb the Earth (Melosh, 1989), which brings us back to the arguments presented by Napier and others regarding the control of Earth impacts by Galactic events.

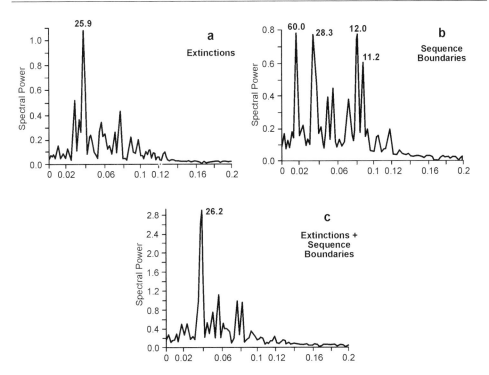

Figure 8.4 (a) Frequency and spectral power of extinctions (b) Frequency and spectral power of sequence boundaries (c) Power spectrum of combined extinction events and sequence boundaries (after Rampino *et al.*, 1986).

8.2.2 The environmental effects of impacts

To clarify how some of the cyclical events may be related to impact events, let us consider the wider aspects that result from the impact of a major meteorite into the ocean. Roddy *et al.* (1988) reported the results of computer modelling of impact effects, in which it was assumed that the impacting body had a diameter of 10 km, travelled at a velocity of 20 km s^{-1} and made impact into a 5 km deep ocean, with an energy of 3.0×10^{22} joules. Although this analysis was conducted before it was generally realised that cometary impacts gave rise to the larger impact features, the modelling is nevertheless instructive. (A degree of correlation between Roddy's modelling and that of a cometary body giving rise to a comparable energy of impact can be obtained if, to off-set the density difference of 3.3/1 between the two types of bodies, one assumes the initial cometary velocity to be about 45 km s^{-1}.)

Their model showed that the immediate effect of the impact is to heat a large mass of air, to a peak temperature of 20,000° K, adjacent to GZ, which then moves out rapidly from the impact area. If a landmass were adjacent, such a 'fireball' would consume combustible materials. In the ocean, a devastating tsunami would be generated with an initial wave height determined by the depth of the ocean, so that the initial wave may be several kilometres high. This wave would spread from the impact area at an initial velocity of 0.5 km s^{-1}. The energy of such a wave would be sufficient to scour the sea- or ocean-floor and rework sediments and develop tsunami deposits, particularly in the shallow waters of the continental shelf. Roddy claims that the volume of reworked material is so large that any ejecta material, including mineral or geochemical signatures from the meteorite or comet at the impact area, could be largely, or totally, obscured. Napier (personal comment) makes the point that whatever the type of impacting body, a very large proportion of the body's energy would certainly arrive at the impact site. Such an impact would result in an earthquake in excess of magnitude M = 12 on the Richter scale (see Melosh, 1989 for the method of calculating M), and such energy would contribute to landslides and submarine slumping as well as initiating, or accelerating, volcanic activity and hydrothermal activity.

Roddy *et al.* estimate that somewhat in excess of 100 km^3 is ejected from the crater area and they report that approximately 10^{14} tonnes of vapourised asteroid, or comet, and crust are thrown to high altitudes. It is thought that 80–90 per cent of these ejecta falls back in or near the crater. However, the remaining material is thrown to heights of about 100 km in the form of dust and ultra-small particles, which remain suspended in the upper atmosphere long enough to circle the globe. These particles block out sunlight and cool the Earth's surface (the oceans would be cooled by 3–4°C and the land surface by as much as 40°C). This would, of course, disrupt the food chain by reducing or preventing photosynthesis. It has been estimated that dust from such an event will not settle in less than 6 months. However, these estimates do not include the effects of soot particles, generated by the fireball, which are much smaller and so will settle more slowly. Soot absorbs light better than fine ejecta (Wolbach *et al.*, 1985); nevertheless, it is thought, complete darkness is not likely to persist much beyond 6 months.

Yet another environmental consequence of a major impact is the shock-heating of the lower atmosphere with the ensuing formation of nitrous and nitric acid which, with their compounds, would inhibit photosynthesis by reducing solar radiation, and asphyxiate fauna through exposure to NO_2 (Lewis *et al.*, 1982; Prinn and Fegley, 1987).

It will be clear from these various effects that an impact capable of generating a crater with a diameter of 100–200 km can bring about catastrophic marine and terrestrial results which can give rise to extinction over considerable areas. One may readily imagine that the effects of an impact, several orders of magnitude more energetic than that considered by Roddy, upon Earth's environment must be even more awesome. Thus, the *mass extinction* of many forms of marine life and of many species of terrestrial flora and fauna must result from a really major impact (i.e. an impact which generates a crater with a diameter of several hundreds of kilometres). In addition, moderate size impacts in the shallow seas of the continental margins can easily give rise to such disruption as to be used to define *sequence boundaries*. (See McClaren and Goodfellow, 1990 for evidence of a catastrophic nature relating to the K/T and other boundaries back to the Precambrian.)

As regards more recent events, Stothers (1993) has even related seven stage boundaries to seven specific impact events, which range in diameter from 10 to 45 km. However, these relatively small impact structures are terrestrial, so it is difficult to envisage how they could give rise to stratigraphic stage boundaries. It is reasonable to assume that if these boundaries are indeed related to meteoritic impacts, they would most likely result from quite large, oceanic impact events.

The interrelationships between the various events given in Table 8.2 are statistically intriguing. It is also interesting to note that the then known impact events are not cited in this table. Nevertheless, with the aid of the Roddy scenario, it is evident that three of the classes of events (mass extinction, anoxic conditions and stage boundaries) can be explained in terms of major impacts.

Moreover, we have already argued that, as the result of a moderate-to-large impact, partial melting of the LVZ could result in such reduction of basal resistance to glide that the velocity of *sea-floor spreading* could be influenced. The violent motions of the ocean in the vicinity of an impact would be so intense that the event would cause at least a regional problem, as regards continuity of life forms. Changes of toxicity could also occur in the oceans. The terrestrial atmospheric conditions would be sufficiently toxic to render the life style of large land animals untenable. Despite the critical comments we have cited in an earlier paragraph, we do not doubt that the shock-wave of oceanic water must have a dramatic influence on the sediments and, thereby, would be likely to cause the initiation of a stage boundary. The presence of nodules and stressed minerals at the K/T boundary attest to this conclusion. In addition, Oberbeck *et al.* (1933) point out that the tsunamis associated with such major marine impacts could give rise to sediments which are remarkably akin to glacial, boulder-clay deposits.

Thus, of the important geological events cited in Table 8.2 only *evaporite deposits* and *orogenic events* have not yet been shown to be directly relatable to impact events. Orogenic events are reasonably well explained in terms of conventional plate tectonics, though a major impact could exert a significant trigger-mechanism in bringing about the initiation of an orogeny. As regards the remaining type of event, which relates to the development of salt and other evaporitic deposits, only three examples are cited in Table 8.2. Because such deposits require marine incursions and climatic conditions acting for long periods, we find it difficult to envisage an obvious direct relationship between impacts and the development of such major evaporite sequences.

Let us now consider sudden sea-level changes. We have already indicated how major oceanic plateau basalts (e.g. the Ontong-Java at 119.3 Ma) resulted in the emplacement of many tens of millions of cubic kilometres of basalt on the ocean floor. Let us assume that the area of the deep oceans has remained essentially unchanged in the last 250 Ma, so that we can take the current area of the ocean, not including continental margins, to be 3.1×10^{14} m^2. If we take the average depth of ocean to be 3770 m (after Turcotte and Schubert, 1982), then the 'constant' volume of the ocean is approximately 11.7×10^{17} m^3 or 11.7×10^8 km^3. If we further assume that the volume of lava that erupted to form the Ontong-Java plateau basalt was 60×10^6 km^3, then this represents approximately 5 per cent of the volume of the ocean and would give rise to an increase in sea-level of about 190 m.

However, we noted in Chapter 6 that in the initial stages of development of the Ontong-Java plateau basalt the erupted basalts were, in part, subaerial, so that one would expect this event to cause a sea-level rise perhaps significantly less than 190 m.

Harland *et al.* (1982) show an abrupt rise of sea-level at 116 Ma of approximately 100 m. These authors point out that the exercise they conducted in plotting eustatic changes in sea-level 'is not likely to be construed as presenting a consensus'. However, in light of the difficulty of assessing the precise age of on-set and the magnitude of the change in sea-level, the estimated dates of change of sea-level and also the magnitude of the rises are probably in reasonable agreement with the impact date of this feature as inferred from its track anomaly. This abrupt eustatic increase in sea-level was followed by a period lasting about 20 Ma, in which sea-level decayed towards the 0 mark, which represents the present-day sea-level. We suggest that this gradual lowering of the sea-level can be attributed to downward deflection of the oceanic lithosphere, in response to the regional loading by the basalt extruded onto the sea-floor.

In a more recent publication, Rampino and his co-authors strengthened their original approach to cyclicity and the interrelation between various geological features, and introduced reference to impacts in a thoroughly forceful manner. Rampino *et al.* (1997) in their paper entitled 'A unified theory of impact crises and mass extinction: quantitive tests' state, 'the impact theory of mass extinctions makes several predictions that can be tested with available data'. They assert that the estimated number and the dynamics of Earth-crossing asteroids and comets permit one to predict that impactors greater than a few kilometres in diameter will almost certainly collide with Earth every few tens of millions of years. Such large impacts will release $>10^7$ mt TNT equivalent of explosive energy, which is sufficient to produce global-scale disasters. They maintain that very large impact events, which release $>10^8$ mt TNT, produce additional severe effects, which include a complete loss of photosynthesis, an 'impact winter', as well as a global heat pulse brought about by wildfires caused by re-entering ejecta. They argue that the extinction record in the Phanerozoic, derived from paleontological studies, exhibits five major pulses and about 20 minor pulses, which these authors note are in excellent agreement with the predictions of about five major impact events of $>10^8$ mt yield and 25 +/–5 impacts of 10^{7-8} mt yields. (As we shall see later in this chapter, these figures are completely in agreement with our findings, that during the Phanerozoic we have evidence of 30 large impact events, some of which comprise multiple impacts.)

These authors present what they term the 'kill curve' (Figure 8.5). This is the result of a quantitative analysis which shows the relationship between the crater diameter in km (from which the energy of impact can be inferred) and the percentage extinction that can be expected. Two impacts (Puzech-Katunchi and Chicxulub) fall on the mean (solid line) curve, while the other known large impacts (Popigai and Manicouagan) fall on, or within, the dashed lines which represent the possible limits of error. Ward (1996) recently pointed out that this quantitative kill curve concept represents 'one of the most powerful to emerge from the entire extinction debate'.

Let us leave this aspect for the moment and see what can be gleaned by using data contained in the Atlas 3.3 system, to establish the incidence of abrupt track anomalies and hence, major impacts in the Phanerozoic.

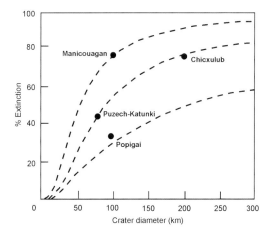

Figure 8.5 The 'kill curve' for Phanerozoic marine species plotted against energy of impacts named.

8.3 More tracks

In previous chapters, we have reported on the behaviour of relatively short sections of tracks that exhibit examples of abrupt anomalous behaviour, which correlate with a variety of effects and structures that are attributed to major impacts. However, when we started on this particular exercise it was not immediately obvious how major impacts could be related to some or all of the various types of geological phenomena which are cyclically related to one another. Using the Atlas 3.3 system, we extended our study to the tracks' specific locations in six new regions. To these six regions we added data for a reference point in the Pacific, discussed in the previous chapter.

As we have already noted, this initial time limit of 255–260 Ma to the present-day has been set for a number of practical reasons, namely:

(1) Beyond about 250 Ma the accuracy with which positions of the plates can be established and the dating of events becomes less precise than one would wish.
(2) The current maximum residence time of an oceanic plate is approximately 150–200 Ma, although occasionally in the past it may have extended to 200–250 Ma. A track change between 150 and 260 Ma may lead to ambiguity, in that a 'track' event may be attributed to an OPB that has been subducted or (what is far less likely) a CFB that has occurred on a continent but has not yet been identified (e.g. an event possibly hidden beneath the Antarctic or Greenland Ice Sheet).
(3) The oldest known continental flood basalt (Siberian event) occurred at about 250 Ma ago. (The most recent dating put this event at 248 Ma.) Hence, by arbitrarily setting the limit at 255 or 260 Ma, tracks in the areas of the world we have selected for study will show whether or not this Siberian, and other such large impact events, left a significant record.

The maps relating to the path or track of a specific, easily identified coastal feature were compiled by taking the present-day position of the reference point O and tracing it back through time. This usually required several hundred operations before the main features of, for example, the Hawaian track could be reasonably clearly defined. Each of the important events which caused direction and/or velocity change of track is noted, with the appropriate age for all six new localities.

We presented the track for a point in the Pacific, relating to the Mariana Trench, in Chapter 7. However, in this example, we omitted details of the track for the period 0 to 100 Ma. The data for this period for this track will be presented later in this chapter. Hence, the total number of tracks from 260 or 255 Ma to the present day is seven.

8.3.1 Details of tracks from seven localities

In this study we set out to define track anomalies for each area studied in which a particular geographical marker abruptly changes its rate and/or its direction of motion.

We very soon discovered that minor track anomalies could be found every million years or so. These extremely small track anomalies, which gave rise to a very small deflection of track and/or a minuscule change in speed, may possibly be attributed to relatively minor impact events which created craters with diameters of perhaps 20 km. Alternatively they possibly could be 'noise' inherent in the Atlas program.

We have attempted to distinguish the larger impact events, which we assume gave rise to craters of at least 100 km diameter, from this 'background noise'. However, as we have noted, there are problems in attributing a definitive 'magnitude' to a specific event. For example, it has been pointed out that an impact event, which we invoked to explain the generation of the Paraña plateau basalt, caused a dramatic change in rate and direction of motion of what is now S America, 135 Ma ago. However, Africa exhibited no detectable disturbance to its track, as measured at the current E coast of that continent. Clearly, one can infer that the response to an impact is not only influenced by the energy of that event. In some circumstances, the effects of the impact may be masked by the resistance to movement of the mass, and by the boundary conditions of the plate in (or adjacent to) which the impact event took place. The estimated magnitude of a specific impact may appear to be very large in one of the areas under discussion, whereas in another area the track anomaly, of the same date, would lead one to suggest that the magnitude of the event was significantly smaller, while in a third area, evidence of a sudden change in track at this time may be entirely lacking.

Consequently, we have been forced to adopt a subjective method of assessment for the various specific impact events. We consider that all the major events in this class of impact were capable of causing a crater with a diameter of at least 100 km. Within this 'large class' we take *large* (L) to give rise to craters possibly in excess of 400 km in diameter, *moderate* (M) sized events, we suggest, result in craters in the range 200–400 km in diameter, while *small* (S) events are likely to have craters with diameters in the 100–200 km range. The letters L, M and S are used in Table 8.3 to designate the inferred magnitude of the various events. In areas where the signal of a distant major event is extremely small or completely lacking this condition is represented by a bar (–).

The six new sets of tracks were prepared for specific, easily recognised, points in the different chosen areas. In order of presentation these areas are (a) Hawaii, (b) Borneo, (c) Madagascar, (d) Japan, (e) S America and (f) N America. To this set of six areas, we add the track obtained for a point adjacent to the Mariana Trench which we considered in Chapter 7. (Only the track for the period 0 to 100 Ma for this example is given in this chapter (see Figure 8.26). The reader should refer back to Chapter 7 for details of this track from 100 to 260 Ma.)

Of the seven tracks noted above only that for Hawaii is discussed in any detail. We have done this so that the reader will understand how we assessed the order of magnitude of small, medium and large impact structures.

Hawaii

As the overall track for Hawaii is relatively extensive, it was decided to present the generalised track for this area in two diagrams, namely Figure 8.6a for the period from 265 Ma to 100 Ma and Figure 8.6b for the period from 100 to 0 Ma. We have used an insert in Figure 8.6a to show the track for the period from 265 to 248 Ma. This is on a sufficient scale to show that from 265 to 250 Ma the track is linear, but at 250 Ma it exhibits a sharp angular change of 55° and an increase in velocity by a factor of 7.8. This can clearly be classified as a very major impact event (L), which coincides in time with the initiation of the Siberian plateau basalt and is also within 2 Ma of the Permo/Trias boundary.

The next events occurred at 238.24, 230, 220 and 208 Ma, all of which are represented in Figure 8.7a. At 238.24 Ma, the track is deflected by 28° and subsequently slows by a few per cent. We classified this as a moderately large event (M). The 230 Ma event is clearly a

Table 8.3 Impact and stratigraphic events inferred from tracks.

	Age	Hawaii	Borneo	Mada-gascar	Japan	America S	America N	Mariana	Events
1	5	L	L	M	M	L	L	L	Plio/Mio (5.1 Ma)
2	8	–	–	S	M	S	S	–	Mes/Tort (?)
3	19	–	S	–	S	S	S*	–	Col. Riv. PB*
4	35	S	M	–	M	S	S	–	1 Popagai*, Banda Arc
5	40	–	–	–	–	–	–	S	Afar*
6	42xx	S	–	–	M	S	S	–	Priabon/Barton (42)
7	50	S	S	M	S	–	S	S	Lutet/Ypre (50.5)
8	59	M	M	–	S	S	S	S (60)	Thanet/Dan. (60.2)
9	65–67	S	M	S	M	L	S	S	2 K/T – 4+Imps. DT*
10	73.5 (Antilles Arc Event)								Maas/Camp 73
11	98	M	M	M	M	S	S	S	Cenom/Alb (97.5)
12	110	–	M	–	S	M	S	S	Alb/Apt ? (113)
13	119	M	M	L	M	L	S	S	Apt/Barr (119)
14	124	M	–	L	M	S	–	S	Barr/Haut (125)
15	128	M	–	M	–	–	M	S	Haut/Val (131)
16	135	–	M	–	–	L	–	S	Paraña*
17	156	M	–	–	–	–	M	S	Kim/Oxf (156)
18	164	L	–	–	–	–	–	–	Oxf/Call (163)
19	170	M	–	S	–	M (171)	–	–	3 Callov/Bath (169) AI*
20	175	L	L	–	L	–	–	L	Bath/Bajoc (175)
21	180	S	S	S	M	S	L	–	4 Bajoc/Aal (181) AII*
22	190	M	M (193)	S	–	S (188)	–	–	Dogger/Lias (188)
23	195	M	–	L	M	L	L	L	5 Karoo*Toar/Plein(194)
24	200	L	S	–	–	–	–	–	Pleins/Sinem (200.0)
25	208	L	S	L	–	L	L	L	6 Sinem/Het (206.0) Man*
26	220	L	M	L	M	L	M	L	Rhaet/Nor (219)
27	225	–	–	–	M	–	–	–	Nor/Carn (225)
28	230	S	S	–	–	S	–	S	Late.Tr/Mid.Tr. (231)
29	238	L	L	–	L	L	L	M	Ladin/Anis. (238)
30	250	L	S	M	L	S	L	L	Trias/Perm. (248) Sib*

Continental impacts designated with *. Popagai and Manicouagan are continental craters. The others are associated with CFBs and are as named or, DT = Deccan Traps, AI = Antarctic I, AII = Antarctic II and Sib = Siberian. Multiple impacts ocean and continental are indicated as 1–6. The Hawaian-Emperor Chain 'kink' is indicated by 42xx. All other events are marine impacts and their association with specific stratigraphic junctions are as indicated.

Note
The number of impact events cited in Table 8.3, we take to be in the 100–500 km range of crater diameters. These events have been inferred from tracks in seven areas of the world. Although the combined areas of these seven maps are large, a significant proportion of the surface area of the Earth is still not represented here. It is reasonable to infer, therefore, that the actual number of major impact events, world-wide, in the Upper Phanerozoic is likely to be at least 10 per cent larger than the total presented in this table.

small (S) event, while the 220 and 208 Ma event, because of the marked change in the direction of the track, we would also classify as (M). Three main events are shown in Figure 8.7b, namely at 200, 194.9 and 190 Ma. At 200 Ma, the track shows a 40° change in direction and a significant increase in rate of movement, so that we classify this event as at least an (M). The events of 195 and 190 Ma we classify as relatively low intensity (S) (but still major) impacts.

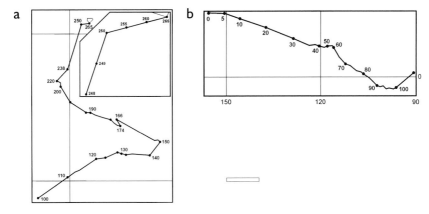

Figure 8.6 (a) and (b) Upper Phanerozoic track of 'Hawaii' from 265 to 0 Ma.

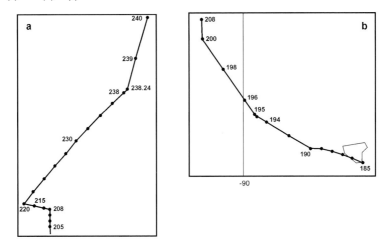

Figure 8.7 Details of 'Hawaii' track from (a) 240 to 205 Ma and (b) 208 to 185 Ma.

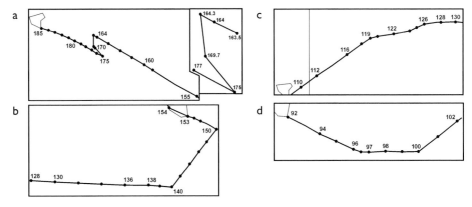

Figure 8.8 (a–d) Details of track of 'Hawaii' from 185 to 92 Ma.

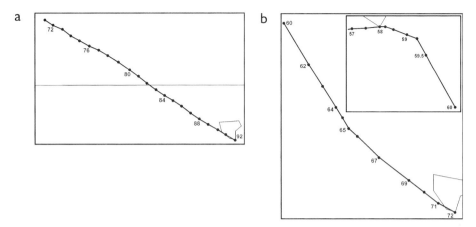

Figure 8.9 (a, b and inset) Details of 'Hawaii' track from 92 to 60 Ma.

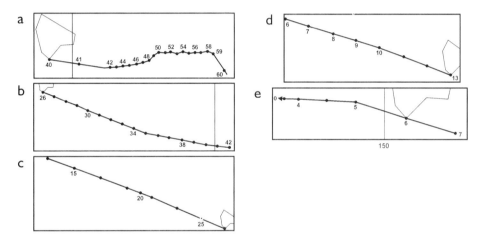

Figure 8.10 (a–e) Details of 'Hawaii' track from 60 to 0 Ma.

The 175, 169.7 and 164.3 Ma events are shown in Figure 8.8a (with details in the inset diagram). Because of the large changes (almost reversals) in the direction of track, we classify as large (L) events. Events at 150 and 140 Ma (Figure 8.8b) we would also classify as (L), while the 138 Ma event is obviously an (S). As regards the tracks shown in Figures 8.8c and 8.8d, the events at 124 Ma, we would classify as (S) while those at 119, 100 and 96.5 Ma we would put at the low end of the (M) range. As regards the track events in Figures 8.9 and 8.10, all the remaining track changes are (S) except for the two events with the marked change in track at 59.8 and 58.2 Ma which are (M).

The classification we have used here is clearly approximate and, of necessity, subjective, but it does enable one to indicate approximate magnitudes within this range of large events (26 major track anomalies).

It would be tedious for the reader if we classified each abrupt track change for the remaining areas. Accordingly, we present the ages of the track changes and the estimated magnitude of the event which gave rise to them, for Hawaii and the other locations, in Table 8.3.

The track data and the various ages at which abrupt track changes take place and which we now refer to as major impact events are given in the maps and captions relating to the specific areas. These are:

Borneo

It will be seen from Figures 8.11 to 8.13 there are marked changes in speed and/or direction of plate motion at 5, 35, 47, 47.5, 54, 71.3, 77.2, 96, 110, 118, 121, 135, 151.3, 175, 193, 212, 220, 242 and 250 Ma. These changes we classify as relating to moderate/large impact events. The 250 Ma point may at first sight seem of very minor importance. However, there was a track change of about 1° and an increase in velocity of approximately 15 per cent. These various points are interspersed by relatively minor events such as 10, 50, 66.4, 68, 90, 150, 180 and 187.5 Ma (27 anomalies).

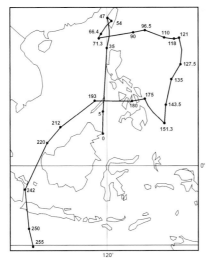

Figure 8.11 General track of 'chicken-beak' of Borneo for the period 255 to 0 Ma.

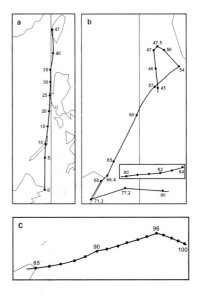

Figure 8.12 Details of track of Borneo for the period 0 to 100 Ma.

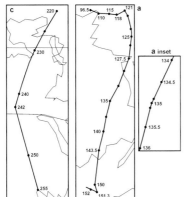

Figure 8.13 (a, a inset, b and c) Details of track from 96.5 to 255 Ma.

Madagascar

Moderate to large anomalies occurred at 5, 28, 77.3, 88, 96, 110, 119, 124.5, 127.7, 149, 160, 170, 180, 190, 195, 208, 220, 238 and 250 Ma. Smaller events occurred at 7.5, 28, 51 and 135 Ma (23 anomalies) (Figures 8.14 to 8.16).

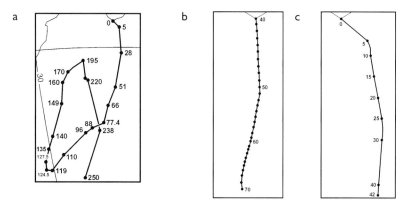

Figure 8.14 (a) General track for Madagascar. (b) and (c) Details of track from 0 to 70 Ma.

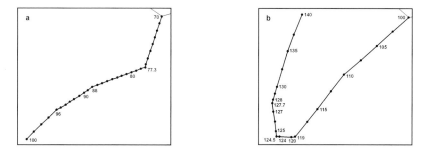

Figure 8.15 Track of the southernmost tip of Madagascar for the period (a) 70 to 100 Ma and (b) 100 to 140 Ma.

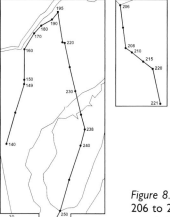

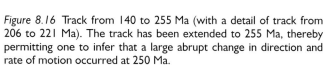

Figure 8.16 Track from 140 to 255 Ma (with a detail of track from 206 to 221 Ma). The track has been extended to 255 Ma, thereby permitting one to infer that a large abrupt change in direction and rate of motion occurred at 250 Ma.

Japan

The track of Japan is quite intricate. From Figures 8.17 to 8.19 it can be inferred that large (L), or moderate (M) anomalies occurred at 250, 238, 225, 220, 210, 195, 193, 187.8, 180.6, 175, 152, 151.25, 149.25, 135, 124.5, 121, 119, 110.4, 104, 96, 77.4, 76, 73.2, 71.5, 66.5, 57, 54, 50.5, 48, 41, 34.5, 27, 19.5, 10, 8 and 5 Ma. (Japan has 36 anomalies. We suspect that these include impacts which are less than 'minor'.)

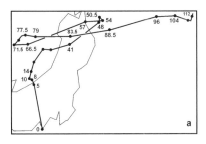

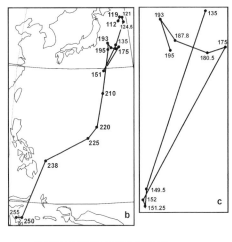

Figure 8.17 Maps (a) and (b) show the tracks for a cape on the S coast of Japan from 0 to 255 Ma.
(c) Detail of major, abrupt track changes between 195 and 135 Ma.

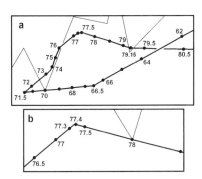

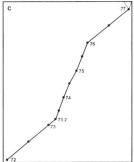

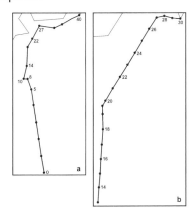

Figure 8.18 (a) Details of the 'goose-head' track from 80.5 to 62 Ma. (b) and (c) Details of track in this period.

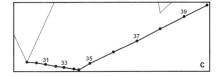

Figure 8.19 (a) General track for the period 40 to 0 Ma. (b) and (c) Details of track.

S America

From these tracks one can infer that large or moderate changes occurred at 238.2, 220, 208, 195, 188, 171.5, 152, 138, 135, 133.5, 127.5, 119, 116, 110.3, 63.35, 50, 43, 11, 7.8 and 5 Ma. Minor track changes occurred at 180, 124, 96, 72, 58.3, 27 and 21 Ma (27 anomalies) (Figures 8.20 to 8.23).

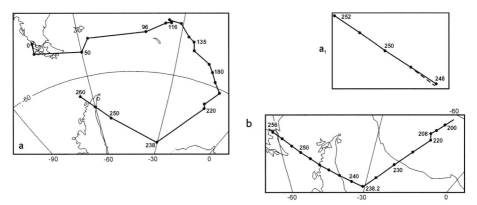

Figure 8.20 (a) General track of S America from 260 to 0 Ma of a point in southern S America is shown. A map of the track from 252 to 248 Ma, (a_1) records a very minor track change of about 2° at 250 Ma, with no detectable alteration in velocity. This is the only location for a small reaction to what elsewhere (with the possible exception of Borneo) is recorded as a major event. (b) Track from 256 to 200 Ma shows abrupt changes at 238.2, 220 and 208 Ma.

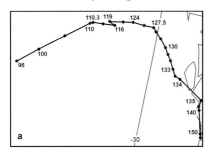

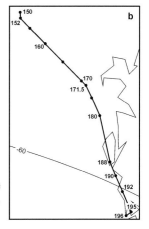

Figure 8.21 (a and b) The track from 196 to 96 Ma. Moderate or large track anomalies are shown at 195, 188, 171.5, 152, 142, 135, 133.5, 127.5, 119, 116 and 110.3 Ma.

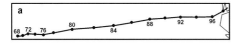

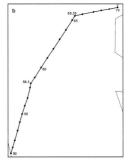

Figure 8.22 The track from 98 to 50 Ma. (a) shows minor deflections in a generally westward track, which continues in (b) until 65.35 Ma, where a major (K/T boundary) track change takes place which continues until 58.3 Ma, and thereafter continues with only minor changes to 50 Ma.

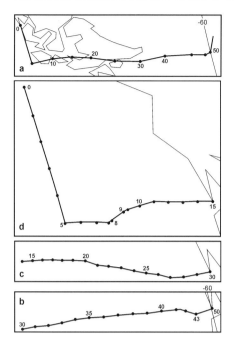

Figure 8.23 (a) The general track from 50 to 0 Ma, starting with an abrupt change at 50 Ma. (b) to (d) This track is shown on a larger scale. The most dramatic change, in the period from 50 to 0 Ma, occurred 5 Ma ago, all other changes in track could be classified as modest, minor or trivial.

N America

The track for the period 0 to 48 Ma (Figure 8.24b) is approximately straight with only minor deviations. The most recent major event at 5 Ma is marked more by the change in rate of movement than a change in the direction. There is a minor change in track at 19 Ma. It is, therefore, important to note that this event has been correlated with the emplacement of the Columbia River plateau basalt. The main deflection occurs at 48 Ma. The track, which is continued in Figure 8.24c, continues smoothly from 48 to 62.5 Ma. There is a very minor change in track at this point, which continues to 67.5 Ma, which we have suggested relates to one of the events associated with the K/T boundary. Thereafter, the track gradually changes to a south-easterly direction, until a moderately large change occurred at 95 Ma.

The track from 95 to 119 Ma (Figure 8.25a) was easterly, but from 119 to 180 Ma it changes to a more south-easterly trend. There are, however, relatively minor changes at 130, 150 and 155 Ma. At 180 Ma there is a more marked change to W of S, and this trend is seen to continue, in Figure 8.25b, to 195 Ma. The track changes to E of S from 195 to 208 Ma. From 208 to 220 Ma it turns abruptly, once again to E of S, and from 220 to 236.75 Ma swings back to SE. At 236.65 Ma the track swings abruptly anticlockwise through 150° and keeps a steady heading until 250 Ma when it swings abruptly clockwise through 113°. The track changes of Figure 8.25b are dramatic relative to those exhibited in the previous figures.

This presents obvious difficulties in assessing the magnitudes of the effects with which we are here concerned. Accordingly, we shall consider the size of effect starting from the older ones. Large or moderate track changes occurred at 250, 236.75, 220, 208, 195, 180, 155, 119, 95, 67.5, 48 and 5 Ma. Minor, or apparently minor, events occurred at 130, 73, 62, 42, 35, 27, 19 and 10 Ma (20 anomalies).

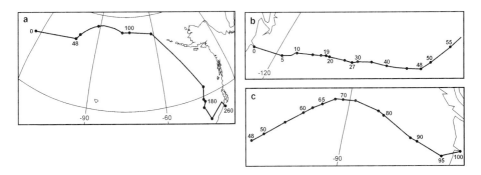

Figure 8.24 (a) General track for marker on the west coast of the USA from 0 to 260 Ma. (b) Detail of track from 0 to 55 Ma and (c) from 48 to 100 Ma.

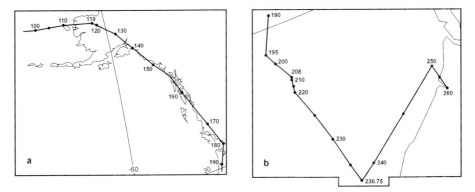

Figure 8.25 Track (a) from 95 to 195 Ma and (b) from 190 to 260 Ma.

Mariana

We have dealt with the tracks of points representing the Mariana in Chapter 7. The track from 100 to 0 Ma for this marker is only shown schematically in Figure 7.34. A more detailed representation of the track for Mariana can be seen in Figures 8.26a and 8.26b. It will be seen that the track with deflections occurring at 96, 80, 67, 40, 27 and 8 Ma with additional changes in track at 60 and 50 Ma.

The track record to the Mariana is shown in Figures 7.34 to 7.35, in which it will be seen that the Mariana track shows moderate or large changes at 250, 238, 220, 207.8, 195, 175, 156, 135, 128, 124, 119.2 and 110 Ma. Smaller changes in the Mariana track from 250 to100 Ma include those at 160 and 150 Ma, as well as all those cited relating to the 100 to 0 Ma period. The reader should refer back to Chapter 7, Figures 7.34 and 7,35 for deatails of the Mariana Track.

We have collected sufficient data from the tracks represented in Figures 8.6 to 8.26 to enable us to indicate the main events which we consider can reasonably be attributed to the major impacts that have struck Earth in the last 250 Ma, as well as many smaller impact events. We have noted that it is difficult to ascertain the magnitude of minor, and even moderately large, impacts when movement is constrained by large continental masses, which often possess deep lithospheric roots. When impacts can be related to the generation of major continental and oceanic flood, or plateau, basalts, one may infer that the energy of

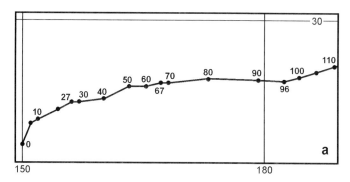

Figure 8.26   Track of Mariana reference point from 0 to 100 Ma.

the impact will be related to the volume of melt produced. However, the volume of melt will also be influenced by the thickness of the lithosphere, so that even in these specific events a degree of ambiguity exists.

We suggest that the simplest and most reliable criterion we can use in estimating the importance of an impact event is the number of regions where a particular event leaves a clear and marked track anomaly. However, even this criterion may be masked or enhanced if there are multiple coeval impacts. Correlation of the various data relating to the track in the seven areas cited above is not readily assimilated by simple visual inspection of the maps. We have, therefore, prepared Table 8.3 (see p. 291), which helps one to comprehend the importance of these events, set out in chronological order.

Many of the seven areas showed a record of track events, in the various regions, as being large, moderate or small. However, it was also obvious that when the 'impact', at a specific age was recorded in all seven areas, this was either the mark of a very large single event, or else it represented a multiple coeval set of impacts.

The 5 Ma event, which is such an obvious feature in all the seven areas, immediately presented a problem in that there is no known large impact crater, of this age, currently recorded anywhere in the world. It seemed likely to us, therefore, that the track data related to one or more major oceanic impacts. It has already been noted that an impact could give rise to a change in oceanic environments. However, such events were related to small, continental impacts. Here, we suggest that the 5 Ma events represented in the tracks and Table 8.3, could conceivably be marked by a major stratigraphical change. In this table, it will be seen that the 5 Ma event occurred only 0.1 Ma before the Pliocene/Miocene Epoch boundary (and in addition could have triggered the recent phase in Alpine deformation). It was immediately obvious, therefore, that all events listed in Table 8.3 should be tested for correlation with stratigraphical boundaries.

Of the 30 events listed in Table 8.3, eight are related to continental plateau basalts (indicated by * in Table 8.3). To these we add Popagai and Manicouagan, making a total of 10 continental impacts which probably had little, or only modest, effects upon the oceans. There are six events (numbered 1–6 in Table 8.3) which we infer to be multiple impacts, which landed in a continental and also in a marine environment. Hence, we can infer that, in the Upper Phanerozoic, there were probably at least 28 major marine impacts. That is, of the total number of impacts in this period, about one third struck continental areas and two thirds were oceanic impacts. It will be noted that this ratio is a reasonable representation of the relative areas of the two types of lithospheres.

Of the 26 inferred oceanic events, all are closely related in time to an Era, Sub-era, Epoch or Age, stratigraphical boundary.

This is a most remarkable correlation. The obvious conclusion, we suggest, is that stratigraphy, in the Upper Phanerozoic, at least, is largely determined by major impacts.

8.4 Peridocity and impact-flux changes in the Phanerozoic

We have noted that Seyfert and Sirkin (1979) (Figure 8.1) have suggested that there is evidence of periodicity as regards the maxima of impacts. However, Grieve *et al.* (1977) have argued that, at this time, the number of known 'certain' impacts and the lack of precision in dating of these impact events are such that it is not possible to demonstrate, with any degree of certainty, that periodicity of impacts exists. Accordingly, we took the dates of the impact events that are represented in Table 8.3, which, it will be recalled, relate to the tracks of seven different reference points, but date back only 255 to 260 Ma.

We shall now look briefly at impact events that may have developed in the age range from 260 to 600 Ma as inferred from the tracks shown in Figures 8.27 to 8.29, relating to Fennoscandia and the Pacific.

As before, we selected impacts cited by Hodge (1994) for this period. There were 21 events which could be clearly ascribed to this period, of which two had a diameter of 54 and 55 km respectively. Three craters had diameters respectively of 32, 15 and 12 km. The remaining 16 events had craters which were less than 10 km in diameter.

All these impacts contribute to the known flux. However, it is probable that none, except perhaps the 54 and 55 km craters, could be classified as likely to give rise to a detectable global or regional disturbance. These impacts, aged 357 and 368 Ma, occurred in Quebec and Sweden respectively. Both these structures are now situated at relatively low altitude. Only one half of the Charlevoir, Quebec, structure is exposed, the other half is buried beneath the St Lawrence River. The Siljan structure in Sweden is currently exposed at a height of about 500 feet. It is possible that one or both of these localities were covered by sea-water at the time of impact. In any event, it is interesting to note that the Siljan event, dated at 368 Ma, occurs within 1 Ma of the assigned Famennian/Frasnian stratigraphic boundary at 367 Ma. The Charlevoir event is dated at 357 Ma, which is within 1 Ma of the Ivorian/Hastarian stratigraphic boundary at 356 Ma, or possibly the Devonian/Carboniferous sub-era boundary at 360 Ma.

8.4.1 Correlation between impacts and other events in the Lower Phanerozoic

Earlier, we used the data from the tracks of the seven regional reference points, relating to events in the Upper Phanerozoic, to ascertain whether we could correlate the abrupt changes in track motion with specific geological features such as continental flood basalts, oceanic plateau basalts, small arcuate subduction zones and stratigraphic horizons. Based, as it was, on data in the last 250 Ma, this study showed good correlation between the dates of the sudden track changes and the various aspects cited above.

For this study we chose to introduce a new reference point at the northern limit of the Baltic, in order to plot the track of the Fennoscandian Shield. This is an old Precambrian continental mass which has been in existence for well over a thousand million years. The tracks we obtained for this shield area, from 250 to 600 Ma are shown in Figure 8.27.

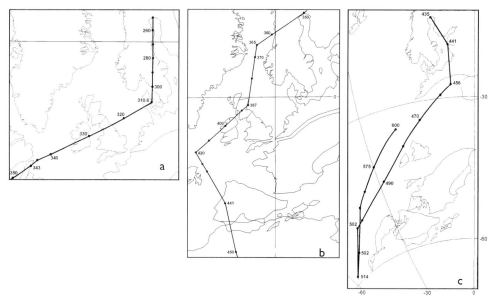

Figure 8.27 Track of Fennoschandia from 250 to 600 Ma. (a) Track from 250 to 350 Ma. (b) Track from 350 to 450 Ma. (c) Track from 435 to 600 Ma (Note change in viewing direction as inferred from latitude and longitude.) (9 anomalies).

Throughout the period of study, this landmass has been in relatively close association with other landmasses. Hence, it would not be surprising if it exhibited a track record, as regards the number of major inferred impact events, which was markedly different from the second area of study which is, again, the Pacific plate area (Figures 8.28 and 8.29).

In this latter study, we extended the track for Hawaii back to 600 Ma and revealed nine more major track changes between 600 and 250 Ma. The ages of these inferred impacts and the associated stratigraphic boundary are listed in Table 8.4.

It will be recalled that the track changes listed in Table 8.4 were based on the track of an island that has not been in existence more than about 30 Ma. Hence, to chase this image back to the base of the Cambrian, without mention of possible changes of plates and other problems is to invite criticism. Indeed, one does not need to be excessively sceptical to suggest that the changes in track are the result of glitches in the program and have no real significance. Nevertheless, it will be seen from Table 8.4 that almost all the sudden track changes are associated with specific stratigraphic boundaries. Also, the 365 Ma event coincides with the inferred impact in the S China Sea and the contemporaneous Central Nevada Circular Feature.

It will be seen that the dates of the changes in track of the Fennoscandian shield occurred at times which are, in the main, very similar to, or exactly the same as, those in the Pacific. There are the same number of sudden changes in track, and most of the matching events occurred close in time. The only real difference between the two lists is that the Pacific tracks show a change at 526.5 Ma (which we correlate with the St David/Merioneth stratigraphic boundary at 523 Ma on the Harland time-scale), while the Fennoscandian track shows an abrupt change at 514 Ma which relates to the Maentwrog/Dolgellian boundary at 514 Ma.

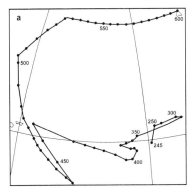
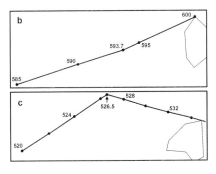
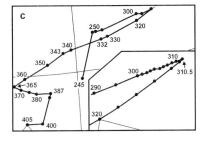

Figure 8.28 (a) General track of 'Hawaii' from 600 to 245 Ma. (b) Detail of track from 600 to 585 Ma. (c) Detail of track from 530 to 520 Ma.

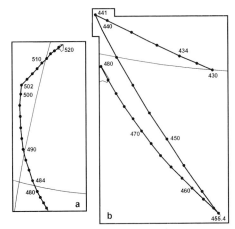

Figure 8.29 (a) General track of 'Hawaii' from 520 to 474 Ma. (b) Track from 480 to 430 Ma. (c) General track from 405 to 245 Ma with detailed insert track (9 anomalies).

These two areas are widely separated. The Pacific area is mainly devoid of continental rocks, while the Fennoscandian shield is closely associated with other continental blocks at the other side of the world. Both sets of tracks are of equal validity and any differences merely reflect differences in geology and the effects of impact events which are remote from each other. We can infer from Table 8.4 that there were at least nine major events in the Lower Phanerozoic. Moreover, if tracks for the Lower Phanerozoic for six or seven other areas were to be constructed, it is probable that several more impact events could be inferred. If four such events of comparable magnitude were to exist in other areas, it would bring the total number of major impacts to 14, which would be in accord with the number that should develop in 250 Ma, predicted by Oberbeck *et al.* (1993)

As regards the position of plates at 300–550 Ma, the accuracy and knowledge regarding the precise geometry and position of plates at specific times in the Phanerozoic is less perfect than subsequent times. Indeed, dating of events within the Phanerozoic is in a continual process of refinement. Thus, within the past few years about 50 Ma have been lopped off the lower limit of the Phanerozoic. Radiometric dating techniques are being continually

Table 8.4 Dates of impacts and stratigraphic boundaries in the Lower Phanerozoic.

Dates (Ma)	Strategic boundary (after Harland et al. 1982)	Ages (Ma)
A Fennoscandia		
550	Base of Cambrian	545*
514	Maentwrogian/Dolgellian	514
502	Cambrian/Ordovician	505
456	Llandeilo/Caradoc	458
441	Ordovician/Silurian	438
420	Wenlock/Ludlow/	421
387	D2/D1 Devonian	387
365	Frasnian/Famennian	367
310.5	Bashkirian/Moscovian	310
B Pacific		
550	Base of Cambrian	545*
526.5	St David/Merioneth	523
502	Cambrian/Ordovician	505
455.4	Llandeilo/Caradoc	458
441	Ordovician/Silurian	438
400	Silurian/Devonian	408
387	D_2/D_1 Devonian	387
365	Frasnian/Famennian	367
310.5	Mississip./Pennsylvanian	320

* Base of Cambrian 'moved' from 600 Ma to 545 Ma.

improved, but once we become involved with events prior to 150 Ma ago, it is still necessary to be aware of the error bars which may be 5 per cent, or even more, of the estimated date. In the light of these remarks, we suggest that the degree of correlation between the dates for the track changes and the ages of the various stratigraphic boundaries as cited by Harland *et al.* (1982) are surprisingly close. Moreover, all but one of the inferred impact events (i.e. that at 365 Ma) relate to major stratigraphic boundary changes. However, it will be recalled (Chapter 6) that the Central Nevada Circular Feature and the S China Sea inferred impact are also dated at close to 365 Ma (i.e. at 364.8 Ma). The exercise of tracing the Hawaiian track back to 600 Ma was also primarily to establish the number of major impacts in the Lower Phanerozoic. That these events could be correlated with important stratigraphic boundaries was a bonus.

We can only conclude that major impacts were controlling, driving, or otherwise determining the geological history of Earth throughout the Phanerozoic – and hence, by inference, throughout most, if not all of the history of Earth.

It should be noted that although a sharp change in track at a specific location provides clear evidence that a major impact has occurred at a specific time, we emphasise that such sharp track changes are not generally directly related to the points of impact. Obviously, a track can be directly related to a known continental impact site. It is reasonable to assume that a detailed geophysical survey of major continental flood basalts would eventually reveal the impact site beneath the basalts. (By 'back-tracking' to the time of impact, as inferred from the date of origin of the basalts, it would be possible to ascertain the point of impact that gave rise to the eruptive event, if for any reason this seemed necessary.) We have noted that the eruption of the Ontong-Java plateau basalt can be inferred to be the result of two separate impact events. However, the precise sites of these two impact events may be difficult to identify. This is also likely to hold true for any other of the major oceanic plateau basalts which are possibly the result of a major impact. Moreover, oceanic plates have a limited residence time, so that direct evidence of such impacts is likely to disappear well within a period of 250 Ma. The only record of such events are, therefore, the effects (i.e. the sharp changes in track) they had on the tracks of continental masses.

We have noted that when following the tracks of continental masses, it is convenient to assume that they maintain their geometry back through time. However, from time to time, continents collide so that they become deformed. In such tectonically deformed regions, evidence of sharp changes in track could then be destroyed.

Nevertheless, although the record of tracks is likely to be progressively less well defined as one regresses in time, it is the sharp changes in direction or speed of movement of continental units which best records the 'ancient' impacts that date back before the Jurassic.

8.4.2 Impact flux in the Lower Phanerozoic

In this section, we are mainly concerned with the question of the rate of flux of impacts in the Phanerozoic. To this end, we prepared tracks of a marker in Fennoscandia and also took the previously used 'Hawaii' locality back through time from 240 to 600 Ma, with the object of determining what change in flux of impacts, if any, can be demonstrated from the sharp track anomalies in plate movement, which we have demonstrated are the results of major impact events (Figures 8.27 to 8.29).

To these we added those certain major impacts, as catalogued by Hodge (1994). However, we rejected reference to every crater with a diameter less than 1 km and also those for which the age is quoted as 'less than' some specific time. Also, we ignored the small number of recorded impacts with ages greater than 610 Ma. Together, data from tracks and details of known certain craters came to a total of 135.

These data are plotted in Figure 8.30. We grouped together impact data in 10 Ma 'bins', and plotted these as a histogram. We then plotted the zig-zag line, so that the probable maxima could be more readily seen. The plot reveals 16 maxima, which are labelled A–P, with A as the most recent of these maxima. This maxima, with a peak assumed to have occurred at 5 Ma, contains by far the largest number of recorded events.

This number reflects the fact that recent impacts, even if they are relatively small, are more likely to resist the agents of weathering and erosion, and are also less likely to be covered by sediments than similar size structures which are many tens of millions of years older. However, one should note that this 'bin' contains at least one major 5 Ma event, of unknown location, which is recorded in all the tracks recorded and presented in Figures 8.6 to 8.26. The inferred ages of the maxima A–P in Figure 8.30 are listed in Table 8.5 with the interval between adjacent peaks. Using, as we have, bin widths of 10 Ma, it is difficult to be sure of the position of a maxima, when two adjacent bins (e.g. 40 and 50 Ma) exhibit the same number of events. Nevertheless, despite the possible errors in locating the position of individual maxima, when the spread of intervals between adjacent maxima range from 20 to 70 Ma, it is difficult to support the suggestion put forward by Seyfert and Sirkin (1979). The argument put forward by Grieve et al. (1977) that data are not sufficent in number or accuracy, as regards the age of individual impacts, to support any conclusions regarding the cyclical nature of impacts on Earth appears currently to be well founded.

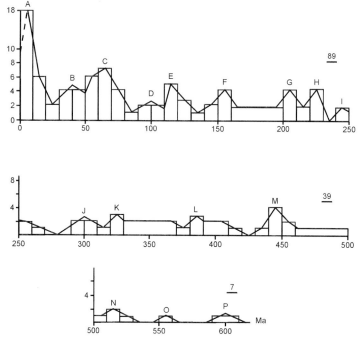

Figure 8.30 Maxima and minima of impact intensity (in 10 Ma bins) during the Phanerozoic.

Table 8.5 Time between maxima.

Adjacent maxima	Time between maxima (Ma)
A – B	35
B – C	25
C – D	35
D – E	15
E – F	40
F – G	50
G – H	20
H – I	20
I – J	55
J – K	25
K – L	60
L – M	60
M – N	70
N – O	40
O – P	45

The number of events represented in each of the three sections in Figure 8.30 is indicated, and underlined, as 89 for the period 0 to 250 Ma, 39 for the period 250 to 500 Ma and 7 for the period 500 to 610 Ma. If we take the division between the Upper and the Lower Phanerozoic to occur at the base of the Mesozoic (approximately 250 Ma), then the number of events which occurred in this period is 89, while that in the slightly longer Lower Phanerozoic (i.e. back to 550 to 600 Ma) is given as 46. This is almost exactly half the number of events for the Upper Proterozoic, and, therefore, lends support to the findings of McEwen *et al.* (1997), that the flux of impacts has approximately doubled in the Upper Phanerozoic relative to the flux in the earlier Lower Phanerozoic period.

We have already noted that old impact craters are likely to be more readily lost by erosion or covered by sediments. In addition, many craters older than about 150 to 200 Ma, which have developed in oceanic lithosphere, will have been subducted. Hence, one would perhaps expect that the number of impacts will exceed the cited number. Of course, this is true, as regards relatively small impact events (i.e. those with crater diameters of less than about 50 km.) However, the large impact events, which leave distinctive track anomalies, even if they are ancient oceanic events which have long been subducted, can still be inferred.

Consequently, we suggest that these records of track anomalies throughout the Phanerozoic support the conclusions reached by the authors cited above, that the impact flux on Earth during the last 250 Ma is approximately twice that which existed from 600 to about 250 Ma.

One can understand why various authors have included every known certain impact when attempting to ascertain the periodicity of impacts. Indeed, we followed the same process in compiling Figure 8.30. However, we now suggest that such a practice may well obscure the periodicity of major impacts.

For every major impact with a known, or inferred, crater diameter in excess of 100 km, there will be hundreds or even thousands of smaller impact events. Most of the smaller impact features have been eroded, buried or remain unidentified. In Table 8.3 we have cited 30 major events in the last 250 Ma, some of which represent multiple impacts. These impacts are shown in Figure 8.31a where the date of impact is shown for each event. An attempt has also been made to indicate the diameter of crater likely to be associated with each specific major event, although, of course, this is largely conjectural.

A histogram compiled from Table 8.3 and Figure 8.31a is shown in Figure 8.31b. This figure shows that there are events separated by 3–4 Ma in the time range 12–25 Ma, which can be loosely correlated with the lower end of the maxima cited in Table 8.1. What is important is that the majority of impact activity appears to have occurred at 5, 8 and 12 Ma. The concentration of 8 impacts at 5 Ma intervals could be disturbing, for the last really major impact took place 5 Ma ago!

It would appear that there are unidentified periodicity inducing mechanisms, other than those involved with the position of the Solar System relative to the median plane of our Galaxy, which are influencing the flux of major impacts!

8.5 Impacts, a hazard to humanity

Although no relatively 'destructive impact' (by which we mean a feature with a crater diameter of 1–5 km) has been definitively recorded in historical times, this does not mean that such events have not occurred. Thus, from Table 5.2, it can be argued that Earth should have received at least four impacts capable of producing 1.0 km diameter craters during the period of recorded history.

As the continents occupy only a little more than 25 per cent of the Earth's surface, statistically, one would expect three out of the four impacts to have occurred in the oceans. The remaining impact need only have hit a continental area remote from then existing centres of population to have escaped general notice. Nevertheless, in the light of the facts and concepts outlined in this and earlier chapters, it is not surprising that impact events are now being used to estimate the risk of individuals and communities dying as the result of a

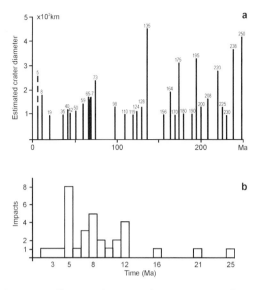

Figure 8.31 (a) Age and 'estimated' crater diameter of major impacts (from Table 8.3) in the Upper Phanerozoic. (b) Histogram compiled from the data shown in (a) and Table 8.3, which reveals important impact peaks at 5, 8 and 12 Ma.

moderate to large meteoritic or cometary impact (Chapman and Morrison, 1994). Chapman and Morrison's paper reiterates several points already made in this chapter. Their opening statement is that there is 1 : 10,000 chance of a 2 km diameter meteorite hitting the Earth in the next century. (Such an impact could cause a crater with a diameter of up to 20 km, so we are dealing with a relatively small impact event, but one which would still have the potential of destroying life over a wide area.) Indeed, these authors state that such an impact would disrupt the ecosphere and kill a large fraction of the world's population.

Obviously, the number of initial fatalities resulting from such an impact would depend on the impact site. A hit in central Sahara, or other desert regions, including the tundra areas and Antarctica, would directly deprive relatively few people of their lives. However, the effects on the ecosphere would be far from trivial.

At the other extreme, if the impact fell in the central area of the North Sea, where the water depth exceeds 150 m, so that a tsunami wave of at least 150 m high would be initiated, the results would be devastating to the east coast of Scotland and England, destroying towns and cities from the Shetlands to Kent. On the mainland of Europe, the west coast would receive comparable devastation, from southern Norway to northern France. Holland and most of Belgium would cease to exist as communities. The impact could ignite fires over a wide land area and the accompanying air blast and ground quake would add to the death toll which would be measured in tens of millions; and these would only be the immediate casualties. A comparable scenario could be presented for an impact off the coast of the 'Bos-Wash' conurbation, USA, particularly if it occurred in deep water, some 100 or so miles off the coast where it could generate an extremely large tsunami. (These comments were written long before various 'impact' films with their excellent scale-model tsunami and the attendant destruction of cities were released.)

In their estimates of hazard, Chapman and Morrison (1994) present their conclusions in graphical and list form. It is the latter which we consider most effective, and accordingly present their conclusions of fatality rates for impacts, and the chances of dying in the USA from selected causes in Tables 8.6 and 8.7 in a slightly modified and/or abbreviated form.

Obviously, the conclusions reached by these authors are debatable, some may even say contentious. Nevertheless, the table of chances of dying from selected causes is especially thought provoking. One's chance of being killed in a passenger aircraft crash and by a relatively small impact event is cited, in Table 8.7, as being the same. If it is any solace to someone killed in an air-crash, he or she is likely to be accompanied by as many as 300 passengers suffering the same fate. Although the chances are the same, the number of deaths caused by a 'small impact' may be about 300,000, i.e. a thousand times worse than a particularly disastrous air-crash.

Chapman and Morrison may be in error regarding their estimate of the *average number* of deaths for a 1–2 km impact by as much as an order of magnitude. Nevertheless, this article presents the type of data already dealt with in this chapter in a way that has yard-sticks which can immediately be related to human experience. It is a line of thought that should completely dispel any false optimism as regards the occurrence of impacts, large and small, and their chances of affecting the continuing existence of major communities and even of humanity itself. Moreover, it will be recalled that the flux of impact in the last 250 Ma has recently been inferred to be twice that of the period earlier than 250 Ma. Hence, we can conclude that currently, the risk of death as the result of a meteoritic/comet impact is likely to be 1 : 5000 rather than 1 : 10000 cited in Table 8.7; so that the chances of being killed by a meteorite/comet impact, with a diameter of 1–1.5 km, are twice as likely as that of being killed in a crash of a commercial airliner.

Table 8.6 Type of event and estimated death-toll.

Type of event	Diameter of impactor	Energy	Typical interval	Deaths
High atmospheric break up	<50 m	<9 Meg.T	N.A.	0
Tunguska-like event	50–300 m	9–2000	250 a	5×10^3
Large Sub-global event	300–600 m	2000–15000	35×10^3a	3×10^5
	0.6–1.5 km	$0.15–2.5 \times 10^5$	..	5×10^5
	1.5–5.0 km	0.025×10^7	..	1.2×10^6
High-Global threshold	> 5.0 km	$> 10^7$	6×10^6	1.5×10^9
Rare K/T scale event	> 10.0 km	$> 10^8$	10^8	5×10^9

Table 8.7 Chances of dying from selected causes (USA, 1990).

Cause of death	Probability
Motor vehicle accident	1 : 100
Murder	1 : 300
Fire	1 : 800
Firearms accident	1 : 2500
Asteroid/comet impact (1–2 km diameter)	*1 : 20000*
Passenger aircraft crash	1 : 20000
Flood	1 : 30000
Tornado	1 : 60000
Venomous bite or sting	1 : 100000
Asteroid/comet impact (>10 km diameter)	1 : 250000
Fireworks accident	1 : 1000000
Food poisoning by botulism	1 : 3000000

8.5.1 Comment and conjecture relating to impact flux

We have noted that Oberbeck *et al.* (1993) derived an expression regarding the flux of impacts in the last 2000 Ma, which they tacitly take to represent a mean or possibly constant flux. When expressed in terms of a 250 Ma period, they have estimated that there should be about 14 such impacts.

Absolutely reliable impact data which form the basis of the relationship derived by Oberbeck *et al.* are far from plentiful. However, we suggest that the relationship between sharp changes in track and their relationships in time to a variety of features, which range from known 'certain' impacts, through continental flood basalts and a major oceanic plateau basalts, and small arcuate or semicircular subduction arcs, have been sufficiently well established in earlier chapters to convince the reader that major impacts are the prime cause of most and probably all the cited features. Hence, the record of these tracks which for two areas have been traced back 600 Ma, permits one to infer that the meteoric and/or cometry flux has doubled during the Upper Phanerozoic. From the tracks of Fennoscandia and 'Hawaii', in the period from 550 to 250 Ma it can be inferred that there were 10 major impact events (Figures 8.27 and 8.28) and if a larger number of track sites were examined, this could easily be increased to about 14. Such a number is in reasonably close agreement with the 'average flux' estimated by Oberbeck *et al.*

Current estimates of future impacts tend to be based on tenuous dating with large potential errors in the periodicity of the recurrence of such events. For example, it has been estimated that the Tunguska event in 1908 may be associated with the return of a similar event between 100 and 300 years of that 1908 event. Hence, the next comparable event may occur between 8 and 208 years from now.

As far as is known, no-one died as the result of the Tunguska event and the next piece of ice that explodes in the atmosphere may be much smaller. Hence, the danger could be minor and 200 years in the future, so easily ignored!

In March 1998, the situation may have changed. Scientists of Spacewatch discovered that a stony asteroid, with a diameter of about 1 mile, was in an orbit that could possibly cause it to collide with Earth on 26 October 2028. From the mass and speed of the body, it was estimated that the asteroid could give rise to a crater with a diameter of 20 km. However, the asteroid may only approach Earth to within a few tens of thousands of miles. The astronomers of Spacewatch, at the time of writing, considered there to be little risk of impact, but that the most likely eventuality was that the asteroid will pass Earth at a safe distance. Past records are currently being searched and future observations will be directed towards determining the orbit of this particular body with the greatest possible accuracy. If a near-miss, fly-by is assured, mankind will be vastly relieved.

However, should such an impact occur, the results will not be trivial.

That the Moon has been bombarded since its earliest times is plain to see. The Earth too has been bombarded for billions of years. Indeed, it became a larger 'target' as the mass of the proto-planet accumulated, through gravitational attraction of the smaller chunks of matter that wandered through the proto-Solar-System. It is now generally considered that the Moon was originally part of the proto-Earth and was generated by a body about the size of Mars, which collided with proto-Earth.

In the Hadean period, it must have been truly 'hellish', with a bombardment rate far higher than has been experienced since the late Archean. Thereafter, in the Proterozoic and into the Phanerozoic to the present day, the mean flux of impacting bodies has probably been approximately constant. However, recent work has shown that an increase in flux has occurred in the Upper Phanerozoic by a factor of about two, relative to the value of flux for the Lower Phanerozoic.

Such are the huge numbers of ice bodies in the Oort cloud, that this state of bombardment will almost certainly continue unabated until the Sun expands to engulf the inner planets, possibly including Earth itself, some 3000 Ma hence.

This bombardment has been responsible for a range of geological effects, some of which have been considered in the text, and doubtless there are effects and by-products of periodic major impacts that are as yet unrecognised. The general demise of the dinosaurs and other life forms some 65 Ma ago has been largely or wholly attributed to one or more such major impacts. Large marine and terrestrial life forms are particularly sensitive to the shocks and environmental changes induced by major oceanic impacts. Different species have become extinct in the history of the Phanerozoic. It is a process that is probably driven in part by a cosmic clock, coupled or modified by a process which is perhaps somewhat akin to 'Russian roulette'.

Despite the evidence that supports the importance of impacts in the evolution of Earth and the continuing activity of this mechanism throughout the Earth's history and into the future, many geologists appear to exhibit a natural antipathy to the concept (despite the evidence) that life forms, especially in the Phanerozoic, can suddenly be extinguished. It is to be hoped that the arguments presented in this book and by Rampino and his collaborators in a series of papers published in the last decade or so, will do much to counter the dominantly inward-looking tendencies of many geologists. Certainly, there are periods, measured in millions of years, which approximate to the steady-state conditions proposed by Hutton 200 years ago. However, these steady-state periods in which evolution of the Earth continues slowly are punctuated by brief catastophic events, which give rise to massive extinctions and often the generation of major stratigraphic boundaries.

It is understandable that most geological studies have been directed to the 'steady-state' periods in Earth's history. However, it is our thesis that the catastrophic events are of vital importance. A catch-phrase, attributed to Hutton, is that 'The past is the key to the present'. We would like to extend this vision to 'The past is the key to the future'. To which one could add 'The future is bleak'.

The extinctions that occurred at or close to the K/T boundary were spectacular. There have been other, less spectacular events since then. One or more impact events occurred 5 Ma ago and is marked by the Pliocene/Miocene boundary. This event is to be seen in every track of the regions studied. Since that time, the impact record has been relatively insignificant. Mankind has evolved in this 'quiet period'. How soon will a major catastrophic impact occur on Earth? We have seen that there is a rough periodicity as regards major impacts on Earth, which appears to be largely controlled by a 'galactic clock'. However, there may be other controlling but as yet unidentified mechanisms which ensure a major impact as frequently as every 5 or 8 Ma.

From the past record, we can expect to be free of a major impact (i.e. one which will give rise to a crater diameter of 300–400 km) for perhaps 3–20 Ma. With our present or foreseeable technology, there is nothing that can be done to prevent such an event. However, before then, we must expect several thousands of smaller impacts which are likely to be extremely dangerous and devastating over large areas of the Earth. Thus, it is highly probable that mankind will experience extinction from this galactic controlled 'smoking gun'.

However, we now have the ability to detect and forecast whether at least some bodies are likely to strike Earth. Nevertheless, we cannot yet be sure that some relatively small, but still highly destructive, bodies may escape detection.

8.6 Synopsis

World War II was brought to an end by exploding two atomic devices of modest energies. The end of WW II rapidly changed to the Cold War and a race by various nations to acquire atomic and nuclear devices of ever-increasing power. As a legacy of this recent Cold War we now have the beginnings of a technology which one day may be able to destroy or redirect an incoming asteroid or cometary body before it destroys us. So there is cause for hope.

During the early part of the Cold War, it became apparent that testing of high energy atomic or nuclear devices was both extremely expensive and, in peace-time, extremely antisocial. To obviate these problems, series of explosive experiments were conducted in the USA and Canada. Similar experiments have certainly been conducted in the former USSR, China, Australia and in the Pacific by the French. However, only the USA and Canada have a 'Freedom of Information' Act, which permits access to at least some of the test results.

Some of the explosive tests conducted in the USA were of sufficient energy that they produced shatter cones and high-pressure polymorphs and shock-induced features within individual grains, from which it was possible to correlate the structures and minerals with the stress levels generated in the explosion. This very rapidly led to the identification of these high-stress minerals and shatter cones associated with natural craters. Indeed, it was the existence of such stressed minerals in association with a structure that permitted the title of 'certain' impact.

The experimental work in these two countries also enabled the relationship between crater diameter and the magnitude of the energy of the explosion to be established, and theoretical studies of the mechanics of cratering, the magnitude of impact stresses and the manner in which these stresses decayed away from GZ laid the foundation for estimating the effects of an impact explosion, including the level of stress at which rocks vaporised, liquified or deformed in a plastic or brittle manner.

In other ways, however, the experiments carried out in the USA must have been disappointing, if not frustrating, for they only generated simple craters. But in Canada, things were scientifically more exciting. In the mid-1950s and 1960s, a whole series of explosive experiments were conducted at Suffield, Alberta. These experiments, ranging from 25 to 500 ton TNT explosions, were conducted mainly on very weak, lacustrine sediments with a water-table at a shallow depth. Such water-logged, non-indurated sediments were perhaps only one thousandth the strength of well-indurated sediments, so that these explosive experiments were 'scale-models'.

The 500 ton Snowball and Prairie Flat explosions, which gave rise to a central uplift crater and a ringed uplift crater respectively, caused a great deal of interest among Canadian and US astronomers, for such features had been seen on the Moon (see the quotation on the Dedication page). As a consequence, natural examples on Earth were also readily identified. Hence, based on the geometry of the craters alone and the high-stress minerals and shatter cones which developed, real impacts could be identified with complete certainty. Had the explosive experiments conducted in the USA and Canada not been carried out, we would be little further forward today than we were in 1950.

However, the larger Suffield craters yielded further vital information. Outside the craters, three different forms of fractures were found. Firstly, there was the circumferential graben. Even when a crater in basement rocks had been deeply eroded and later covered by flat-lying, younger sediments, the trace of the underlying graben, or single circular fault, with diameters of 50–250 km, revealed the existence of a covered impact structure.

The second feature comprised the vertical fractures which developed radially from below the rim of the crater which, like the graben faults, also brought water, mud and sand to the surface. Radiating dykes were known to occur on the Moon and also occur around the Central Nevada Circular Feature. (We have inferred that such super-large dykes could give rise to the splitting of continents.) Both these types of fractures were instantly recognised, when the surface outside the crater rim was surveyed immediately after the explosion.

The third and last set of fractures were series of low angle fractures which dipped inward towards GZ. These were only discovered by deep trenching along survey lines, and often gave the impression of being low-angle, normal faults. However, it was found that some of the fractures were thrusts, so that it was reasonable to assume that the normal fault movement was induced by inward slumping towards the crater, subsequent to the outward blast of the explosion. If it were not for these thrusts, any explanation regarding the development of relatively tight-radius arcuate, subduction zones would, of necessity, need to be conjectural.

Relatively few 'unclassified' reports, regarding this work, were published while the research program was in progress at Suffield, Alberta. It is to his credit that Gareth Jones spent the last couple of years of his life presenting an overview of the cratering program (carried out from 1960 to 1970 at Suffield) based on his memories, notes and personal records.

Early theoretical modelling was, of necessity, simplistic. However, as more powerful computers became available in the final decade or so of the 20th century, the modelling became progressively more complex and more realistic. For example, such a model forecast with remarkable accuracy the morphology of the back-jet of ejecta that would result from the impacts of the various bodies which made up the Shoemaker-Levy 9 that struck Jupiter in 1994 (see Figure 1.12). This computer program was also used to show that a back-jet must also have developed following the 1908 Tunguska air-burst.

The field studies of natural large 'certain' craters, coupled with theoretical analyses, enabled the quantity of melt rock for an event of specified size and energy to be estimated. The greater the energy of impact, the greater the quantity of melt rock generated. Melosh (1989) argued that, for large energy impact events the quantity of melt which developed could fill a crater to the brim. It required only a small intellectual step further to realise that impacts of even higher energy could form quantities of melt that would overflow the crater, and produce volumes of extrusive melt which could completely hide the obvious evidence of such a major impact event.

Although one cannot claim that the mechanisms of cratering are completely understood, we suggest that the sequence in the order of magnitude of cratering set out in this book requires very little speculation. The step, which postulates that impacts of very high energy could generate volumes of melt which were so large that not only does the crater become filled with erupted melt rock, but that the volume of melt is sufficient to generate a carapace forming a continental flood basalt, is an extremely reasonable hypothesis. Indeed, Negi *et al.* (1993) have already suggested that such a major impact initiated the Deccan Traps.

We resurrected this concept, introduced a mechanistic reason for the development of the eruption and demonstrated how such events can be recognised by changes in the rate of plate motion and/or changes in the direction of plate motion. This presents a model for explaining abrupt changes in plate movements, which is absent from the analyses of plate motions reported by Lithgow-Bertelloni and Richards (1998) and other workers in this field.

The next step was to establish that the mechanistic hypothesis was viable and supported by plate movement records. This was made possible by the Atlas System 3.3 which permitted the movements of continents, or smaller areas, to be traced back through the Phanerozoic. This program represented the second of our 'foundation stones'.

We tested the hypothesis against three known certain impact features and two near-certain impacts, by plotting the track of these events for the period at which it was known that the impacts took place. The correlation between the abrupt change in track and the age of the impact as determined by normal dating procedure was impressive. We followed this with similar studies of eight continental flood basalts and the Ontong-Java plateau basalt. All nine of these events showed a track change which coincided closely with the age of the event as determined by radiometric dating. We also showed that five arcuate, relatively small radius, subduction zones were also associated with abrupt track changes.

We continued this study by comparing the tracks of seven areas for the period 0–260 Ma and found that there were 30 major impact events in this period, with several of them being multiple impacts. Also, in this group of events, 25 coincide exactly, or closely (+/– 1 Ma), with a stratigraphic boundary.

We also traced the tracks for two areas back from about 250 to 600 Ma, and found that the major track changes also coincided with stratigraphic boundaries. Moreover, tracks in both areas showed the 365 Ma event which coincides with the Central Nevada Circular Feature and the coincident S China Sea event. It was also apparent that the impact flux in the Lower Phanerozoic is approximately half the flux in the Upper Phanerozoic, as was predicted by McEwen *et al.* (1997). We have been able to indicate that the volume of melt rock generated is determined not only by the energy of the impact event, but that the strength and thickness of oceanic and/or continental lithosphere plays a role in determining the volume of the flood or plateau basalt which is erupted.

It has recently been realised that the vast majority of impacts are caused by large cometary bodies. The type of impact that such comets engender when they hit a continental surface, compared with a meteoritic impact of comparable energy, will not be identical. The density of a meteorite may be 3–6 times greater than a comet of comparable size. However, a comet is likely to have an impact velocity of at least twice that of a meteorite. Consequently, the energy of impact of these disparate bodies, when they strike continental lithosphere, are not likely to be too great. However, when a meteor strikes in deep ocean, we argued that, compared with a meteoritic strike on 'dry-land' of equal potential energy, the difference in morphology of the impact structure that is engendered will be marked. A large meteorite will plunge through the deep ocean to impact upon the ocean floor with a relatively small degree of retardation. A friable comet may not only begin to break up in the Earth's atmosphere but will be seriously retarded by deep oceanic water. We have suggested that the model of the meteorite may be akin to a rocky sphere, whereas a cometary body will rapidly evolve into an oblate spheroid, striking the deep ocean floor at a relatively low impact velocity. The change in cross-sectional area of the impacting body coupled with a slow impact velocity, will result in a relatively low stress impact at GZ, combined with a low rate of stress attenuation away from GZ. Such a stress environment will probably result in a modest, or even undetectable crater, but can give rise to the development of circumferential thrusts at considerable distances from GZ. These conclusions are based on a simplistic 'shotgun' analogy, so cannot be regarded as rigorous analyses. However, when we applied this model to explain the initiation of five relatively small subduction zones, we found that all five could reasonably be related to changes in track of rate and/or direction which occurred at an appropriate time.

We have found track anomalies for three certain and two probable impact craters, eight continental plateau basalts, and the largest major oceanic plateau basalts as well as five small arcuate subduction zones.

In this final chapter, we have been addressing an even more fundamental concept. Namely, the analysis of the occurrence of a whole series of geological phenomena. Dating of geological events is not yet perfect. However, we have seen that several of these different phenomena appear to exhibit a series of maxima and minima. Although distinct periodicity has yet to be established, peaks appear to vary about a mean of approximately 26–30 Ma. However, from the data presented in this final chapter, it would appear that there are shorter-term major impact periodicities of 5 and 8 Ma which have not so far been explained.

The several periodicity investigations have been mainly concerned with the relationship of certain dramatic geological events. Recently, work by Rampino *et al.* (1997) has resulted in the presentation of a unified theory of impact-induced crises which have brought about major mass-extinction events. These geological data have been subjected to quantitative tests, based on new data as regards extinction intensities and the impact data of several well-dated, relatively large impact events. They identify five major events and about 25 moderate-size events which have occurred within the Phanerozoic.

This information became available only after we had completed the analysis which enabled us to prepare Table 8.3. It will be noted that we identify 30 separate events (some of which include multiple impacts) which have occurred within the Phanerozoic. Thus, the number of events in the two studies are close to being identical. We suggest that these two fundamentally different approaches to the impact history of the Phanerozoic constitute well founded, mainly factual, analyses which point strongly to the control of major impacts in geological history.

There is ample evidence, to be seen on the Moon, that our satellite has an approximately linear flux and accumulation curve of craters over at least the past 2000–250 Ma. Moreover, it has been demonstrated that a similar accumulation curve is likely to hold for the Earth for a comparable period. This flux can reasonably be extrapolated to show that in the last 2000 Ma there have been, on Earth, well over 100 major impacts with diameters greater than 100 km. Moreover, it can be inferred that in the Phanerozoic about five impacts have resulted in craters with diameters of several hundred kilometres. It has also been shown that the number of impacts in the last 250 Ma has doubled and, therefore, must have had an enhanced influence on the crustal evolution of the Earth, as well as upon the fauna and flora that, from time to time, flourish on its surface and in its oceans.

When viewed from space, our Earth is beautiful to behold, but it is also vulnerable. Probably, for the first time in its history, its surface dwellers have the potential of guarding this life-supporting 'space capsule' against external threats of damage and human extinction. Cooperation between governments and scientific and military bodies in the world is urgently needed to establish means of defending Earth against such potentially distructive bombardment.

We have already noted that in March 1998 it was announced by members of Spacewatch that there is an asteroid in an Earth-crossing orbit that could approach closely to, or even collide with, Earth on 26 October 2028. Moreover, the search for further information regarding bodies which may collide with Earth is being intensified.

Recently, the American Space Agency (NASA) has set up a project for 'Near-Earth Asteroid Tracking' (NEAT), the object of which is to locate those bodies which are a hazard to Earth. At the time of writing, the NEAT project has established about 25,000 objects near Earth, including two comets and 30 hitherto unknown asteroids. The NEAT team has also discovered a new group of asteroids with a tight orbit around the Sun, called 'aten', (ancient Egyptian symbol for Sun) which are also 'Earth-crossers'. Because of their relatively tight orbit they return at relatively short intervals, and so are extremely hazardous. Also, because, like fighter pilots, the aten approach 'out of the Sun', they are extremely difficult to detect.

The NEAT program is obviously worthy of the most urgent and significant financial support. The future of mankind will depend upon the efforts of this and the Spacewatch project in the USA and other programs currently being conducted by astronomers of many nations around the world.

Sooner or later, unless we are vigilant and prepared and also extremely lucky, we and most of the major life forms on Earth will follow the dinosaurs into extinction.

Appendix

Let us now briefly comment on the various events listed in Table 8.3.

1   The most recent major event occurred 5 Ma ago, and shows dramatic changes in speed and/or direction in all six of the tracks studied. Unfortunately, no major impact event of this date has so far been identified, or inferred. However, it may be noted that the event almost exactly coincides with the Miocene/Pliocene boundary and could also be associated with a phase of development of the Alps. This could be a multiple impact event. One or more impacts could have occurred in the ocean, thereby generating a stratigraphical boundary and one (or more) impacts could have initiated mountain building.

2   This 8 Ma event is recorded in four tracks, mainly as a 'small' event. Its location is unknown, but may be associated with the Messinian/Tortonian boundary.

3   The event, at 19 Ma, attributed to the development of the Columbia River flood basalts, is recorded in only two of the tracks studied. This event is likely to be related to the Yellowstone Park hotspot. Dating of the basalt flows is ambiguous and ranges from 6 to 17.5 Ma and 15.7 to 17.2 Ma. (Coffin and Eldholm, 1994).

4   A reasonable amount of wide-spread impact activity occurred 35 Ma ago. For example, tektites and stress metamorphic minerals of this age have been found in the Caribbean area. The 100 km diameter Popigai impact feature in Siberia and the initiation of the Banda Arc also occurred at about this time. Hence, one may infer that there were probably at least two separate impacts associated with this event, one continental and the other, marine, However, it does not appear to be related to a stratigraphic boundary. This, could possibly be related to the land-locked position of the Banda Arc which would minimise the effect of the impact.

5   A relatively minor track anomaly (dated at 40 Ma) for a location on the Red Sea coast can be inferred as being the result of a continental impact of sufficient size to induce considerable eruptive activity in the adjacent areas. The fact that the track anomaly is relatively small, can be attributed to the difficulty of causing Africa to change its direction

and rate of movement, rather than indicating that the impact was of relatively low energy.

6   A relatively small (42 Ma old) event, of unknown location, caused anomalies in the tracks of four of the areas studied. The age of the impact coincides with that of the Priabonian/Bartonian stratigraphic boundary, so that the impact is almost certainly marine.

7   This 50 Ma event (or events) is/are recorded in the tracks of six of the seven areas studied. The location(s) is/are not known, but the dating of the impact is within 0.5 Ma of the boundary between the Early and Middle Eocene, so the impact(s) is/are almost certainly marine.

8   This event is recorded in six of the seven areas studied. The ages fall between 59 and 60 Ma. We suggest that this marks a multiple impact event. The ages of impact are a little earlier than Thanetian/Danian boundary at 60.2 Ma, which also marks the boundary between Early and Late Paleocene, so that the impact event is almost certainly oceanic.

9   The period of about 2 Ma prior to, and up to, the K/T boundary (65 Ma) saw the development of at least four major events, namely the Deccan Trap 3, Amirante Arc, the Scotia Arc and Chicxulub. Two of the impacts are marine and two are borderline ocean/continent impacts. Chicxulub is, of course, blamed for the extinction of the dinosaurs. However, we have suggested that this has always seemed somewhat unlikely. Indeed, Robin *et al.* (1993) conclude from the study of meteoritic material that the evidence can best be explained by multiple impact events rather than by a single large impact. Certainly Chicxulub made a significant contribution. However, the emplacement of the Deccan Traps over a period of 1 Ma would have caused widespread atmospheric pollution. In addition, the submarine emplacement of basalts, as the result of the initiation of the Scotia and Amirante Arcs would have influenced marine temperatures, and hence weather patterns, over a long period.

10   This 73 Ma event is associated with the initiation of the Antilles Arc and subduction trench, which in the south, is now largely obliterated by more recent events. This was certainly a marine impact, which coincides with the Maastrichtian/Campanian stratigraphic boundary.

11   This event is recorded in all of the track areas over the period 98 to 100 Ma and may have resulted in the development of the Hess Rise. It is certaily linked in time to the 97.5 Ma Cenomanian/Albian boundary, which also marks the junction between the Early and Late Cretaceous.

12   It is probable that at least two major impacts occurred at this time, and are recorded in four of the track areas, within the age range (110 to 114). This is coincident with the development of the Kerguellen large igneous province and the Elan Bank, both dated at 109.5 to 114 Ma (Coffin and Eldholm, 1994). This age-range also embraces the junction between the Albian and Aptian which is dated at 113 Ma.

13   This 119 Ma event is recorded in all of the track areas and can be correlated with the Ontong-Java Plateau Basalt and the three named Basin Basalts. Coffin and Eldholm (1994) note that there is ambiguity regarding the dates of these large igneous provinces with dates that range a) 117.7 to 118.2, or b) 121 to 124 Ma. More recently, these lavas have been dated at 121.6–123.2. The track-kink would fit closely to any of these dates, within the probable limits of uncertainties in dating. We would favour a date intermediate

between these two ranges of about 119 Ma, for this would coincide with the Aptian/ Barremian stratigraphic boundary.

14   The 124 to 125 Ma event(s) is/are recorded in six of the seven track areas and has an age which is in reasonable agreement with the cited age of the Cape Verde hotspot. As it also coincides with the Barremian/Hauterivian stratigraphic boundary, it is in all probability a marine impact.

15   This event of unknown location, dated at 128 Ma, is recorded in four of the seven areas. The date of this event is earlier than, but only 2 per cent different from, the Hauterivian/Valanginian boundary of 131 Ma. The possible error in age for this boundary is approximately 3 to 4 Ma. This is also likely to be a marine impact.

16   The Paraña event at 135 Ma is recorded in four of the seven track areas. This event is marked by an extremely large increase in the rate of plate movement, for S America; and we have argued, resulted in the opening of the S Atlantic. This very major impact was continental.

17   This oceanic event (at 156 Ma) is of modest proportion and is associated with the Kimeridgian/Oxfordian stratigraphic boundary at 156 Ma.

18   This 164 Ma event is well seen in the Hawaiian track and is attributed to a marine impact that gave rise to the junction between Oxfordian and Callovian at 163 Ma.

19   This 170 Ma event is recorded in four of the seven tracks. It is attributed to a marine impact that gave rise to the Callovian/Bathonian stratigraphic boundary at 169 Ma.

20   An event, which is a feature of the Mariana track reference point, occurs at 175 Ma, which we have associated with the initiation of the Mariana arcuate feature and is also associated with the Bathonian/Bajocian junction.

21   This event which is recorded as 'large' in five of the seven areas, and is attributed to a marine impact age of 180 Ma, coincides closely with the designated age of the Bathonian/ Bajocian stratigraphic boundary, set at 181 Ma.

22   These inferred marine impact events, with ages which range between 188 to 193 Ma are within the designated age-range of the boundary about the Dogger/Lias boundary at about 188 Ma.

23   This inferred marine impact event, which is recorded in all seven areas, is dated at 195 Ma, is within 1 Ma of the mean date of 194 Ma given for the Toarcian/Pliensbachian stratigraphic boundary.

24   This inferred marine impact event is only recorded in two of the seven areas. However, the date of the event at 200 Ma, coincides exactly with the designated age for the Pliensbachian/Sinemurian stratigraphic boundary.

25   This inferred marine impact is recorded in all seven track areas, with six of the events being termed 'large'. The date of the inferred impact event is 208 Ma. The nearest stratigraphical boundary, in time, is the Sinemurian/Hettangian boundary at 206 Ma. However, because the track anomalies indicate that the event was of large energy, we suggest that events could better be attributed to the Jurassic/Triasic boundary at 213 +/– 7 Ma.

26   The 220 Ma event is evident in all seven sets of tracks. This age can be compared with the the date of 219 Ma for the Rheatian/Norian statigraphic boundary.

27   The 225 Ma event shows only in a single track, but as it compares exactly with the Norian/Carnian stratigraphic boundary, it is assumed to be a marine impact.

28   The 230 Ma event, seen in four of the seven areas, occurred close to the age of 231 Ma, and is nominated for the Carnian/Ladimian stratigraphic boundary.

29   The 238 to 240 Ma event(s) is/are seen in all seven areas, and range from medium to large. The date range is in keeping with the nominated value of 238 for the Landinian/Anisian stratigraphic boundary. However, we note that the likely energy of the impact event(s) is very high. Accordingly, we suggest that this event is a marine impact and may be correlated with the Scythian/Middle Jurassic boundary, age 243 Ma.

30   The final example of this list occurs at 248 to 250 Ma. This event registers in all seven areas, with five of the track changes indicating a high energy event. We have identified this as the continental impact event which gave rise to the Siberian Plateau Basalt. It may also have been coeval with a marine impact because it marks the junction between the Permian and the Triassic; a fitting event to mark the end of the Lower Phanerozoic and the entry to the Upper Phanerozoic.

References

Abrams, L., Larson, R., Shipley, T. and Lancelot, V. (1993) 'Cretaceous volcanic sequences and Jurassic oceanic crust in the E Mariana Pigafetta basins of the W Pacific'. In: *The Mesozoic Pacific* (Eds Pringle, M and Sager, W.). Amer. Geophys. Union Monograph Series.

Ahrens, T.J. and O'Keefe, J.D. (1977) 'Equations of state and impact induced shock wave attenuation on the Moon'. In: *Impact and Explosive Cratering* (Ed. Roddy, D.J.). Pergamon, Oxford.

Albriton, C.A. Jr. (1989) *Catastrophic Episodes in Earth History. Topics in the Earth Sciences*. Chapman and Hall, London.

Allen, R.T. (1969) 'Equations of state of rocks and minerals'. Interim Rept. to DASA under contract DA49-146-XZ-462. General Dynamics Report GA MD 7834.

Allegre, C.J. (1997) 'Limitation on the mass exchange between the upper and the lower mantle: the evolving convection regime of the Earth', *Earth and Planetary Scinece Letters* 150, 1–6.

Alvarez, L.W. (1980) 'Extraterrestrial causes for the Cretaceous-Tertiary extinction'. In: *Geological Implications of the Impact of Large Asteroids or Comets on Earth* (Eds Silver and Schulz). Geol Soc. Am. Special Paper, 190, 304–315.

Alvarez, L.W., Alvarez, W., Asaro, F. and Michel, H.V. (1980) 'Extraterrestrial causes for the Cretaceous-Tertiary extinction'. *Science*, 208, 1095–1108.

Alvarez, W. and Muller, R.A. (1984) 'Evidence from crater ages for periodic impacts on Earth'. *Nature*, 308, 718–720.

Alvarez, W., Asaro, F., Michel, H.V. and Alvarez, L.W. (1982) 'Iridium anomaly approximately synchronous with terminal Eocene Extinctions'. *Science*, 216, 886–888.

Anderson, D., Yu-Shen Zhang and Tanimoto, T. (1992) 'Plume-heads, continental lithosphere, flood basalts and tomography'. In: *Magmatism and the Causes of Continental Break-up* (Eds Storey, B., Alabaster, T. and Pankhurst, R.). Geol. Soc. London Spec. Publ., 68, 99–124.

Anderson, D.L. (1994) 'Lithosphere and flood basalts'. *Nature*, 367, No. 6460, 226.

Anderson, D.L. (1998) *The EDGES of the Mantle*. Geodynamics Series, American Geophyiscal Union, 28, 255–271.

Angelier, J., Tarantola, A. and Valette, B. (1982) 'Inversion of field data in fault tectonics to obtain the regional stress – I. Single phase fault populations: a new method of computing the stress tensor'. *Geophys. J. R. Astron. Soc.*, 69, 609–621.

Archambeau, C., Flinn, E. and Lambert, D.G. (1969) 'Fine structure of the upper mantle'. *J. Geophys. Res.*, 74, 5825–5865.

Arctowski, H. (1895) 'Observations sur l'interêt qui présente l'exploration géologique des Terres Australes'. *Bull. Soc. Geol. France*, 3e series XXIH 589–591.

Arkani-Hamed, J. (1993) 'On the tectonics of Venus'. *Phys. Earth and Planet Inters.*, 76, 75–96.

Asudeh, I., Green, A. and Forsyth, D. (1988) 'Canadian expedition to study the Alpha Ridge complex: results of seismic refraction survey'. *Geophys. Journ.*, 92, 283–301.

Atwater, T. (1970) 'Implications of plate tectonics for the Cenozoic tectonic evolution of western North America'. *Geol. Soc. Amer. Bull.*, 81, 3513–36.

Audley-Charles, M.G. (1976) *J. Geol. Soc. Lond.* 132, 179-198.B

Audley-Charles, M.G. (1986) 'Rates of Neogene and Quaternary tectonic movements in the southern Banda Arc based on micropalaeontology'. *Journ. Geol. Soc.*, 143, 161–175.

Audley-Charles, M.G. (1991) 'Tectonics of the New Guinea Area'. *Annual Rev. Earth Planet. Sci.*, 19, 17–41.

Babuska, V., Plomerova, J. and Granet, M. (1990) 'The deep lithosphere of the Alps: a model inferred from P residuals'. *Tectonophysics*, 176, 137–165.

Bailey, M. and Emil'yanenko. (1997) 'On comets'. *Geoscientist*, 1, 29.

Bailey, M.E., Clube, S.V.M., Hahn, G., Napier, W.M. and Valsecchi, G.B. (1994) 'Hazards due to giant comets: climate and short term catastrophes'. In: *Hazards Due to Comets and Asteroids*, (Ed. Gehrels, T.) Tucson, Univ. Arizona Press.

Baksi, A. (1990) 'Timing and duration of Mesozoic-Tertiary flood-basalt volcanism'. *EOS*, 71, 1835–1836.

Baldwin, R.B. (1949) *The Face of the Moon*. University of Chicago Press, Chicago.

Barazangi, Isacks *et al.* (1973) *Nature*, 242, 98.

Barber, A.J., Audley-Charles, M.G. and Carter, D.J. (1977) *J. Geol. Soc. Australia*, 24, 51–62.

Barker, P., Kennett, J. *et al.* (1988) Proceedings of the Ocean Drilling Programme, Initial Reports, 113. Ocean Drilling Program, College Station, TX.

Barker, P.F., Barber, P.L. and King, E.C. (1984) 'An early Miocene ridge crest-trench collision on the South Scotia Ridge near 36W'. *Tectonophysics*, 102, 3132.

Barker, P.F., Dalziel, I. and Storey, B. (1984) 'Tectonic development of the Scotia Arc region'. 215–248.

Barrell, J. (1914) 'The strength of the Earth's crust'. *J. Geol.*, 22, 729–741.

Barrow, J. (1830) 'Account of the island of deception, one of the New Shetland Isles'. *Journal Roy. Geographical Soc.*, 1, 62–66.

Barton, P. and Wood, R. (1984) 'Tectonic evolution of the North Sea Basin: crustal stretching and subsidence'. *Geophys. Journ. Roy. Astron. Soc.*, 79, 987–1022.

Bemmelen, van R. (1949) *The Geology of Indonesia*, vol. 1A. Government Printing Office, The Hague.

Benioff, H. (1954) 'Orogenesis and deep crustal structures – Additional evidence from seismology'. *Geol. Soc. Am. Bull.*, 66, 385–400

Benka, S.G. (1994) 'Astronomers are poised for the crash of 1994: boom or bust', *Physics Today*, 47, 19–21.

Bergeron, L. (1997) 'Deep waters'. *New Scientist*, 155, 2097, 22–26.

Bergman, E.A. and Solomon, S.C. (1984) 'Source mechanisms of earthquakes near mid-ocean ridges from body waveform inversion – implications for the early evolution of oceanic lithosphere'. *J. Geophys. Res.*, 89, B13, 1415–1441.

Bergman, E.A. and Solomon, S.C. (1985) 'Earthquake source mechanisms from body-waveform inversion and intraplate tectonics in the northern Indian Ocean'. *Physics of the Earth and Planetary Interiors*, 40, 1–23

Bindschadler, D.L., Schubert, G. and Kaula, W.M. (1992) 'Coldspots and hotspots – Global tectonics and mantle dynamics of Venus. *J. Geophys. Res.*, 97, E8, 13 495–13 532.

Bodine, J.H., Steckler M. and Watts, A. (1981) 'Observations of flexure and the rheology of the oceanic oithosphere'. *J. Geophys. Res.*, 86, B5, 3695–3707.

Boslough, M., Chael, E., Trucano, T. and Crawford, D. (1995) 'Axial focusing of energy from hypervelocity impact on Earth'. *Int. J. Impact. Engng.*, 17, 99–108.

Bosworth, W., Strecker, M. and Blisnuik (1992)' Integration of E African paleostress and present-day stress-data: implications for continental stress field dynamics'. *J. Geophys. Res.*, 97, B8, 11 851–11 866.

Bott, M. and Dean, D. (1973) 'Stress diffusion from plate boundaries'. *Nature Phys. Sci.*, 243, 339–341.

Bott, M.H.P. (1993) 'Modelling the plate-driving mechanism'. *Journ. Geol. Soc.*, 150, 5, 941–951.

Bouysse, P. and Westercamp, D. (1990) 'Subduction of atlantic aseismic ridges and late cenozoic evolution of the lesser Antilles island-arc'. *Tectonophysics*, 175, 349–380.

Bowland, C. and Rosencrantz, E. (1988) 'Upper crustal structures of the western Colombian Basin, Caribbean Sea'. *Geol. Soc. Amer. Bull.*, 100, 534–546.

Brewer, T.S., Hergt, J.M., Hawkesworth, C.J., Rex, D. and Storey, B.C. (1992) 'Costs Land dolerites and the generation of Antarctic continental flood basalts'. In: *Magmatism and the Causes of Continental Break-up* (Eds Storey, B.C., Alabaster, T. and Pankhurst, R.J.). Geol. Soc. Special Publication, 68, 185–208.

Brown, E. and Hoek, E. (1978) 'Trends in relationships between measured in situ stress and depth'. *Int. J. Rock Mech. Min. Sci. and Geomech. Abstr.*, 15, 4, 11–15.

Bullard, E.C. (1936) 'Gravity measurements in East Africa'. *Phil. Trans. Roy. Soc. London*, 235A, 445–531.

Burke, K. (1988) 'Tectonic evolution of the Caribbean'. *Ann. Rev. Earth. Planet. Sci.*, 16, 201–130.

Burke, K., Fox, P. and Sengor, A. (1978)'Buoyant ocean floor and the evolution of the Caribbean'. *J. Geophys. Res.*, 83, 3949–3954

Burns, K., Rickard, M., Belbin, L. and Chamelaun, F. (1980) 'Further palaeomagnetic confirmation of the Patagonian orocline'. *Tectonophysics*, 63, 790.

Caldwell, J. and Turcotte, D. (1979) 'Dependence of the thickness of the elastic oceanic lithosphere on age'. *J. Geophys. Res.*, 84, 7572–7576.

Campbell, I.H. and Griffiths, W.R. (1990) 'Implications of mantle plume structure for the evolution of flood basalts'. *Earth and Planetary Science Letters*, 99, 79–93.

Cardwell and Isacks (1978) *J. Geophys. Res.*, 83, 2825–2838.

Carey, E. (1979) 'Recherece des directions principales de contraintes associés aujeu d'une population de faillles'. Rev. Geol. dyn. Geogr. Phys., 21, 57–66

Carey, S. (1958) 'The orocline concept in geotectonics'. *Proc. R. Soc. Tasmania*, 89, 2588.

Carpenter, B. and Carlson, R. (1992) *Oklahoma Geol. Surv.*, 52, 206–223.

Carter, D.J., Audley-Charles, M.G. and Barber, A.J. (1976) 'Stratigraphical analysis of island-arc/ continental collision in eastern Indonesia'. *J. Geol. Soc. Lond.*, 132, 179–198.

Case, J. and Holcombe, T. (1980) Geologic-tectonic Map of the Caribbean Region. U.S. Geol. Surv. Misc. Invest. Ser. Map I-1100, scale 1:2,500,000.

Catchings, R.D. and Mooney, W.D. (1988) 'Crustal structure of the Columbia plateau – evidence for continental rifting'. *J. Geophys. Res.*, 93, B1, 459–474

Chapman, C. and Morrison, D. (1994) 'Impacts on the Earth by asteroids and comets: assessing the hazard'. *Nature*, 367, 33–40.

Chappell and Veeh (1978) *Bull. Geol. Soc. Am.*, 89, 356–368.

Chase, C.G. (1978) 'Plate kinematics of the Americas, East Africa, and the rest of the World'. *Earth Planet. Sci. Lett.*, 37, 355–368.

Chatelain, J.-L., Molnar, P., Prevot, R. and Isacks, B. (1992) Detachment of part of the downgoing slab and uplift of the New Hebrides (Vanuatu) Islands'. *Geophys. Res. Letts.*, 19 (14), 1507–1510.

Christensen, U.R. (1989) 'Models of mantle convection – one or several layers'. *Phil. Trans. Roy. Soc. Lond.*, A328, 417–424.

Christie, D., Duncan, R., McBirney, A., Richards, M., White, W., Harpp, K. and Fox, C. (1992) 'Drowned islands downstream from the Galapagos hotspot imply extended speciation times'. *Nature*, 355, 246–248.

Christiensen, R. and McKee, E. (1978) 'Late Cenozoic volcanic and tectonic evolution of the Great Basin and Columbia intermontane regions'. *Geological Society of Amer. Mem.*, 152.

Chyba, C.F. (1993) 'Explosions of small Spacewatch objects in the Earth's atmosphere'. *Nature*, 363, 701–703, June.

Cloetingh, S. and Wortel, R. (1985) 'Regional stress-field of the Indian plate'. *Geophysical Research Letters*, 12, 77–80

Cloetingh, S. and Wortel, R. (1986) 'Stress in the Indo-Australian plate'. *Tectonophysics*, 132, 1–3.

Clube, S.V.M. and Napier W.M.. (1982) *The Cosmic Serpent*. Faber and Faber, London.

Clube, S.V.M. and Napier W.M. (1984) 'Comet capture from molecular clouds'. *Monthly Notices of the Royal Astrom. Soc.*, 208, 575–588

Clube, S.V.M. and Napier W.M. (1986) 'Comets and the galaxy: implications of the terrestrial record'. In: *The Galaxy and the Solar System*, (Eds. Smoluchowski, R. Bahcall, J.N. and Matthews, W.S.). Tucson, Univ. Arizona Press.

Clube, S.V.M. and Napier W.M. (1996) 'Galactic dark matter and terrestrial periodicities' *Quart. Journ. R. Astron. Soc.*, 37, 617–642.

Clube, S.V.M., Hoyle, F., Napier, W.M and Wickramasinghe, N.C. (1997) 'Giant comets, evolution and civilization'. *Astrophysics and Space Science* 245, 43.

Coblentz, D. and Richardson, R. (1995) 'Statistical trends in the intraplate stress field'. *J. Geophys. Res.*, 100, B10, 20, 245–255.

Coffin, M. and Eldholm, O. (1992) 'Volcanism and continental break-up: a global compilation of large igneous provinces'. In: *Magmatism and the Causes of Continental Break-up*. (Eds Storey, B., Alabaster, T. and Pankhurst, R.). Geol. Soc. London Spec. Publ., 68, 17–30.

Coffin, M. and Eldholm, O. (1994) 'Large igneous provinces: crustal structures, dimensions, and external consequences'. *Rev. of Geophys.*, 32, 1, 1–36.

Coleman, P., Michael, P. and Mutter, J. (1982) 'The origin of the Naturaliste Plateau, SE Indian Ocean: implications from dredged basalts'. *Journ. Geol. Soc. of Australia*, 29, 457–468.

Condie C. and Kent, C. (1976) *Plate Tectonics and Crustal Evolution*. Pergamon Press, Oxford.

Courtillot, V. (1990) 'Deccan volcanism of the K-T boundary: past climatic crises as a key to the future. Paleogeograph'. *Paleoclimat. Paleoecol.*, 189, 291–299.

Courtillot, V.R. and Besse, J. (1987) 'Magnetic field reversals, polar wandering and core-mantle coupling'. *Science*, 237, 1140–1147.

Courtney, R. and White, R. (1986) 'Anomalous heat flow and geoid across the Cape Verde Rise: evidence for dynamic support from a thermal plume in the mantle'. *Geophys. Journ. Roy. Astro. Soc.*, 87, 815–867.

Cowgill, M. (1994) 'The orientation of crustal stresses in the North Sea Basin and their geological origin'. Ph.D. Thesis, Univ. London.

Cox, K. (1988) 'The Karoo Province'. In: *Continental Flood Basalts* (Ed. MacDougall, J.L.). Kluwer, Dordrecht, The Netherlands, 239–271.

Crough, S.T. (1978) 'Thermal origin of mid-plate hot-spot swells'. *Geophys. J. R. Astron. Soc.*, 55, 4451–4469.

Cunningham, D., Klepeis, K., Wulf, G. and Dalziel, I. (1991) 'The Patagonian Orocline: New paleomagnetic data from the Andean Magmatic Arc, in Tierra del Fuego, Chile'. *J. Geophys Res.*, 96, B10, 16061–16067.

Dalziel, I. and Elliot, D. (1973) 'The Scotia Arc and Antarctic Margin'. In: *The Ocean Basins and Margins 1. The South Atlantic* (Eds Nairn and Stehli). Plenum, New York, 171–246.

Darwin, G.H. (1882) 'On the stress caused in the interior of the earth by the weight of continents and mountains'. *Phil. Trans. Roy. Soc.*, 173, 187–230.

Davies, G.F. (1988) 'Role of the lithosphere in mantle convection'. *J. Geophys. Res.*, 93, 10451–10466.

Davies, G.F. (1992) 'On the emergence of plate tectonics'. *Geology*, 20, 963–966.

Davis, L.K. (1967) 'Effects of a near-surface water table on crater dimensions'. U.S. Army Engineer Waterways Experimental Station. Corps of Engineers, Vicksburg, MS.

Davey, C. and Stephens, W. (1992) 'Deccan-related magmatism west of the Seychelles-India rift'. In: *Magmatism and the Cause of Continental Break-up* (Eds Storey, B., Alabaster, T. and Pankhurst, R.). Geol. Soc. London Spec. Publ., 68, 271–291.

DeMets, C., Gordon, D., Argus and Stein, S. (1990) 'Current plate motions'. *Geophys. J. Int.*, 101, 425–478.

Den, N., Ludwig, W. *et al.* (1969) 'Seismic refraction measurements in the NW Pacific Basin'. *J. Geophys. Res.*, 74, 1421–1434.

Den, N., Ludwig, W. *et al.* (1971) 'Sediments and structure of the Eauripik New Guinea Rise'. *J. Geophys. Res.*, 76, 4711–4723.

Dence, M., Grieve, R. and Robertson, D. (1977) 'Terrestrial impact: principal characteristics and energy considerations'. In *Impact and Explosive Cratering* (Ed. Roddy, D.). Oxford, Pergamon.

Detrick, R.S., and Crough, S. (1978) 'Island subsidence, hot spots and lithospheric thinning'. *J. Geophys. Res.*, 83, 1236–1244.

Detrick, R.S., Vonherzen, R.P., Parsons, B., Sandwell, D. and Dougherty, M. (1986) 'Heat flow observations on the Bermuda Rise and thermal models of mid-plate swells'. *J. Geophys. Res.*, 91, 3701–3723.

Devey, C.W. and Stephens, W.E. (1992) 'Deccan-related magmatism west of the Seychelles–India rift' In: *Magmatism and the Cause of Continental Break-up* (Eds. Storey, Alabaster and Pankhurst) Special Publication 68, Geol. Soc. London.

Dewey, J. (1988) 'Extensional collapse of orogens'. *Tectonics*, 7, 6, 1123–1139.

Dewey, J. and Pindell, J. (1986) 'Reply to Amos Salvador'. *Tectonics*, 5(4), 703–705.

Diament, M. and Goslin, J. (1987) 'Emplacement of the Marion, Dufresne, Lena and Ob seamounts (S Indian Ocean) from a study of isostacy'. *Tectonophysics*, 121, 252–262.

Dickin, A. (1988) 'The N Atlantic Tertiary Province'. In: *Continental Flood Basalts* (Ed. McDougall, J.L.). Kluwer, Dordrecht, The Netherlands, 111–149.

Dietz, R.S. (1960) 'Meteorite impact suggested by shatter cones in rock'. *Science*, 131, 1781–1784.

Dietz, R.S. (1963) 'Cryptoexplosion structures: A discussion'. *Am. Journ. Sci.*, 261, 650–664.

Doutch, F. (1981) 'Plate-Tectonic Map of the Circum-Pacific Region, Southwest Quadrant'. American Association of Petroleum Geologists, Tulsa, OK.

Duncan, R. (1984) 'Age progressive volcanism in the New England Seamounts and the opening of the central Atlantic Ocean'. *J. Geophys. Res.*, 89, 9980–9990.

Duncan, R. (1990) 'The volcanic record of the Reunion Hotspots'. Proceedings of the Ocean Drilling Program, Scientific Results, 115, Ocean Drilling Program, College Station, TX.

Duncan, R. and Clague, D. (1985) 'Pacific plate motions recorded by linear volcanic chains'. In: *The Ocean Basins and Margins*, Vol. 7A, The Pacific Ocean (Eds Nairn, A *et al.*). Plenum, New York, 89–121.

Duncan, R. and McDougall, I. (1976) 'Linear volcanism in French Polynesia'. *Journ. Volcanology and Geothermal Research*, 1, 197–227.

Duncan, R.A. (1981) 'Hotspots in the southern oceans – an absolute frame of reference for motion of the Gondwana continents'. *Tectonophysics*, 74, 29–42.

DuToit, A.L. (1933) *Our Wandering Continents*, Edinburgh, Oliver and Boyd.

Dziewonski, A.M. and Woodhouse, J.H. (1987) 'Global images of the Earth's interior'. *Science*, 236, 37–48.

Earetani, D.G. and Richards, M.A. (1995) 'Thermal entrainment and melting in mantle plumes'. *Earth and Planetary Science Letters*, 136, 251–267

Eaton, A., Wahl, H., Prostka, H., Marby, D. and Kleinkopf, M. (1978) 'Regional gravity and tectonic pattern: their relationship to late Cenozoisc epeirogeny and lateral spreading in the Western Cordillera'. *Geol. Soc. Amer. Mem.*, 152.

Engebretson, D., Cox, A. and Gordon, R. (1984) 'Relative motions between ocean plates of the Pacific basin'. *J. Geophys. Res.*, 89, 10291–10310.

England, P. (1996) 'The mountains flow', *Nature*, 381, 23–24.

Fastovsky and Gartner (1996) 'The Cretaceous-Tertary event and other catastrophies in Earth history'. Special Paper, 307, Geol. Soc. Am.

Feighner, M.A. and Richards, M.A. (1995) 'The fluid-dynamics of plume-ridge and plume-plate interactions – an experimental investigation'. *Earth and Planetary Science Letters*, 129, 171–182.

Fischer, K., McNutt, M. and Shure, L. (1987) 'Thermal and mechanical restraints on the lithosphere beneath the Marquesas Swell'. *Nature*, 322, 733–736.

Fisher, R.L., Engel, C.G. and Hilde, T.W.C. (1968) 'Basalts dredged from the Amirante Ridge, western Indian Ocean'. *Deep-Sea Res.*, 15, 521–534.

Fornari, D. and Gallo, D. (1989) 'Structure and topography of the Siqueiros transform fault system: evidence for the development of intra-transform spreading centre's. *Mar. Geophys. Res.*, 11, 263–299.

Forsyth, D.W. and Uyeda, S. (1978) 'On the relative importance of the driving forces of plate motion'. *Geophys. J. Roy. Astron. Soc.*, 43, 163–200.

Fox, P. and Gallo, D. (1986) 'The geology of North American transform plate boundaries and their aseismic extension'. In: *The Geology of North America. M – The Western North Atlantic Region* (Eds Vogt, P. and Tucholke, B.). Geol. Soc. Am., Boulder, CO, 157–172.

French, B. and Short, N. (Eds) (1968) *Shock Metamorphism of Natural Materials.* Mono Book Corp., Baltimore.

Freund, R. (1965) 'A model of the structural development of Israel and adjacent areas since Upper Cretaceous times'. *Geol Mag.*, 102, 189–205.

Fyfe, W. and Leonardos, O. (1977) 'Speculations on the causes of crustal rifting and subduction with applications to the Atlantic margin of Brazil'. *Tectonophysics*, 42, 29–36.

Fyfe, W., Price, N.J. and Thompson, A.B. (1978) *Fluids in the Earth's Crust.* Elsevier, Amsterdam.

Gallagher, K. and Hawkesworth, C. (1992) 'Dehydration melting and the generation of continental flood basalts'. *Nature*, 358, 57–59.

Gallet, Y., Weeks, R., Vandamme, D. and Courtillot, V. (1989) 'Duration of Deccan Trap volcanism: a statistical approach'. *Earth and Planet. Letts.*, 93, 273–82.

Gamboa, L. and Rabinowitz, P. (1984) 'The evolution of the Rio Grands Rise in the SW Atlantic Ocean'. *Marine Geology*, 58, 35–58.

Garfunkel, Z. (1986) 'Review of oceanic transform activity and development'. *J. Geol. Soc. London*, 143, 775–784

Gill, R., Pedersen, A. and Larsen (1992) 'Tertiary picrites in W. Greenland: melting at the periphery of a plume?'. In: *Magmatism and the Causes of Continental Break-up.* (Eds Storey, B., Alabaster, T. and Pankhurst, R.) Geol. Soc. Spec. Publ. No. 68. 335–348.

Girdler, R.W., Taylor, P.T. and Frawley, J.J. (1992) 'A possible impact origin for the bangui magnetic anomaly (Central Africa)'. *Tectonophyscis*, 212, 45–58.

Girling, C.M. (1992) 'Hydrocarbon habitat of the Seychelles microcontinent related to plume tectonics and palaeogeography'. In: *Proceedings of the Indian Ocean Petroleum Symposium* (Ed. Plummer P.S.) Victoria, Seychelles, 223–231.

Gladczenko, T.P., Coffin, M.F. and Eldholm, O. (1997) 'Crustal structure of the Ontong Java Plateau: Modeling of new gravity and existing seismic data'. *J. Geophys. Res.*, 102, 22711–22729.

Goertz, C. and Evans, B. (1979) Stress and temperature in the bending lithosphere as constrained by experimental rock mechanics. *Geophys. J. Roy. Astron. Soc.*, 59, 463–478.

Goslin, J. and Diament, M. (1987) 'Mechanical and thermal isostatic response of the Del Cano Rise and Crozet Bank (S Indian Ocean) from altimetry data'. *Earth and Planetary Sci. Letters*, 84, 285–294.

Govers, R., Wortel, J., Cloetingh, S. and Stein, C. (1992) Stress magnitude estimates from earthquakes in oceanic plate interiors. *J. Geophys. Res.*, 97, B8, 11, 749–760.

Gowd, T., Srirama Roa, S. and Gaur, V. (1992) 'Tectonic stress fields in the Indian subcontinent'. *J. Geophys. Res.*, 97, B8, 11, 867–888.

Grady, M.M. (1997) *Geoscientist*, 7 (1), 8–12.

Gradstein, F.M. and Ogg, J. (1996) 'A Phanerozoic time scale'. *Epsidosdes*, 19, 3–5

Grieve, R.A.F. (1991) 'Terrestrial impact: the record in the rock'. *Meteoritics*, 26, 175–194.

Grieve, R.A.F. (1993) 'When will enough be enough? News and Views'. *Nature*, 363, 670–671, June.

Grieve, R.A.F. (1994) 'The impact cratering rate in ancient times'. *J. Geophys. Res.*, 89, 403–408.

Grieve, R.A.F. (1997) *Geoscientist*, 7(1), 30.

Grieve, R.A.F. and Dence, M.R. (1979) 'The terrestrial cratering record, II. The Crater Production Rate'. *Icarus*, 38, 230–242.

Grieve, R.A.F. and Dence, M.R. (1980) 'Impact bombardment and its role in protocontinental growth on the early Earth'. *Precamb. Res.*, 10, 217–247.

Grieve, R.A.F. and Pilkington, M. (1996) 'The signature of terrestrial impacts'. *AGSO Journ.*, 16, 4, 399–420.

Grieve, R.A.F., Dence, M.R. and Robertson, P.B. (1977) 'Cratering processes: As interpreted from the occurrence of impact melts'. In: *Impact and Explosion Cratering* (Eds Roddy, D.J., Pepin, R.O. and Merrill, R.B.). Pergamon Press, Oxford, 791–814.

Grieve, R.A.F., Sharpton, V.L., Rupert, J.D. and Goodacre, A.K. (1988) 'Detecting a periodic signal in the terrestrial cratering record'. *Proceedings of the 18th Lunar and Planetary Science Conference*, 375–382.

Griffiths, R.W. and Campbell, I.H. (1990) 'Stirring and structures in the mantle starting plumes'. *Earth and Planetary Science Letters*, 99, 66–78.

Griffiths, R.W. and Campbell, I.H. (1991) 'On the dynamics of long-lived plume conduits in the convecting'. *Earth And Planetary Science Letters*, 103, 214–227.

Griggs, D. (1939) 'A theory of mountain building'. *Amer. Journ. Science*, 237, 703–720 (see also pp. 721–730).

Griggs, D., Turner, F. and Heard, H. (1960) 'Deformation of rocks at 500°C to 800°C, in Rock Deformation'. *Geol. Soc. Am. Mem.*, 79, 39–104.

Grunow, A.M. (1993) 'New palaeomagnetic data from the Antarctic Peninsula and their tectonic implications'. *J. Geophys. Res.*, 98, 13815–13833.

Grunow, A.M., Kent, D.V. and Dalziel, I.W.D. (1987) 'Mesozoic evolution of west Antarctica and the Weddell Sea basin – new paleomagnetic constraints'. *Earth And Planetary Science Letters*, 86, 16–26.

Grünthal, G. and Stromeyer, D. (1992) 'The recent crustal stress field in central Europe'. *J. Geophys. Res.*, 97, B8, 11, 821–828.

Gudlaugsson, S.T. (1993) 'Large impact crater in the Barents Sea'. *Geology*, 21, 291–294.

Gudmundsson, A. (1995) 'Stress fields associated with oceanic transform faults'. *Earth And Planetary Science Letters*, 136, 603–614.

Gunn, R. (1944) 'A quantitative study of the lithosphere and gravity anomalies along the Atlantic Coast'. *J. Franklin Inst.*, 237, 139–154.

Hales, A.L. (1969) 'Gravitational sliding and continental drift'. *Earth And Planetary Science Letters*, 6, 31–34.

Hamilton W. (1988) 'Plate tectonics and island arcs'. *J. Geol. Soc. Amer.*, 100, 1503–1527.

Hanks, T.C. (1971) 'The Kuril Trench-Hokkaido Rise system and simple models of deformation'. *Geophys. J. R. Astron. Soc.*, 23, 173–189.

Haq, B.U., Hardenbol, J. and Vail P.R. (1987) 'Chronology of fluctuating sea levels since the Triassic'. *Science*, 235, 1156–1167.

Haq, B.U., Hardenbol, J. and Van Eysinga (1987) *Geological Time Table*. Elsevier, Amsterdam.

Harland, W.B., Cox, A.V., Llewellyn, P.G., Picton, C.A.G., Smith, A.G. and Walters, R. (1982). A *Geologic Time Scale*. Cambridge University Press, Cambridge.

Harris, R. (1989) 'Processes of allochthon emplacement with special reference to the Brooks Range ophiolite, Alaska and Timor, Indonesia'. Ph.D Thesis, University of London, England.

Hawkes, D.D. (1962) 'The structure of the Scotia Arc'. *Geol. Mag.*, 99, 891.

Hawkesworth, C.J. and Glallagher, K. (1992) 'Mantle hotspots, plumes and regional tectonics as causes of intraplate magmatism'. *Terra Nova*, 5, 552–559.

Haxby, W., Turcotte, D. and Bird, J. (1976) 'Thermal and mechanical evolution of the Michigan Basin'. *Tectonophysics*, 36, 57–75.

Heard, H. (1963) 'Effects of large changes in strain rates in the experimental deformation of Yule Marble'. *J. Geol.*, 71, 162-95.

Heezen, B. and Matthews, J. (1973). 'Western Pacific guyots'. In: *Initial Reports of the Deep Sea Drilling Project, 20* (Eds Heezen *et al.*). Government Printing Office, Washington, 653–724.

Heimann, A., Fleming, T.H., Elliot, D.H. and Foland, K.A. (1994) 'A short interval of Jurassic continental flood-basalt volcanism in Antarctica as demonstrated by Ar-40/Ar-39 geochronology'. *Earth And Planetary Science Letters*, 121, 19–41.

Herrick, D.L. and Parmentier, E.M. (1994) 'Episodic large-scale overturn of 2-layer mantles in terrestrial planets. *J. Geophys. Res.*, 99, 2053–2062.

Hess, H. (1962) 'History of ocean basins'. In: *Petrologic Basins* (Eds Engel, James and Leonard). Geol. Soc. Amer., 599–620.

Hildebrand, A.R., Penfield, G.T., Kring, D.A., Pilkington, M., Carmargo, A., Jacobsen, S.B. and Boynton, W.V. (1991) 'Chicxulub Crater: a possible Cretaceous/Tertiary boundary impact crater on the Yucatan Peninsula, Mexico'. *Geology*, 19, 867–871.

Hirth, G. and Kohlstedt, D.L. (1995) 'experimental constraints on the dynamics of partially molten upper mantle' *J Geophys Res.* 100, 1981–2001

Hodge, P. (1994) *Meteorite craters and impact structures of the Earth.* Cambridge University Press, Cambridge.

Holm, P., Hald, N. and Nielsen, T. (1992) 'Contrasts in composition and evolution of Tertiary CFBs between West and East Greenland and their relations to the establishment of the Icelandic mantle plume'. In: *Magmatism and the Causes of Continental Break-up* (Eds Storey, B., Alabaster, T. and Pankhurst, R.). Geol. Soc. Spec. Publ., 68, 349–364.

Holmes, A. (1927) *The Age of the Earth: An Introduction to Geological Ideas*, Benn, London.

Holmes, A. (1965) *Principles of Physical Geology*, 2nd edn. Nelson, London.

Holsapple, K.A. (1993) 'The scaling of impact processes in planetary sciences. *Ann. Rev. Earth Planet. Sci.*, 21, 333–373.

Hooper, P.R. (1988) 'The Columbia River basalt'. In: *Continental Flood Basalts* (Ed. MacDougall, J.D.). Kluwer Academic, Amsterdam, 1–33.

Hooper, P.R. (1990) 'The timing of crustal extension and the eruption of continental flood basalts'. *Nature*, 345, 246–249.

Houtz, R., Hayes, D. and Markl, R. (1977) 'Kerguelen Plateau bathymetry, sediment distribution, and crustal structure'. *Marine Geology*, 25, 95–130.

Hubbert, M.K. and Willis, D.G. (1957) 'Mechanics of hydraulic fractures Parts I and II'. *Trans A.I.M.E.*, 210, 153–168

Hunt, C. (1953) 'Geology and geography of the Henery Mountains region of Utah'. U.S. Geol. Survey, Professional Papers, 228.

Hussong, D., Wipperman, L. and Kroenke. (1979) 'The crustal structure of the Ontong Java and Manihiki plateaus'. *J. Geophys. Res.*, 84, 6003–6010.

Illies, J.H. (1981) 'Mechanism of graben formation – epilogue'. *Tectonophysics*, 73, 249–266

Isaaks, B., Oliver, J. and Sykes, L. (1968) 'Seismology and the new global tectonics'. *J. Geophys. Res.*, 5855–5899.

Ivanov, B.A. (1986) 'Cratering mechanics'. English translation of Russian original. NASA Tech. Memorandum 88477 (N87-15662).

Iwabuchi, V. (1984) General Bathymetric Chart of the Oceans (GEBCO). Canadian Hydrographic Service, Ottawa, 5.06.

Jacoby, A. (1980) 'Plate sliding and sinking in mantle convection and the driving mechanism'. In: *Mechanisms of Continental Drift and Plate Tectonics* (Eds Davies, P. and Runcorn, K.). Academic Press, New York, 159–172.

Jackson, I. (1998) *The Earth's Mantle.* Cambridge University Press, Cambridge.

Jaeger, J.C. (1962) *Elasticity, Fracture and Flow*, 2nd edn. Methuen, London.

Jaeger, J.C. and Cook, N. (1969) *Fundamentals of Rock Mechanics*. Methuen, London.

Jeffreys, H. (1942) 'On the mechanics of folding'. *Geol. Mag.*, 79, 291–295.

Johnson, A. (1970) *Physical Processes in Geology*. Freeman, Cooper and Co., San Francisco.

Jones, G.H.S. (1963) 'Strong motion Seismic effects of the Suffield explosions'. Ph.D. Thesis University of Alberta and Suffield Rpt. 208 Part I and II.

Jones, G.H.S. (1965) 'A scale model study of the Bosumtwe Crater'. Proc. Astron. Soc. and in *Sky and Telescope* 30, 1.

Jones, G.H.S. (1977) 'Complex craters in alluvium'. In: *Impact and Explosive Catering* (Eds Roddy et al.). Pergamon, Oxford.

Jones, G.H.S. (1978) 'Coherent flaps surrounding craters'. *Nature*, 273, 211–213.

Jones, G.H.S. (1995) 'The Suffield Craters as analogues of impact structures'. *Suffield Special Publ.* No. 177, W7703-2-R332, Department of National Defence, Canada.

Jones, G.H.S. and Diehl (1964) 'On the prediction of the crater diameter for the 500 ton shot'. *Suffield Special Publ.*, 4.

Jouannic, C. (1985) 'Holocene coral reef emergence in the Sunda Strait'. Proc. Fifth International Coral Reef Congress, Tahiti, 3, 193.

Jouannic, C., Hoang, C.T., Hantoro, W.S. and Delinom, R.M. (1988) 'Uplift rate of coral-reef terraces in the area of Kupang, West Timor – preliminary results'. *Palaeogeog., Palaeoclimat., Palaeoecology*, 68, 259–272.

Jurdy, D.M., Stefanick, M. and Scotese, C. (1995) 'Palaeozoic plate dynamics'. *J. Geophys. Res.*, 100, B9, 17965–17975.

Kearey, P. and Vine, F. (1990) *Global Tectonics*. Blackwell, Oxford.

Kenyon, C. (1974) 'Stratigraphy and sedimentology of the late Miocene to Quaternary deposits of Timor', PhD thesis, London University.

Kiefer, W.S. and Hager, B.H. (1991) 'A mantle plume model for the equatorial highlands of Venus'. *J. Geophys. Res.*, 96, 20947–20966

King-Hubbert, M. (1937) 'Theory of scale models as applied to the study of geological structures'. *Geol. Soc. Amer. Bull.*, 48, 1459–1520.

Kirby, S.H. (1980) 'Tectonic stresses in the lithosphere: Constraints provided by the experimental deformation of rocks'. *J. Geophys. Res.*, 85, B11, 6353–6363.

Kirschner, C.E., Grantz, A. and Mullen, M.W. (1992) 'Impact origin of the Avak Structure, Arctic Alaska, and genesis of the Barrow Gas-Fields'. *American Association of Petroleum Geologists Bulletin*, 76, 651–679

Kohlstedt, D.L., Evans, B. and Mackwell S.J. (1995) 'Strength of the lithosphere: constraints imposed by laboratory experiments'. *J Geophys Res.* 100, 17587–17602.

Kuenen, P. and de Sitter, L.U. (1938) *Leidsche Geol. Mag.*, 10, 217–240.

Kumar, N. (1979) 'Origin of "paired" aseismic ridges: Ceara and Sierra Leone rises in the equatorial and the Rio Grands Rise and Walvis Ridge in the S Atlantic'. *Marine Geology*, 30, 175–191.

Lamé. (1852) *Leçons sur la Théorie Mathématique de l'Elasticité des Corps Solides*, Paris.

Larsen, R., Mutter, J. et al. (1979) 'Cuvier Basin: a product of ocean crust formation by early Cretaceous rifting off Western Australia. *Earth and Planetary Science Letters*, 45, 105–114.

Lawvers, L., Sclater, J. and Meinke, L. (1985) 'Mesozoic and Cenozoic reconstructions of the South Atlantic'. *Tectonophysics*, 114, 233–254.

Leary, P.N. and Rampino, M. (1990) 'A multi-causal model of mass extinctions: Increase in trace metals in the oceans'. In: *Extinction Events in Earth History*. Springer, Berlin, 44–55.

Lenardic, A, Kaula., W.M. and Bindschadler, D.L. (1993) 'A mechanism for crustal recycling on Venus'. *J. Geophys. Res.*, 98, 18697–18705.

Lepersonne, J. (1961) 'Quelques problèmes de l'histoire géologique de l'Afrique, au Sud de Sahara, dupuis la fin du carbonifère. *Annales de la Société Géologique de Belgique*, 84, 21–85.

Lewis, J.S. (1996) *Rain of Iron and Ice*. Addison-Wesley, New York.

Lewis, J.S., Watkins, G.H., Hartman, H. and Prinn, R. (1982) 'Chemical consequences of major impact events on Earth'. In: *Geological Implications of Impacts of Large Asteroid and Comets on the Earth* (Eds Silver and Schulz). Geol. Soc. Am. Spec. Paper 190.

Lister, C.R.B. (1992) 'The volcano-tectonic setting of oceanic lithosphere generation'. In: *Orphiolites and their Modern Oceanic Analogues*, (Eds. Parson *et al*) Geol. Soc. London, Special publication no. 60.

Lithgow-Bertelloni, C. and Richards, M.A. (1998) 'The dynamics of Cenozoic and Mesozoic plate motions'. *Reviews of Geophysics*, 36, 27–78.

Lonsdale, P. (1988) 'Geography and history of the Louisville hotspot chain in the SW Pacific'. *J. Geophys. Res.*, 93, 3078–3104.

Loper, D. and McCartney, K. (1986) 'Mantle plumes and the periodicity of magnetic field reversals'. *Geophys. Res. Lett.*, 13, 1524–1528.

Loper, D., McCartney, K. and Buzyna, G. (1988) 'A model of correlated episodicity in magnetic field reversals, climate and mass extinction'. *J. Geol.*, 96, 1–15.

Love, A.E.H. (1944) *The Mathematical Theory of Elasticity*. Dover, New York.

MacKenzie, K. (1984) 'Crustal structure and realistic seismic data'. Ph.D. Thesis, University of California, San Diego.

Maddox, J. (1994) 'Sun's encounter with star predicted'. *Nature*, 368, 395.

Mahoney, J.J. (1988) 'Deccan traps'. In: *Continental Flood Basalts* (Ed. MacDougall, J.D.). Kluwer, Dordrecht, The Netherlands, 151–194.

Mahoney, J.J., MacDougall, J.D., Lugmair, G.W. and Gopalan, K. (1983) 'Kerguelen hot spot source for Rajmahal Traps and Ninetyeast Ridge?'. *Nature*, 303, 385–389.

Mann, P. and Burke, K. (1984) 'Neotectonics of the Caribbean'. *Rev. Geophys. Space Phys.*, 22, 309–362.

Markov, A.V, (1966) 'Relative depths of Lunar ring mountains and craters in the Mare Nubium'. *Dok. Akad. Nauk SSSR*, 167, 1, 63–64.

Martin, A.J. (1969) 'Possible impact structure in southern Cyrenaica, Libya'. *Nature*, 223, 940–941.

Marzocchi, W. and Mulargia, F. (1993) 'Patterns of hot-spot volcanism'. *J. Geophys. Res.*, 98, B8, 14029–14039.

Mattey, D. (1982) 'Minor and trace element geochemistry of volcanic rocks from Truk, Ponape, and Kusaie, E Caroline Is.: evolution of a young hot spot trace across old oceanic crust'. *Mineral. and Petrology*, 80, 1–13.

McAdoo, D.C. and Sandwell, D.T. (1985) 'Folding of oceanic lithosphere'. *J. Geophys. Res.*, 91, 8373–8386.

McCaffrey, R. (1988) Active deformation in the New-Guinea fold-and-thrust belt – seismological evidence for strike-slip faulting and basement-involved thrusting'. *J. Geophys. Res.*, 93, 13323–13354.

McCaffrey, R., Molnar, P., Roecker, S.W. and Joyodiwiryo, Y.S. (1985) 'Microearthquake seismicity and fault plane solutions related to arc-continent collision in the Eastern Sunda Arc, Indonesia'. *J. Geophys. Res.*, 90, 4511–4528.

McCord, T.B., Morris, J., Persing, D., Tagliaferri, E., Jacobs, C., Spalding, R., Grady, L. and Schmidt, R. (1995) 'Detection of a meteorite entry into the Earth's atmosphere on Feb. 1st, 1994'. *J. Geophys. Res.*, 100, 3245–3250.

McEwen, A.S., Moore, J.M. and Shoemaker, E.M. (1997) 'The phanerozoic impact cratering rate: evidence from the far side of the Moon'. *J. Geophys. Res.*, 102, E4, 9231–9242.

McKenzie, D. (1969) 'Speculation on the consequences and causes of plate motions'. *Geophys. J. R. Astr. Soc.*, 18, 1–32.

McKenzie, D. (1977) 'The initiation of trenches: A finite amplitude instability'. In: *Island Arcs, Deep Sea Trenches and Back-Arc Basin* (Eds Talwani, M. and Pitman W.). Maurice Ewing Series 1, Am. Geophys. Union, Washington DC, 57–61.

McKenzie, D. (1978) 'Some remarks on the development of sedimentary basins'. *Earth and Planetary Sci.*, 40, 25–32.

McKenzie, D. and Bickle, M. (1988) 'The volume and composition of melt generated by extension of the lithosphere'. *Journ. of Petrology*, 29, 625–679.

McKenzie, D. and Sclater, J.G. (1971) 'The evolution of the Indian Ocean since the late Crataceous'. *Geophys. J. R. Astr. Soc.*, 24, 437–528.

McKenzie, D., Ford, P.G., Johnson, C., Parsons, B., Sandwell, D., Saunders, S. and Solomon, S.C. (1992) 'Features on Venus generated by plate boundary processes'. *J. Geophys. Res.*, 97, 13533–13544.

McKerrow, W.S., Scotese, C.R. and Brasier, M.D. (1992) 'Early Cambrian continental reconstructions', *Journ. Geol. Soc*, 149, 599–606.

McLaren, D.J. and Goodfellow, W. (1990) 'Geological and biological consequences of giant impacts'. *Annu. Rev. Earth Pln. Sci.*, 18, 123–171.

McSween, H.Y. and Weissman, P.R. (1989) 'Cosmochemical implications of the physical processing of cometary nuclei. *Geochim. Cosmchim. Acta*, 53, 3628.

Meijer, P.T. and Wortel, M.J.R (1992) 'The dynamics of motion of the South-American plate'. *J. Geophys. Res.* 97(B8), 11915–11931.

Melosh, H.J. (1989) 'Impact cratering: A geologic process'. *Oxford Monographs on Geology and Geophysics*, 11, Oxford University Press, Oxford.

Menard, H. (1969) 'The deep ocean floor'. *Scient. Amer.*, 221, 127–142.

Meyerhoff, A. and Teichert, C. (1971) 'Continental drift III: late Palaeozoic glacial centers and Devonian-Eocene coal distributiond. *Journ. Geology*, 79, 285–321.

Milsom, J. and Audley-Charles, M.G. (1986) Geol. Soc. Spec. Pub. 219, London, 353–364.

Milton, D.J., Glickson, A. and Brett, R. (1996a) 'Gosses Bluff, Part 1: geological structure'. *AGSO Journ. Austral. Geol. and Geophys.*, 16, 453–486.

Milton, D.J., Glickson, A. and Brett, R. (1996b) 'Gosses Bluff, Part 2. seismic, magnetic and gravity studies'. *AGSO Journ. Austral. Geol. and Geophy.s*, 16, 487–527.

Minster, J. and Jordan, T. (1978) 'Present-day plate motions'. *J. Geophys. Res.*, 83, 5331–5354.

Mohr, P. and Zanettin, B. (1988) 'The Ethiopian flood basalt province'. In. *Continental Flood Basalts* (Ed. McDougall, J.). Kluwer Dordrecht, The Netherlands, 63–100.

Molnar, P. and Atwater, T. (1973) 'Relative motion of hot spots in the mantle'. *Nature*, 246, 288–291.

Molnar, P. and Stock, J. (1987) 'Relative motions of hotspots in the Pacific, Atlantic and Indian Oceans since late Cretaceous times'. *Nature*, 327, 587–591.

Monahan, D., Falconer, R. and Tharp, M. (1984)'General Bathymetric Chart of the Oceans (GEBCO), Canadian Hydrographic Service, Ottawa, Canada, 5,10.

Moody, J.D. and Hill, M.J. (1956) 'Wrench fault tectonics'. *Geol. Soc. Am. Bull.*, 67, 1207–46.

Morgan, W.J. (1965) *J. Geophys. Res.*, 70, 6175–687.

Morgan, W.J. (1971) 'Convection plumes in the lower mantle'. *Nature*, 230, 42–43.

Morgan, W.J. (1972) 'Plate motions and deep mantle convection'. *Geol. Soc. Amer. Mem.*, 132, 7–22.

Morgan, W.J. (1975) 'Heat flow and vertical movements in the crus't. In: *Petroleum and Global Tectonics* (Eds Fisher, A. and Sheldon, J.). Princeton University Press, 23–43.

Morgan, W.J. (1983) 'Hotspot tracks and the early rifting of the Atlantic'. *Tectonophysics*, 94, 123–139.

Morris, E., Detrick, R., Minshull, T., Mutter, J., White, R., Wusi Su and Buhl, P. (1993) 'Seismic structure of oceanic crust in the western N Atlantic'. *J. Geophys. Res.*, 98, B8, 13879–13903.

Muller, R.A. and Morris, D.E (1986) 'Geomagnetic reversals from impacts on the Earth'. *Geophys. Res. Let.* 13, 1177–1180.

Müller, B., Zoback, M., Fuchs, K., Mastin, L. Gregersen, S., Pavoni, N., Stephansson, O. and Ljunggren, C. (1992) 'Regional patterns of tectonic stresses in Europe'. *J. Geophys. Res.*, 97, B8, 11783–117804.

Nakiboglu, S. and Lambeck, K. (1985) 'Thermal response of a moving lithosphere over a mantle heat source'. *J. Geophys. Res.*, 90, 2985–2994.

Napier, W.M. (1997) *Geoscientist*, 7(1), 30.

Nataf, H.-C. and VanDecar, J. (1993) 'Seismological detection of a mantle plume?'. *Nature*, 364, 115–120.

Negi, J., Agrawal, P., Pandey, O. and Singh, A. (1993) 'A possible K-T boundary bolide impact site offshore Bombay and triggering of rapid Deccan volcanism'. *Phys. of Earth and Planetary Interiors*, 76 189–197.

Norman, J.W. (1980) Regional tectonic structure of the Ccover rock. 2nd Prog. Report. Prep. by Imp. Coll. Div. of Min. Geol. for Min of Petr. and Mineral Resources, Jeddah, Saudi Arabia.

Oberbeck, V.R., Marshall, J. R. and Aggarwal, H. (1993) 'Impacts, tillites and the breakup of Gondwanaland'. *Journ. Geol.*, 101, 1–19.

Obert, L. and Duvall, W. (1967) *Rock Mechanics and the Design of Structures*. Wiley, New York.

O'Keefe, J. and Ahrens, T. (1982) 'Impact mechanisms of large bolides interacting with Earth and their implication to extinction mechanisms'. In: *Geological Implications of Impacts of Large Asteroids and Comets on Earth* (Eds Silver and Schultz). Spec. Pap. Geol. Soc. Am., 190, 103–120.

Olson, P. and Nam, I.S. (1986) 'Formation of sea-floor swells by mantle plumes'. *J. Geophys. Res.*, 91, 7181–7191.

Olson, P., Schubert, G., Anderson, C. and Goldman, P. (1988) 'Plume formation and lithosphere erosion: a comparison of laboratory and numerical experiments'. *J. Geophys. Res.*, 93, 15065–15084.

Pal, P.C. and Creer, K.M (1986) 'Geomagnetic reversal spurts and episodes of extraterrestrial catatrophism'. *Nature*, 320, 148–150.

Parsons, B. and Richter, F.M. (1988) 'A relation between the driving force and geoid anomaly associated with mid-ocean ridges'. *Earth and Planetary Science Letters*, 55, 445–450.

Parsons, B. and Sclater, J.G. (1977) 'An analysis of the variation of ocean floor bathymetry and heat flow with age'. *J. Geophys. Res.*, 82, 803–827.

Patriat, P. and Achache, J. (1984) 'India-Eurasian collision chronology has implications for crustal shortening and driving mechanism of plates'. *Nature*, 311, 615–621.

Peel, F. (1982) 'The driving mechanism of plate motion and deformation'. M.Sc. dissertation, University of London.

Peirce, C. and Barton, P. (1991) Crustal structure of the Madeira-Tore Rise, eastern N Atlantic - results of DOBS wide-angle and normal incidence seismic experiment in the Josephine Seamount region'. *Geophysical Journal International*, 106, 357–378.

Peltier, W.R. (1984) 'The thickness of the continental lithosphere'. *J. Geophys. Res.*, 89, 1303–1316

Petford, N., Kerr, R. and Lister, J. (1993) 'Dyke transport of granitoid magmas'. *Geology*, 21, 845–848.

Piccorillo, E., Melfi, A., Comin-Chiaramonte, P., Bellieni, G., Ernesto, M., Margues, L., Nardy, A., Pacca, I., Roisenberg, A. and Stolfa, D. (1988) 'Continental flood volcanism from the Parana Basin (Brazil)'. In: *Continental Flood Basalts* (Ed. McDougall). Kluwer, Dordrecht, The Netherlands, 192–238.

Pierce, J. and Weissel, J., *et al.* (1989) Proceedings of the Ocean Drilling Program, Initial Reports, 121, Ocean Drilling Program, College Station, TX.

Pike, R.J. (1974) 'Depth/diameter relations of fresh lunar craters: revision from spacecraft data. *Geophys. Research Letters*, 1, 291–294.

Pilkington, M. and Grieve, R. (1966) 'The signature of terrestrial impacts'. *AGSO Journal*, 16, 399–420.

Pindell, J. and Barrett, S. (1988) 'Geological evolution of the Caribbean region: a plate tectonic perspective'. In: *The Geology of North American, Vol H: The Caribbean Region* (Eds. Dengo, C. and Case, C.E.) Geol. Soc. Amer., Boulder, CO. 405–432.

Plawman, T.L. and Hager, P.I. (1983) 'Seismic expression of structural styles'. In: *A Picture and Work Atlas* (Ed. Bally, A.W.), Vol 1, Am. Assoc. Petrol Geol.

Price, N.J. (1966) *Fault and Joint Development in Brittle and Semi-brittle Rock*. Pergamon, Oxford.

Price, N.J. (1975) 'Rates of deformation'. *J. Geol. Soc. London.*, 131, 553–575.

Price, N.J. and Audley-Charles, M.G. (1983) 'Plate rupture by hydraulic fracture resulting in overthrusting'. *Nature*, 306, 572–575

Price, N.J. and Audley-Charles, M.G. (1987) 'Tectonic collision processes after plate rupture'. *Tectonophysics*, 140, 121–129.

Price, N.J. and Cosgrove, J.W. (1990) *Analysis of Geological Structures*. Cambridge University Press, Cambridge.

Price, N.J., Price, G.D. and Price, S.L. (1988) 'Gravity glide and plate tectonics'. In: *Gondwana and Tethys* (Eds Audley-Charles, M.G. and Hallam, A.). Geol. Soc. London, Special Publ., 37, 5–21.

Prinn, R.G. and Fegley, B. (1987) 'Bolide impacts, acid rain, and biospheric trauma at the K/T boundary'. *Earth Planet Sci. Lett.*, 83, 1–15.

Rabinowitz, D., Bowell, E., Shoemaker, E. and Muinmonen. (1994) 'The population of Earth-crossing asteroid's. In: *Hazards Due to Comets and Asteroids* (Ed. Gehels, T.). University of Arizona Press, Tucson, AZ, 285–312.

Raff, R. and Mason, R. (1961) 'Magnetic survey off the west coast of N. America, 40°N to 52.5°N'. *Geol. Soc. Amer. Bull.*, 72, 1259-1265.

Rampino, M.R. (1988) 'Flood basalt volcanism during the past 250 million years'. *Sciences*, 241, 663–668.

Rampino, M.R. and Caldeira, K. (1992) 'Antipodal hotspot pairs on Earth'. *Geophys. Res. Letts.*, 19, 2001–2014.

Rampino, M.R. and Caldeira, K. (1993) 'Major episodes of geological change: correlations, time structure and possible causes'. *Earth and Planetary Letters*, 114, 214–227.

Rampino, M.R. and Stothers, R.B. (1984) 'Geological rhythms and cometary impacts'. *Sciences*, 226, 1427–1431.

Rampino, M.R. and Stothers, R.B. (1986) 'Geologic periodicities and the Galaxy'. In: *The Galaxy and the Solar System* (Eds Smoluchowski *et al.*). University of Arizona Press, Tucson, AZ, 241–259.

Rampino, M.R., Haggerty, B. and Pagano, T. (1997) 'A unified theory of impact crises and mass extinction: quantitative tests'. *Annals New York Acad. Sci.*, 822, 403–431.

Raup, D.M. (1991) 'A kill curve for Phanerozoic marine species'. *Palaeobiology*, 17, 37–48.

Raup, D.M. and Sepkosski, J.J. (1984) 'Periodicity of extinction in the geologic past'. *Nat. Acad. Sci.*, U.S. Proc. 81, 801–805.

Reidel, S.P. and Hooper, P. (1989) 'Volcanism and tectonism in the Columbia River flood-basalt province'. *Geol. Soc. Amer. Special Paper*, 239.

Reidel, S.P., Fecht, K.R., Hargood, M. and Tolan, T.L. (1989) *Geol. Soc. Am. Spec. Paper*, 239, 247–264.

Ribe, N. and Christensen, U. (1994) 'Three-dimensional modeling of plume-lithosphere interaction'. *J. Geophys. Res.*, 99, B1, 669–682.

Richardson, R. (1987) 'The origin of the intraplate stress field of the Indo-Australian plate'. *Eos Trans. AGU*, 68, 1466.

Richardson, R. (1989) 'The origin of the intraplate stress field'. Proceedings of the 28th Int. Geol. Conf., Washington DC, International Union of Geology, Washington DC.

Richardson, R. (1992) 'Ridge forces, absolute plate motions and the intraplate stress field'. *J. Geophys. Res.*, 97, 11739–11749.

Richardson, R., Solomon, S. and Sleep, N. (1979) 'Tectonic stresses in the plates'. *Rev. Geophys.* 11, 391.

Robertson, O.B. and Grieves, R.A.F. (1977) 'Shock attenuation in terrestrial impact structures'. In: Impacts and Explosion Cratering (Ed. D.J. Roddy). Pergamon Press, New York, 687–702.

Robin, E., Froget, L., Jehanno, C. and Rocchia, R. (1993) 'Evidence for a K/T impact event in the Pacific Ocean'. *Nature*, 363, 615.

Roddy, D. (1968) 'The Flynn Creek crater, Tennessee'. In: *Shock Metamorphism of Natural Material* (Eds French and Short). Mono Book Corp., Baltimore, 1304–1317.

Roddy, D.J., Schuster, S.H., Rossenblatt,M., Grant, L.B., Hassig, P.J. and Kreyenhagen, K.N. (1988) 'Computer modelling of large asteroid impacts into continental and oceanic sites: atmospheric, cratering and ejecta dynamics'. In: *Proceedings of the Lunar Planet. Inst. Natl. Acad. Sci.*, Washington DC, 158–159.

Rodean, H. (1971) Nuclear-Explosion Seismology. Lawrence-Livermore Lab., University of California. Atomic Energy Commission.

Romanowicz, B. (1991) 'Seismic tomography of the Earth's mantle'. *Ann. Rev. Earth Planet Sci.*, 19, 77–99.

Ronov, A.B. (1983) 'The Earth's sedimentary shell'. Am. Geol. Inst., Falls Church, VA.

Rosenblatt, P., Pinet, P.C. and Thouvenot, E. (1994) 'Comparative hypsometric analysis of Earth and Venus'. *Geophys. Res. Letts.*, 21, 465–468.

Rowan, L.C. and Wetlaufer, P.H. (1978) 'Iron-adsorption band analysis for the discrimination of iron-rich zones'. U.S.G. Type III Final Report, Contract S-70243-AG.

Runcorn, S.K. (1980) 'Mechanics of plate tectonics: mantle convection currents, plumes, gravity gliding or expansion'. *Tectonophys.*, 63, 297–307.

Ryder, G., Fastovsky, D. and Gartner, S. (1996) *The Cretaceous–Tertiary Event and Other Catastrophes in Earth History*. Geol. Soc. Amer. Special Paper 307

Salmon, E. (1952) *Materials and Structures*, Vol. 1. Longmans, Green and Co., London.

Sandwell, D. and Renkin, M. (1988) 'Compensation of swells and plateaus in the N Pacific: no direct evidence for mantle convection'. *J. Geophys. Res.*, 93, 2775–2783.

Saul, J.M. (1978) 'Circular structures of large scale and great age in the Earth's crust'. *Nature*, 271, 343–345.

Saunders, A., Storey, M., Kent, R. and Norry, M. (1992) 'Consequences of plume-lithosphere interaction'. In: *Magmatism and the Causes of Continental Break-up* (Eds Storey, B., Alabaster, T. and Pankhurst, R.). Geol. Soc. Spec. Publ., 68, 41–60.

Schlanger, S., Campbell, J., Haggerty, J. and Premoli-Silva, I. (1981) 'Cretaceous volcanism and Eocene failed atolls in the Radak Chain: Implications for the geological history of the Marshall Is.'. *EOS Transactions of the Amer. Geophys. Union*, 62, 1075.

Schwan, W. (1980) 'Geodynamic peaks in alpinotype orogenies and changes in ocean-floor spreading during Late Jurassic-Late Tertiary time'. *Am. Assoc. Petrol. Geol. Bull.*, 64, 359–373.

Sclater, J. and Fisher, R. (1974) 'Evolution of the east central Indian Ocean with emphasis on the tectonic setting of the Ninetyeast Ridge'. *Geol. Soc. Amer. Bull.*, 85, 683–702.

Searle, R. (1992) 'The volcano-tectonic setting of oceanic lithosphere generation'. In: *Ophiolites and their Modern Oceanic Analogues* (Eds Parson, L., Murton, B. and Browning, P.) Geol. Soc. London Spec. Pub. 60, 65–79.

Sepkoski, J.J. (1989) 'Periodicity in extinction and the problems of catastrophism in the history of life'. *J.Geol.Soc. London.*, 146, 7–19.

Sepkoski, J.J. (1994) 'Extinction and the fossil recod' *Geotimes*, 39, 5–17.

Seyfert, C. and Sirkin, L. (1979) *Earth History and Plate Tectonics*, 2nd edn. Harper and Row, New York.

Shaw, H.R. (1994) *Craters, Cosmos and Chronicles*. Stanford University Press, Stanford, CA.

Sheridan, R.E. (1983) 'Phenomena of pulsation tectonics related to the break-up of the eastern North American continental margin'. Int. Report Deep-Sea Drilling Proj., 76, 897–909.

Shipley, T., Abrams, L., Lancelot, V. and Larson, R. (1993) 'Later Jurassic-Early Cretaceous ocean crust, mid-Cretaceous volcanic sequences and seismic sequences of the Naura Basin, W Pacific'. In: *The Mezozoic Pacific* (Eds Pringle, M. and Sager, W.). Amer. Geophys. Union, Monograph Series.

Shoemaker, E. and Shotts. (1968) 'Pseudo-vulcanism and lunar impact craters, *Eos Trans. AGU*, 49, 457–461

Shoemaker, E.M. (1960) 'Penetration mechanics of high velocity meteorites, illustrated by Meteor Crater, Arizona'. In: Report of the Int. Geol. Congress, XXI Session Norden, Part XVIII, 418–434, Copenhagen.

Shoemaker, E.M. (1983) 'Asteroid and comet bombardment of the Earth'. *Ann. Rev. Earth Planet. Sci.*, 11, 461–494.

Shoemaker, E.M. (1997) 'Long-term variations in the impact cratering rate on Earth'. *Geoscientist*, 1, 29.

Shoemaker, E.M. and Wolfe, R.F (1986) 'Mass extinctions , crater ages and comet showers'. In: *The Galaxy and the Solar System*, (Eds. Smoluchowski, R. Bahcall, J.N. and Matthews, W.S.). Tucson, Univ. Arizona Press.

Shoemaker, E.M., Wolfe, R. and Shoemaker, C.S. (1990) 'Asteroid and comet flux in the neighbourhood of Earth'. In: *Global Catastrophes in Earth History* (Eds Sharpton and Ward). Spec. Pap. Geol. Soc. Am., 247, 155–170.

Shotts, R. (1968) 'Pseudo-volcanism and Lunar impact craters'. Trans. A.G.U., 49, 457–461.

Sibson, R.H. (1975) 'Generation of pseudotachylite by ancient seismic faulting'. *Geophys. J. R. Astro. Soc.*, 43, 774–789.

Sinha, H., Louden, K. and Parsons, B. (1981) 'The crustal structure of the Madagascar Ridge'. *Geophys. J. R. Astro. Soc.*, 66, 351–377.

Skogseid, J., Pedersen, T., Eldholm, O. and Larsen, B. (1992) 'Tectonism and magmatism during NE Atlantic continental break-up: the Voring Margin'. In: *Magmatism and the Causes of Continental Break-up* (Eds Storey, B., Alabaster, T. and Pankhurst, R.). Geol. Soc. Spec. Publ., 68, 305–320.

Sleep, N.H. (1994) 'Martian plate tectonics'. *J. Geophys. Res.*, 99, 5639–5655.

Sleep, N.H. and Snell, N.S. (1976) 'Thermal contraction and flexure of mid-continent and Atlantic marginal basins'. *Geophys. J. R. Astr. Soc.*, 45, 125–154.

Smith, A.D. (1993) 'The continental mantle as a source for hot spot volcanism'. *Terra Research*, 452–460.

Smith, A.G. and Hallam, A. (1970) 'The fit of the southern continents', *Nature*, 225, 139–44.

Smith, A.G., Hurley, A. and Briden, J. (1981) *Phanerozoic Paleocontinental World Maps*. Cambridge University Press, Cambridge.

Solheim, L.P. and Peltier, W.R. (1994) 'Avalanche effects in-phase transition modulated thermal-convection – a model of Earth's mantle'. *J. Geophys. Res.*, 99, 6997–7018.

Solomon, S.C. (1993) 'The geophysics of Venus'. *Physics Today*, 46, 48–55

Speed, R. (1985) 'Cenozoic collision of the Lesser Antilles arc and origin of the El Pilar Fault'. *Tectonics*, 4, 41–69.

Spohr, T. and Schubert, G. (1982) 'Convection thinning of the lithosphere: a mechanism for continental rifting'. *J. Geophys. Res.*, 87, 4669–4681.

Stewart, C.A. and Rampino, M.R (1992) 'Time-dependent thermal convection in the Earth's mantle: theory and observation'. *EOS*, 73, 303.

Stocker, R. and Ashby, M. (1973) 'On the rheology of the upper mantle'. *Rev. Geophys. Space Phys.*, 11, 391.

Stothers R.B. (1986) 'Periodicity of the Earth's magnetic reversals'. *Nature*, 332, 240–242.

Stothers R.B. (1993) 'Impact cratering at geologic stage boundaries'. *Geophys. Research Letters*, 20, 10, 887–890.

Strom, R., Schaber, G. and Dawson, D. (1994) 'The global resurfacing of Venus'. *J. Geophys. Res.*, 99, E5, 10899–10926.

Suess, E. (1909) *Das Atlitz der Erde* (3 Vols). Freytag, Leipzig.

Sullivan, W. (1990) *Continents in Motion*, 2nd edn. American Inst. of Phys., New York.

Supko, P. and Perch-Neilsen, K. (1977) 'General synthesis of central and southern Atlantic drilling results'. In: *Initial Report of the Deep Sea Drilling Project, 39* (Eds Supro and Perch-Nielsen *et al.*). U.S. Government Printing Office, Washington DC, 1099–1132.

Symonds, P. and Cameron, P. (1977). 'The structure and stratigraphy of the Caernavon Terrace and Wallaby Plateau'. *Austral. Petrol. Expl. Assoc. Journ.*, 17, 30–41.

Tackley, P.J., Stevenson, D.J., Glatzmaier, G.A. and Schubert, G. (1993)' Effects of an endothermic phase-transition at 670 km depth in a spherical model of convection in the Earth's mantle'. *Nature*, 361, 699–704

Tagliaferri, E., Spalding, R. *et al.* (1994) 'Detection of meteorite impacts by optical sensors in Earth orbit'. In: *Hazards Due to Comets and Asteroids* (Ed. Gehrels, T.). University of Arizona Press, Tucson, AZ, 199–220.

Tarduno, J. and Gee, J. (1995) 'Large-scale motion between Pacific and Atlantic hot-spots'. *Nature*, 378, 477–480.

Taylor, S.R. (1992) *Solar System Evolution*. Cambridge University Press, Cambridge.

Tolan, T.L. and Reidel, S.P., Beeson, M.H., Anderson, J.L., Fect, K.R., and Swanson D.A. (1989) 'Revisions to estimates of areal extent and volume of the Columbia River Basalt group' In: *Volcanism and Tectonim in the Columbia River Flood Basalt Province,*Geol. Soc. of Amer. Special Paper 239, 1–20.

Tollmann, A. and Tollman, E. (1993) *Und die Sintflut gab es doch*. Droemer-Knaur, Munich.

Tollmann, A. and Tollman, E. (1994) 'The youngest big impact on Earth deduced from geological and historical evidence'. *Terra Nova*, 6, 209–217.

Toon, O.B., Turco, R.P. and Covey, C. (1997) 'Environmental Pertubations caused by the impacts of asteroids and comets'. *Reviews of Geophysics*, 35, 41–78.

Turcotte, D.L. (1983) 'Mechanism of crustal deformation'. *J. Geol. Soc. London*, 140, 701–724.

Turcotte, D.L. (1993) 'An episodic hypothesis for Venusian tectonics'. *J. Geophys. Res.*, 98, 17061–17068

Turcotte, D.L and Schubert, G. (1982) *Geodynamics*. Wiley, New York.

Turcotte, D.L. and Oxburgh, E.E. (1967) 'Finite amplitude convection cells and continental drift', *J. Fluid Mech.* 28, 29–42.

Turcotte, D.L., Haxby, W. and Ockenden, J. (1977) 'Lithospheric instabilities'. In: *Island Arcs, Deep Sea Trenches, and Back-Arc Basins*, Maurice Ewing Series 1 (Eds Talwani, M. and Pitman, W.). Am. Geophys. Union, Washington DC, 63-69.

Turner, S., Regelous, M. *et al.* (1989) Geol. Soc. Amer. Special Paper, 239, 1–20.

Turner, S., Regelous, M., Vail, P.R., Mitchum, R. and Thompson, S. (1977) *Am. Assoc. Petrol. Geol. Mem.*, 26, 83–97.

Upton, B. (1988) 'History of Tertiary igneous activity in the N Atlantic borderlands'. In: *Early Tertiary Volcanism and the Opening of the NE Atlantic* (Eds Morton, A. and Parson, L.). Geol. Soc. London, Special Publ. 39, 429–453.

Vail, P.R., Mitchum, R. and Thompson, S. (1977) 'Seismic stratigraphy and global changes in sea-level. Pt.4 Global cycles of relative changes of sea-level'. *Am. Assoc. Petrol. Geol. Mem.*, 26, 83–97.

Vallier, T., Dean, W., Rea, D. and Thiede, J. (1983). 'Geological evolution of the Hess Rise, central N Pacific Ocean'. *Geol. Soc. Amer. Bull.*, 94, 1289–1307.

Veevers (1962) *Earth Planet. Sci. Lett.* 1, 335–338.

Veevers, J. (1986) 'Break-up of Australia and Antarctica, estimated as mid-Cretaceous (95 +/– 5 Ma) from magnetic and seismic data at the continental margins'. *Earth Planet. Sci. Lett.*, 77, 91–99.

Vening-Meinesz, F. (1941) 'Gravity over the Hawaiian archipelago and over the Madeira area: Conclusions about the Earth's crust'. *Proc. K. Ned. Akad. Wet.*, 44, 1–14.

Veershuur, G. (1966) *Impact! The threat of Comets and Asteroids*. Oxford University Press, Oxford.

Vine, F. and Matthews, D. (1963) 'Magnetic anomalies over oceanic ridges'. *Nature*, 199, 947–949.

Vita-Finzi. C. (1986) *Recent Earth Movements*. Academic Press, London.

Vogt, P. (1974) 'The Icelandic phenomenon: imprints of a hot spot on the ocean crust and implications for flow beneath the plates'. In: *Geodynamics of Iceland and the N Atlantic area* (Ed. Kristjanson, L.). Reidel, Himgham, MA, 105–126.

Von Karman, T. (1911) 'Festigkeitsversuche unter allseitigem Druck'. *Z. Ver. dt Ing.*, 55, 1749–1757.

Walcott, R.I (1970) 'Flexural rigidity, thickness and viscosity of the lithosphere'. J. Geophys. Res.. 75. 3941–3954.

Wang, K, Chatterton, B.D.E, Attrep, M and Irth, C.J.. (1993) 'Late Ordovician mass extinction in the Selwyn basin, Notherwestern Canada: Geochemical, sedimentological and palaeontological evidence' *Canadian Journal of Earth Sciences* 30, 1870–1880.

Wang, K., Geldselzer, H.H.J. and Chatterton, B.D.F (1994) 'A late Devonian extraterrestrial impact and extinction in E. Gondwana'. Geol. Soc. Amer. Special Paper 293, 111–120.

Ward, P.D. (1996). 'After the fall: Lessons and directions from the K/T debate'. *Palaios*. 10, 530–538.

Watts, A.B. (1978) *J. Geophys. Res.*, 83, 5989.

Watts, A.B. and Cochran, J. (1974) 'Gravity anomalies and flexure of the lithosphere along the Hawaiian-Emperor seamount chain'. *Geophys. J. R. Astr. Soc.*, 38, 119–141.

Watts, A.B. and Talwani, M. (1974) 'Gravity anomalies seaward of deep-sea trenches and their tectonic signifcance'. *Geophys. J.R. Astr. Soc.*, 36, 57–102.

Watts, A.B., Bondine, J. and Ribe, N. (1980) 'Observations of flexure and the state of stress in the oceanic lithosphere'. *Nature*, 238, 532–537.

Watts, A.B., Karner, G. and Steckler, M. (1982) 'Lithospheric flexuring and the evolution of sedimentary basins'. *Phil. Trans. R. Soc. Lond.*, A 305, 249–281.

Weertman, J. and Weertman, J.R. (1975) 'High-temperature creep of rocks and mantle viscosity'. *Annu. Rev. Earth Planet. Sci.*, 3, 293–316.

Wellman, P. and McDougall, I. (1974) 'Cainozoic igneous activity in eastern Australia'. *Tectonophysics*, 23, 49–65.

Wen, L.X. and Anderson D.L. (1997) 'Present-day plate motion constraint onf matle rheology and convection',. *Journ, Geol. Res.* 102, 24639–24653.

Wezel, F-C. (1988) 'The origin and evolution of Arcs'. *Tectonophysics*, 146, 380.

White, R.S. (1992) 'Magmatism during and after continental break-up'. In: *Magmatism and the Causes of Continental Break-up* (Eds Storey, B., Alabaster, T. and Pankhurst, R.). Geol. Soc. Spec. Publ. 68, 1–30

White, R.S. and McKenzie, D.P. (1989) 'Magmatism of rift zones: the generation of volcanic continental margins and flood basalts'. *J. Geophys. Res.*, 99, B6, 7685–7729.

White, R.S., Spence, G.D., Fowler, S.R., McKenzie, D.P., Westbrook, G.K. and Bowen, A.N. (1987) 'Magmatism at rift zones: The generation of volcanic continental margins and flood basalts'. *J. Geophys. Res.*, 94, 7685–7729.

Wiens, D.A. and Stein, S. (1983) 'Age dependence of oceanic intraplate seismicity and implications for lithospheric evolution'. *J. Geophys. Res.*, 88, 6455–6468

Wiens, D.A. and Stein, S. (1984) 'Intraplate seismicity and stresses in young oceanic lithosphere'. *J. Geophys. Res.*, 89, 1442–1464.

Willar L.-A. (1991) 'Mines today'. *Col. School Mines*, 5, 3.

Wilson, L. and Head, J.W. (1994) 'Mars – review and analysis of volcanic-eruption theory and relationships to observed landforms'. *Review of Geophysics*, 32, 221–263.

Wilson, M. (1992) 'Magmatism and continental rifting during the opening of the S Atlantic Ocean: a consequence of Lower Cretaceous super-plume activity?'. In: *Magmatism and the Cause of Continental Break-up* (Eds Storey, B., Alabaster, T. and Pankhurst, R.). Geol. Soc. London Spec. Publ. 68, 241–256.

Wilson, M. (1993) 'Plate-moving mechanisms – constraints and controversies'. *J. Geol. Soc. London*, 150, 923–926.

Wilson, T.J. (1963) 'Evidence from islands on the spreading of ocean floors'. *Nature*, 197, 536–538.

Wilson T.J. (1966) 'Are the structures of the Caribbean and Scotia Arc analagous to ice rafting?'. *Earth Planet. Sci. Lett.*, 1, 335–338.

Winterer, E. and Meltzer, C. (1984) Origin and subsidence of guyots in mid-Pacific mountains'. *J. Geophys. Res.*, 89, 9969–9979.

Winterer, E., Lonsdale, P., Matthews, J. and Rosendahl, B. (1974) 'Structure and acoustic stratigraphy of the Manihiki Platform'. *Deep Sea Research*, 21, 793–814.

Wolbach, W.S., Lewis, R.S. and Anders, E. (1985) 'Cretaceous extinctions: evidence of wildfires and search for meteoritic material'. *Science*, 230, 167–170.

Woodhouse, J.H. and Dziewonski, A.M. (1989) 'Seismic modeling of the Earth's large-scale 3-dimensional structure'. *Phil. Trans. R. Soc. Lond.*, A328, 291–308.

Woodward, R.L., Dziewonski, A.M. and Peltier, W.R. (1994) 'Comparisons of seismic heterogeneity models and convective flow calculations'. *Geophys. Res. Letts.*, 21, 325–328

Xu Zhonghuai, Wang Suyun, Huang Yurui and Gao Ajia. (1992) 'Tectonic stress fields inferred from a large number of small earthquakes'. *J. Geophys. Res.*, 97, B8, 11867–11878.

Yamaoka, K. (1988) 'Spherical-shell tectonics – on the buckling of the lithosphere at subduction zones'. *Tectonophysics*, 147, 179–191.

Zahnle, K. (1990) 'Atmospheric changes in chemistry caused by large impacts'. In: *Global Catastrophes in Earth History* (Eds Sharpton and Ward). Spec. Pap. Geol. Soc. Am., 247, 271–288.

Zaychenko, V., Kuznetsov, O. and Popsuy-Shapko, G. (1982) 'Nature of ring-shaped photoanomalies identified by remote sensing surveys'. *Internat. Geol. Rev.*, 24, 1148–1154.

Zhen-Ming Jin, Green, H. and Yi Zhou (1994) 'Melt topology in partially molten mantle peridotite during ductile deformation'. *Nature, Letters*, 372, 164–167.

Ziegler, P.A. (1993) Smith, William Lecture 1992: 'Plate-moving mechanisms – their relative importance'. *J. Geological Society*, 150, 927–940.

Ziegler, W.H. (1975) 'Outline of the geological history of the North Sea'. In: *Petroleum and the Continental Shelf of NW Europe* (Ed. Woodland, A.W.). Geology Applied Science Publications, London, 165–187.

Zoback, M.L. (1992a) 'Stress field constraints on intraplate seismicity in E North America'. *J. Geophys. Res.*, 11761–11782.

Zoback, M.L. (1992b) 'First and second order patterns of stress in the lithosphere: The World Stress Map project'. *J. Geophys. Res.*, 97, 11703–11729.

Zoback, M.L., Nishenko, R., Richardson, H., Howegewa and Zoback, M. (1986) 'Mid-plate stress, deformation and seismicity'. In: *The Geology of North America, Vol. M, The Western North Atlantic Region* (Eds Vogt, P. and Tucholke). Geol. Soc. America, Boulder, CO, 297–312.

Zoback, M.L. and Magee, M. (1991) 'Stress magnitudes in the crust'. *Philos. Trans. R. Soc. London*, A337, 181–194.

Zolotukhin, V. and Al'mukhamedov, A. (1988) 'Traps of the Siberian Platform' In: *Continental Flood Basalts* (Ed. MacDougall, J.). Kluwer, Dordrecht, The Netherlands, 273–310.

Zorin, Y., Kozhevnikov, V. and Turuttanov, E. (1989) 'Thickness of the lithosphere beneath the Baikal rift zone and adjacent regions'. *Tectonophysics*, 168, 327–338.

Zuber, M.T. (1987) 'Compression of oceanic lithosphere – an analysis of intraplate deformation in the central indian basin'. *J. of Geophys. Res.*, 92, 4817–4825.

Index